Community Policing Today

Sara Miller McCune founded SAGE Publishing in 1965 to support the dissemination of usable knowledge and educate a global community. SAGE publishes more than 1000 journals and over 800 new books each year, spanning a wide range of subject areas. Our growing selection of library products includes archives, data, case studies and video. SAGE remains majority owned by our founder and after her lifetime will become owned by a charitable trust that secures the company's continued independence.

Los Angeles | London | New Delhi | Singapore | Washington DC | Melbourne

Community Policing Today
Issues, Controversies, and Innovations

Deborah A. Parsons
California State University, San Bernardino

Janine Kremling
California State University, San Bernardino

Los Angeles | London | New Delhi
Singapore | Washington DC | Melbourne

FOR INFORMATION:

SAGE Publications, Inc.
2455 Teller Road
Thousand Oaks, California 91320
E-mail: order@sagepub.com

SAGE Publications Ltd.
1 Oliver's Yard
55 City Road
London, EC1Y 1SP
United Kingdom

SAGE Publications India Pvt. Ltd.
B 1/I 1 Mohan Cooperative Industrial Area
Mathura Road, New Delhi 110 044
India

SAGE Publications Asia-Pacific Pte. Ltd.
18 Cross Street #10-10/11/12
China Square Central
Singapore 048423

Acquisitions Editor: Jessica Miller
Editorial Assistant: Sarah Wilson
Production Editor: Astha Jaiswal
Copy Editor: Exeter Premedia Services
Typesetter: Exeter Premedia Services
Proofreader: Lawrence W. Baker
Indexer: Exeter Premedia Services
Cover Designer: Gail Buschman
Marketing Manager: Jillian Ragusa

Copyright © 2021 by SAGE Publications, Inc.

All rights reserved. Except as permitted by U.S. copyright law, no part of this work may be reproduced or distributed in any form or by any means, or stored in a database or retrieval system, without permission in writing from the publisher.

All third-party trademarks referenced or depicted herein are included solely for the purpose of illustration and are the property of their respective owners. Reference to these trademarks in no way indicates any relationship with, or endorsement by, the trademark owner.

Printed in the United States of America

ISBN 978-1-5443-3672-5

This book is printed on acid-free paper.

20 21 22 23 24 10 9 8 7 6 5 4 3 2 1

BRIEF CONTENTS

Preface	xvii
Acknowledgments	xix
About the Authors	xxi

PART I • FOUNDATIONS OF COMMUNITY POLICING

CHAPTER 1 • Community Policing and Community Issues	2
CHAPTER 2 • Community Policing in the 21st Century	24
CHAPTER 3 • Identifying and Responding to Problems Within the Community: Three Case Studies	42

PART II • RESPONSES TO SPECIFIC CRIME TYPES

CHAPTER 4 • Community Policing and Terrorism	64
CHAPTER 5 • Community Policing and Gangs	85
CHAPTER 6 • Community Policing and Hate Crime	109
CHAPTER 7 • Community Policing and Militarism	126
CHAPTER 8 • Community Policing and Drugs	148

PART III • STRATEGIES AND TACTICS

CHAPTER 9 • Net Widening and Social Control	170
CHAPTER 10 • Community Policing and the Use of Force	185

PART IV • DEPARTMENT ORGANIZATION AND CHALLENGES

CHAPTER 11 • Police and the Media	203
CHAPTER 12 • Community Policing: The Men and Women in Uniform	223

CHAPTER 13 • Community Policing and Police Administration   240

PART V • WHERE DO WE GO FROM HERE?

CHAPTER 14 • Community Policing, Engagement, and Outreach   260

CHAPTER 15 • The Future of Community Policing   281

Glossary   302
Notes   307
Index   345

DETAILED CONTENTS

Preface	xvii
Acknowledgments	xix
About the Authors	xxi

PART I • FOUNDATIONS OF COMMUNITY POLICING

CHAPTER 1 • Community Policing and Community Issues — 2

Learning Objectives	2
"I Can't Breathe."	2
Introduction	3
Police and Violence	4
Traditional Versus Community Policing	5
Traditional Policing	6
Organizational Structure of Traditional Policing	6
Crime-Fighting Focus of Traditional Policing	7
Cornerstones of Traditional Policing	7
Community Policing	12
COPS Definition of Community Policing	13
Community Partnerships	14
Organizational Transformation	15
Problem Solving	15
Principles of Community Policing	16
Contrast of Traditional and Community Policing Models	18
Police and Community Relations	18
Expanding the Police Mandate	18
Contrasting Problem-Solving, Information Systems, and New Technologies	19
Improvements in Problem-Solving	19
Predictive Policing and Information Systems	19
New Technologies and Policing	19
Community Issues and Community Policing	20
Summary	22
Key Terms	23
Discussion Questions	23

CHAPTER 2 • Community Policing in the 21st Century — 24

Learning Objectives	24
Covid-19: Testing Police and Community Relations	24
Introduction	25
Three Eras of Policing	25

Political Era	27
Reform Era	29
Community Policing Era	31
Sir Robert Peel's Nine Principles	32
Police Community Relations	32
Team Policing	33
Foot Patrol	34
Broken Windows Theory	37
Summary	40
Key Terms	40
Discussion Questions	41

CHAPTER 3 • Identifying and Responding to Problems Within the Community: Three Case Studies — 42

Learning Objectives	42
Introduction	42
SARA Model	43
Case Studies	44
Addressing Alcohol-Related Crimes and Victimization	44
The Problem of Alcohol-Related Incidents in College Towns	44
Pharmacological Effects of Alcohol	46
Alcohol Outlet Density and Violent Crime	47
Reducing Alcohol-Related Violence	48
Police Response to People With Mental Disorders: A Gray Zone	49
Solutions to Chronic Vulnerability	51
Using Local Knowledge for Guided Decision-Making	51
Negotiating Peace With Complainants and Call Subjects	51
Crisis Intervention Teams as a Best Practice	52
Domestic Violence	53
Domestic Violence and Intimate Partner Homicide	55
Domestic Violence and Officer Safety	56
The Problem of Officer-Involved Domestic Violence	57
The Challenges of Community Policing and Domestic Violence	58
Promising Practices in Community Policing of Domestic Violence	59
Coordinated Partnerships	59
Neighborhood-Based Organizing and Problem Solving	60
Community Policing Action Team (CPAT)	60
Access and Collaboration	61
The Safety and Accountability Audit as a Community Problem-Solving Tool	61
Summary	61
Key Terms	62
Discussion Questions	62

PART II • RESPONSES TO SPECIFIC CRIME TYPES

CHAPTER 4 • Community Policing and Terrorism — 64

Learning Objectives	64
Introduction	65
Types of Terrorism	65
Defining Terrorism	65
International and Domestic Terrorism	66
The Role of Community Intelligence in Countering Terrorism	68
Types of Intelligence	70
Intelligence-Led Policing	72
The Importance of Building Partnerships Between Police and Muslim and Arab American Communities	73
Office of Community-Oriented Policing Services	74
Tensions between Police and Muslim Communities, Overcoming Tensions, and Building Trust	75
Distrust Between Muslim and Arab-American Communities and the Police	75
Balancing the Priorities of Intelligence Gathering, Community Engagement, and Trust Building	79
Determining Who to Partner With in the Community	80
Levels of Consultation and Community Input	81
The Importance of Building Resilience to Violent Extremism Within Communities to Right- and Left-Wing Terrorism	81
Bonds	83
Bridges	83
Linking	83
Summary	83
Key Terms	84
Discussion Questions	84

CHAPTER 5 • Community Policing and Gangs — 85

Learning Objectives	85
Introduction	85
What Is a Gang?	86
Classification and Characteristics of Gangs	86
Street Gangs	87
Prison Gangs	87
Motorcycle Gangs	88
Prevalence of Gangs and Gang Violence	89
Gang Violence Against Law Enforcement	90
Why Do Juveniles Join Gangs?	93
Risk Factors and Community Programs	94
The Story of Josephina Ramirez, 20	94
Motivations by Juveniles to Join Gangs	96
The Story of Trei Hernandez, 21	97
White Street Gangs	98
The "Royals"	98
Asian Gangs	99
Gangs and Drugs	100

Recruitment of Children	101
Drug Distribution	102
How Can Community Policing Reduce Gang Activity and Recruitment?	103
A Comprehensive Strategy to Address Gang Violence	103
Community Policing and Juvenile Gang Membership	103
Education and Training	106
Gang Resistance Education and Training (G.R.E.A.T.)	107
Summary	107
Key Terms	107
Discussion Questions	108

CHAPTER 6 • Community Policing and Hate Crime — 109

Learning Objectives	109
Introduction: Defining Hate Crime	110
Hate Crime Data Collection	111
Data Collection From the General Population	111
Data Collection From the College Population	113
Extent and Trends of Hate Crime in the United States	113
The Extent and Purpose of Federal and State Hate Crime Legislation	115
Federal Hate Crime Legislation	115
State Hate Crime Laws	117
Types of Recording Statutes	117
Types of Laws	117
The Problem of Motive in Hate Crime Investigations: Are Hate Crime Laws Effective?	118
Arguments Against Hate Crime Legislation	118
Community Policing Strategies to Respond to Hate Crime Incidents	121
Community Policing Strategies in Response to Hate Crime Incidents	123
Summary	124
Key Terms	125
Discussion Questions	125

CHAPTER 7 • Community Policing and Militarism — 126

Learning Objectives	126
Introduction	127
Death by Swatting	127
The History of Police Militarization	128
The Civil Rights Riots: How It All Began	128
The War on Drugs and Police Militarization	128
The War on Terrorism and the Militarization of Police	131
Police Perceptions of Militarization and SWAT Developments in the 21st Century	132
The Department of Defense Excess Program	133
Militarization as a Threat to Community Policing	135
The Impact of Battle Dress Uniforms on Community Policing	136
The Impact of Military-Style Training of Police Officers on Community Policing	139

The Importance and Use of Military Equipment	141
Uses of the Military Excess Equipment	141
Benefits of the LESO Program	142
Summary	147
Key Terms	147
Discussion Questions	147

CHAPTER 8 • Community Policing and Drugs — 148

Learning Objectives	148
A Deadly Combination: Heroin Laced With Fentanyl	148
Introduction	149
Extent of the Current Drug Problem in the United States	150
Opioids	150
Community–Police Partnerships Help Combat the Opioid Epidemic	153
The Changing Role of Police in the Face of the Current Drug Crisis	156
Marijuana	158
Cocaine	160
Methamphetamine	160
Open-Air Drug Markets	161
Police Strategies to Eliminate Open-Air Drug Markets	164
Problem-Oriented Policing Strategies: The Drug Market Intervention Model	166
Summary	168
Key Terms	168
Discussion Questions	168

PART III • STRATEGIES AND TACTICS

CHAPTER 9 • Net Widening and Social Control — 170

Learning Objectives	170
Scared Straight	170
Introduction	171
The Net Widening Effect	171
What Is Net Widening?	171
Diversion Programs and Net Widening	171
The Negative Effects of Net Widening	172
Loss of Human Capital	169
Group Stigma	174
Quality-of-Life Policing	175
Asset Forfeiture	178
Stop and Seize: The Practice of Asset Forfeiture	181
In Defense of Asset Forfeiture	182
Summary	183
Key Terms	184
Discussion Questions	184

CHAPTER 10 • Community Policing and the Use of Force — 185

- Learning Objectives — 185
- Introduction — 186
- Community Policing and Use of Force — 186
  - Police Legitimacy and Procedural Justice — 187
    - Police Legitimacy — 187
    - Procedural Justice — 187
  - Prevalence of Force — 188
    - Race and Use of Force — 189
    - Police Brutality — 189
    - Legacy of Police Brutality — 190
- Use of Force Policies and Legal Case Precedent — 191
  - Use of Force Policies — 191
    - Legal Precedence and Liability — 192
    - Off-Duty, Current, and Retired Law Enforcement Ofcers — 193
- Law Enforcement Tactical Gear — 193
- Training, Force Option Models, and Officer Discretion — 193
  - Training — 195
  - Force Option Models — 196
  - Officer Discretion and Use of Force — 197
    - Characteristics of the Officer and Use of Force — 198
    - Job Assignments and Use of Force — 198
    - Arrest Activity and Use of Force — 198
  - Misconceptions Regarding Use of Force — 199
- How Can Community Police Reduce Use of Force Issues? — 199
- Summary — 200
- Key Terms — 201
- Discussion Questions — 201

PART IV • DEPARTMENT ORGANIZATION AND CHALLENGES

CHAPTER 11 • Police and the Media — 203

- Learning Objectives — 203
- Introduction — 204
- Police and Media: A Symbiotic Relationship — 204
- Impact of the Media — 205
  - Social Construction Versus Direct Experience — 205
- Police Portrayals in the Media — 206
  - Entertainment Media — 206
  - News Media — 208
    - Investigative Reporting — 208
- Fear of Crime — 210
- Media Role in the War on Police — 212
- Controlling Media Information — 212

Social Media	213
News Media and Social Media Merge	216
Investigation of Social Media Crimes	217
Public Use of Social Media	218
Summary	219
Key Term	222
Discussion Questions	222

CHAPTER 12 • Community Policing: The Men and Women in Uniform — 223

Learning Objectives	223
Introduction	224
Lessons Learned From Early Community Policing Implementations	225
Split Force	226
Officer-Centered Challenges to Community Policing	227
Police Subculture	228
The Hiring Process	229
Police Recruitment	230
The Recruitment Message	231
Who, What, Where of Recruitment	232
Recruitment of Women and People of Color	232
Recruitment of Educated Officers	233
Police Training	234
Soft Skills	236
Summary	238
Key Terms	238
Discussion Questions	239

CHAPTER 13 • Community Policing and Police Administration — 240

Learning Objectives	240
Introduction	241
Top Cop, Who's the Boss?	242
Top Cop Responsibilities	243
Top Cop and Elected Officials	244
Top Cop as a Public Figure and Community Leader	244
Top Cop and the Police Union	244
Police Officer Rights	245
Top Cop as a Visionary	246
Incentives to Adopt Community Policing	247
Coercive Change Motivations	248
Riots and Civil Unrest	248
Race Relations	248
Crime and Social Disorganization Inducements	249
Crime	249
Fear of Crime	249
Financial Enticement for Change	249

Impediments to Change	**250**
Officer Resistance	250
Challenges of Police Culture	251
Institutionalized Values, Norms, and Beliefs	*252*
Background, Training, and Work Experience	*252*
Closed System	*252*
All or Nothing	*253*
Crime and Budgetary Restraints	253
Crime and Disorder Challenges to Implementation	*253*
Budget Allocations for Policing	*253*
Economic Downturns Create Hardships	*254*
Unintended Consequences for Departments	*254*
Impact of Budget Cuts on the Community	*255*
Community Policing Realities	**255**
Summary	**257**
Key Terms	**258**
Discussion Questions	**258**

PART V • WHERE DO WE GO FROM HERE?

CHAPTER 14 • Community Policing, Engagement, and Outreach — 260

Learning Objectives	260
Introduction	261
From Community Policing to Community Engagement	**262**
Community Defined	262
Community Engagement Defined	263
From Community Engagement to Community Outreach	**263**
Program Development: The Who, What, Where, How, Why?	264
Scale and Focus of Outreach	*264*
Planning a Program	*264*
Community Outreach	**265**
Historical Citizen Participation	266
Neighborhood Watch Programs	*266*
Citizen Police Academies	267
Youth Programs	269
Explorer/Cadets Programs	*269*
Teen Academy	*270*
Senior Programs	270
Programs Based on Population Targets	**271**
Outreach to Black Americans	271
Police and Barbers	*272*
Outreach to Immigrant Communities	273
Aurora, Colorado, Immigration Outreach	*274*
Outreach to People Experiencing Homelessness	275
H.E.A.R.T. Program, Santa Ana Police Department	*276*

Mental Health Outreach	276
Crisis Intervention Teams	278
Community Outreach - Honorable Mentions	278
Summary	279
Key Terms	280
Discussion Questions	280

CHAPTER 15 • The Future of Community Policing — 281

Learning Objectives	281
The Story of Flint Town	281
Public Expectations and Satisfaction With Law Enforcement	283
Officer-Involved Fatal Incidents 2015–2019 and Racial Disparities	285
Community Policing as the Foundation of Restorative Justice	285
Restorative Policing	288
Implementation of Restorative Policing	291
Restorative Policing: From Theory to Practice	291
Police-based Restorative Conferencing Programs	291
Nashville Police Department	293
Baton Rouge Police Department	294
National League of Cities: Four Steps to Building Better Working Relationships between Police and Faith-based Community	295
Police Influences on Law Enforcement: The Potential Role of Congress	296
Political Influences on Policing—The Power of the People	297
How Technology May Change Policing in the Future	298
Augmented Reality	299
Drones	300
Summary	301
Key Terms	301
Discussion Questions	301

Glossary	**302**
Notes	**307**
Index	**345**

PREFACE

Community policing is an umbrella term that refers to many innovative programs, such as problem solving, crime prevention, evidence-based policing, artificial intelligence, predictive analytics, and technological advances. The innovative programs fall under the community policing umbrella because "community" is central to all policing endeavors. Instead of reactive policing, it focuses on proactive strategies for long-term solutions of issues and problems that are unique to that community, such as those relating to homelessness; high crime; immigrant populations; and low socioeconomics, racial-ethnic minority, and at-risk juveniles. Community policing is smarter policing using sophisticated methods, technologies, data analytics, and predictive applications.

Many police officials and police scholars thought that community policing was a passing fad, soon to be disposed of for the next program or a return to traditional law and order crime fighting. Nothing could be farther from the truth. Community policing is not a program; it is a validation of what police do every day—they interact with the public and help facilitate problem solving of community issues and problems. It also rewards officers for innovative and creative policing strategies. It could be asserted that once community policing was introduced, it is highly unlikely that there will ever be a return to a type of policing that fails to recognize the importance of the community's voice in how they are policed.

What community policing has done for both the law enforcement community and the public is to open channels of communication and promote a true partnership. It is through community policing efforts that positive and productive relationships can be built. Police cannot solve community problems without the help of the public. Police can have an open dialog with citizens, giving them an opportunity for input. Input involves the process of identifying community concerns, setting priorities, assigning responsibilities, and working together to devise effective strategies and outreach efforts. The community must be part of that process. It also should be noted that while some individuals and groups may continue to be hostile to police, never has there been a better opportunity for healing past wounds. Police officers trained for community policing learn everything through a procedural justice lens so that their decisions and actions are fair and just. Procedural justice promotes police legitimacy so that the officer's authority is respected, and compliance is a product of that trust.

Implementation of community policing is not a simple process. It takes time and commitment to implement community policing effectively. The officers must be trained and educated properly, and they must buy in to the principles of community policing. All stakeholders, such as community members, elected officials, nonprofits, local business, and other agencies, must be part of the process and understand their responsibilities and roles. Despite the popularity of community policing, many police departments do not use community policing to its full potential; for example, either they implement some components and not others, have specialized units rather than embracing it department-wide, or they claim to have it and they do not. Some of the challenges to full implementation of community policing are identified and discussed in this book.

While there are many texts available on community policing, this one offers lessons about some of the distinctive ways that traditional components of policing are reconciled with community policing philosophy. It also offers a refreshing positive perspective on community policing, a concept that has been talked about for many years, but which has defied a common definition and understanding. This book will provide an overview of the history of community policing and how community policing has developed over time. We will discuss community policing practices and challenges, in the context of different topics, including terrorism, gangs, drugs, hate crimes, and militarism. Additionally, we cover issues of net widening and social control, the relationship of media and police, cutting-edge technology and policing, and the future of community policing. We will identify and discuss the many challenges police officers face every day. Real-world examples of community policing programs are shared to give students an operational understanding of community engagement and outreach. Finally, we will ask students to think about the issues surrounding community policing and police in general. Also unique to this book, we engage the students in thoughtful debates about controversial matters through provocative discussion questions throughout and at the end of each chapter. Students will emerge with a better understanding of the complexity and promise of community policing.

ACKNOWLEDGMENTS

Deborah A. Parsons

I would like to thank my coauthor, Janine Kremling, for asking me to be part of this project and for her support, expertise, and friendship. Much thanks to Jessica Miller and Sarah Wilson from SAGE Publishing for the detailed assistance in bringing the book to completion. Thank you to all my academic and police colleagues, family, and friends, both living and deceased, who have given me the motivation to be successful. Thank you to my former students who have taught me more than I have taught them. Finally, my heartfelt thanks and love to my husband, Dan, and son, Zackary without whose support and humor in all things I could not live without.

Janine Kremling

First of all, I would like to thank Jessica Miller and Sarah Wilson from SAGE Publishing for their patience, continuous support, and encouragement over the past two years. They have been wonderful to work with and I am very grateful for their efforts. I would also like to give a huge thank you to my coauthor, Dr. Deborah Parsons, who is not only a retired police officer but also the assistant dean of the College of Social and Behavioral Sciences at California State University, San Bernardino. She has been amazing to work with and has shown great dedication to the success of this book. I would like to thank the chair of my department, Dr. Andrea Schoepfer, for her consistent support for research.

Finally, I have to express my gratitude to my parents, who have always supported me through all my endeavors, believed in me, and showed me their unconditional love. Without them, I would not have been able to accomplish my goals. They have taught me to tackle challenges with a positive attitude, enthusiasm, and determination. They taught me that it is not about being talented or intelligent, but success comes through hard work and a growth mindset.

SAGE and the authors would also like to gratefully acknowledge the reviewers who provided suggestions on the drafts of these chapters. Their feedback helped shape the development of this text and helped ensure that it meets the needs of instructors teaching this course and their students.

LaNina N. Cooke, Farmingdale State College
Darren D. Gil, Southern University at New Orleans and Tulane University
Ronald W. Glensor, University of Nevada, Reno, and Arizona State University
Melissa S. Harrell, Bainbridge State College
Richard C. Helfers, The University of Texas at Tyler
Quentin D. Holmes, Grambling State University
Nerissa James, Miami Dade College
Ivan M. Kaminsky, Mesa Community College

Marcos L. Misis, Northern Kentucky University
Denise Nation, Winston-Salem State University
Timothy W. Roberts, University of Pikeville
R. D. Robertson, Bryant & Stratton College Online
Kelly Sue Roth, Bloomsburg University of Pennsylvania
Jeffrey Schwartz, Rowan University
Mercedes Valadez, California State University, Sacramento
Ericka Wentz, University of West Georgia
Roger Wright, University of Cincinnati

ABOUT THE AUTHORS

Deborah A. Parsons

Deborah A. Parsons, Ph.D., has been a professor at California State University, San Bernardino, since 1996 and is currently an assistant dean for the College of Social and Behavioral Sciences. Dr. Parsons retired from law enforcement after serving 26 years as a police officer. She earned her Ph.D. in social ecology from the University of California, Irvine. Dr. Parsons teaches courses in policing, women and crime, and justice and the media. Her area of expertise is in law enforcement and multicultural issues in the criminal justice system, with a specific focus on women in policing.

Janine Kremling

Janine Kremling, Ph.D., has been a professor at California State University, San Bernardino, since 2008. She received Ph.D. in criminology from the University of South Florida. Dr. Kremling teaches a wide variety of classes. She has also published five books, including *Cyberspace, Cybersecurity, and Cybercrime*; *Homeland Security*; *Drugs, Crime, and Justice*; and *Estimating Drug Use*. Dr. Kremling has been studying issues pertaining to policing, especially criminal procedure and how it impacts policing. In addition, she has also extensively studied drug use and abuse and police responses.

PART I
FOUNDATIONS OF COMMUNITY POLICING

CHAPTER 1

COMMUNITY POLICING AND COMMUNITY ISSUES

Learning Objectives

1. Identify the three cornerstones of traditional policing.
2. Explain the difference between traditional policing and community policing.
3. Explain the core principles of community policing.
4. Identify community issues that broaden the police mandate.

"I CAN'T BREATHE."

The death of a Black man at the hands of a White police officer is in the news again, the names have changed but the story has been told many times. At approximately 8:00 pm on May 25th, 2020, police responded to a call involving a man passing a counterfeit bill at a grocery story. The man identified as George Floyd, a 46-year-old African American, had purchased a pack of cigarettes with a twenty-dollar bill. Employees believed the bill was counterfeit. They followed Floyd out of the store and confronted him as he was getting into the driver's seat of his SUV across the street, demanding the pack of cigarettes be returned. Floyd refused. Another employee of the store called police to report that a very drunk and out-of-control man was fighting with employees and had passed a phony bill. After police arrived and contacted Floyd, he was placed under arrest. What ensued next was captured by various cameras including bystander's cell phones. Although the entire scenario has yet to be completely pieced together, it appears that Floyd was experiencing problems of breathing and claimed to be claustrophobic after initially being put in the back seat of a police unit. It was when Officer Derick Chauvin arrived and took over the scene that things started to fall apart. Chauvin removed Floyd from the back of the unit. When Floyd was extricated from the vehicle, he fell to the pavement, face down. Officer Chauvin, in what is now a well-publicized photo snapshot, is seen kneeling on the man's neck, although Floyd is clearly in custody, handcuffed, and does not appear to be resisting. In the videos and photos, two additional officers are sitting on Floyd's back and legs, while a fourth officer is keeping worried and vocal bystanders at bay. Over a dozen times, Floyd can be heard saying, "I can't breathe."[1] Bystanders were begging the officers to let him up. For eight long minutes and 46 seconds, Chauvin kneeled on Floyd's neck. An ambulance was called and then expedited when Floyd appeared to go unconscious. After Floyd was loaded in the ambulance, attendants called the fire department to respond because they believed Floyd was going into cardiac arrest. Firefighters arrived and found Floyd to be unresponsive. He was transported to Hennepin County Medical Center where he was pronounced dead, just over an hour and a half from the initial call to police.[2] Since that time, protests, rioting, looting, property damage, injuries, and death are erupting in cities across America. Officer Chauvin and the three other officers have been fired, charged with murder, and remain in custody.[3]

Racial bias and police brutality are frequently at the root of demands for police reform. The tragic killing of George Floyd, as illustrated above, has resulted in unrelenting anger and vociferous demands for police reform, which include radical notions to **defund** police or to eliminate them completely. Defunding the police refers to taking away funds allocated to policing and redistributing those funds to other social services, such as mental health care, drug rehabilitation, domestic violence, and homelessness.[4] Public outrage and emotions are running high, and police and city officials are scrambling to respond to the various demands. Some cities, for example New York, Los Angeles, and Minneapolis, have already announced defunding plans.[5] Activists in Seattle, WA, boarded up the city's East Precinct police building when police fled and have set up a "police no-go zone" where there is no police presence. The area was deemed a safe zone for protestors but has morphed into a community of anarchists by erecting barriers against outsiders.[6] Other cities are taking conservative steps by being more strategic about what to do going forward. There has been immediate response by members of Congress to propose a Justice in Policing Act, which will increase oversite and accountability of police including a move to eliminate legal protections for police, form a national database of excessive-force incidents, boost requirements for body cameras, and increase subpoena power of the Justice Department to conduct investigations. While this bill is in the early stages, many are proclaiming such an act would lead to transformative changes in policing.[7]

INTRODUCTION

In this chapter, we compare traditional and community policing models, outlining features of each. Second, we examine the failure of traditional policing to effectively solve crime or heal inimical relations with the public. Third, we explore the definition and key components of community policing. The transformation from traditional to community policing involves sweeping changes in the way police view their role and relationships with the community; comprehensive organizational changes in structure and management; and the adoption of new technology and information systems to find fresh ways of addressing crime and disorder.

Throughout this text, we consider the claim that community policing might be a viable solution to violence, crime, and hostile relations with the public more so than traditional methods. There are many issues facing our communities today. Let us first take a look at another example of the challenges police face in communities across America.

POLICE AND VIOLENCE

On September 5, 2018, just before 7:30 p.m., 19-year-old Delmonte Johnson was brutally gunned down on a South Side Chicago sidewalk outside his brother's basketball practice. The drive-by shooting that claimed the life of Johnson added to Chicago's death toll of 381 and 2,074 people shot so far that year.[8] According to Uniform Crime Reports (UCR) in 2017, approximately 29,737 violent crimes and 653 homicides occurred in Chicago.[9] Johnson was not an ordinary teen; he volunteered his time and effort to fight gun violence in Chicago as part of a group called GoodKids MadCity. It was with sad irony that Johnson had dedicated a good portion of his young life to encourage others to move away from violence. The death of Delmonte Johnson is yet another statistic in a larger story of rampant, unrestrained violence on the streets of Chicago. Each death represents a community of grieving friends and families. Families of those who died and residents who want to protect their own families are begging for help. Community leaders are calling for stricter gun laws and a declaration of martial law to take over the city's law enforcement.[10] Despite pervasive antipolice sentiment among residents impacted by violence, their willingness to have outside military intervention is evidence of their desperation for change. Although martial law has not yet been declared, residents and community leaders continue to ask the Trump administration for federal intervention in the form of boots on the ground.[11] In a highly controversial response, President Trump called for increased stop-and-frisk, an aggressive and invasive law enforcement practice similar to liberal stop-and-frisk practices in New York City.[12]

Mayor Rahm Emanuel said yes to federal help but no to the National Guard. In July 2018, Illinois's attorney general, the city of Chicago, and the Chicago Police Department drafted a **consent decree** agreement that would grant independent federal oversight of the Chicago Police Department. Consent decrees are mutually binding agreements between two or more parties, which allow federal courts to require oversight and enforcement of the agreement. For the most part, we hear of consent decrees when police officers have egregiously crossed the line in terms of serious police misconduct, abuse of force, and civil rights violations. The Chicago decree agreement, following the death of Delmonte Johnson, included federal oversight in cases involving use of force as well as outlining recommendations for police officer supervision, promotions, accountability, and oversight; implementation of community policing, impartial policing, crisis intervention, officer assistance and support, data management, and guidelines for the role of the independent federal monitor.[13] Considering that the public was calling for military response, it is surprising that community policing was one of the recommendations. We might wonder why community policing would be effective in what is essentially a war zone. The sad truth is that whatever the police are doing is not stopping the flow of blood in Chicago and many other communities across America.

The claim that community policing might be a viable solution to the violence illustrated in the Delmonte Johnson killing more so than traditional methods is an important assertion to consider. Violence is just one of many issues facing our communities today. In this chapter and throughout the text, we identify community issues and highlight the advantages of community policing to address problems and find long-term solutions. For example, community problems include the ravages of the opioid epidemic, homelessness, illegal immigration, poverty, unemployment, single-parent households, increasing signs of disorder, fear of crime, quality of life issues, and social disorganization. Under the traditional police model, such concerns were not considered to be the purview of police. Decades of scholarly research into the causes of crime do show, however, that these issues have a correlation to crime and, therefore, should be of concern to police. After all, police have marketed their services to the public as the experts in crime fighting. The public are encouraged to call 911 to initiate the response and then get out of the way so that police can handle the situation. Unfortunately, much of what police do is ineffective at addressing crime and other community problems. Following a brief comparison of traditional and community policing, we identify where traditional policing has failed and why community policing may be the answer to problems associated with traditional policing.

TRADITIONAL VERSUS COMMUNITY POLICING

To appreciate the extent of differences between community policing and traditional policing, we must first understand strategies and philosophies of traditional policing. Traditional aspects of policing continue to persist, both good and bad. Traditional values and practices have contributed to the dissonance between citizens and their police—something community policing was specifically designed to address. In both traditional and community policing models, police fight crime, make arrests, and use lethal force when necessary; however, community police officers do so with intentionality, and specifically, with long-term solutions in mind. Community policing, while still using traditional tactics, is smarter law enforcement.[14]

It could be argued that the death of Delmonte Johnson and the continuing carnage in Chicago demonstrate the failure of police to prevent crime, respond to crime, and solve crime. In the blame game, it is easy to point fingers at police, thus removing any responsibility of citizens and others to address social disorder, dysfunction, violence, and crime. Many believe that crime is the purview of law enforcement and not of the community. Both the police and the public believe that crime is solely a police matter. For many decades, the belief that police alone can resolve all community issues has been challenged. Under the community policing model, police and the public share the responsibility for resolving community issues and must do so in collaboration.

While aspects of traditional policing will likely be forever engrained in policing, the same could be said about community policing. Once community policing was introduced, it is unlikely that there will be a return to the former model without at least some inclusion of community policing principles. Traditional policing, however, does remain the standard of the profession both in organization and operation. Certain crimes and situations require full law enforcement response and that is not likely to change.

The community policing movement is touted as the most comprehensive police reformation in police history. While rising crime rates were worrisome, the most significant and urgent concern was the animus between police and the public, especially in racial-ethnic minority communities. Rebuilding trust and promoting police legitimacy were central tenets of police reform. In this chapter, we explore the definition of community policing and outline some of the key principles. What is community policing and how does it differ from traditional policing? One important difference between traditional and community policing involves the establishment of police and community partnerships, very much absent from traditional policing. Let us now look at traditional policing and outline some of its key attributes and limitations.

Traditional Policing

The traditional policing model followed on the heels of the political era of policing, which witnessed a rift with the corrupting influences of politics and the public. Police had become puppets for politicians' special interests, often accepting bribes and doing their bidding, even campaigning for them. Addressing corrupt practices in local, state, and federal government paved the way for police reform. The professionalization of policing began in earnest, including greater standardization of hiring and training, formalized policies and practices, and the advent of motorized patrol. To a great extent, the police became a respected and professional organization following the reform from the political era, which is why the traditional model is often referred to as the professional model. The downside of the transition, however, was the separation of police from the public they served. That separation formed the impetus for the community policing model.

Despite the move to community policing, public perceptions of policing reflect the traditional model. Moreover, police officers hold similar views. Traditional policing is the model that is depicted in the media, both in entertainment and the news, fortifying the view that police are gun-wielding, badge-heavy, action figures who fight crime and arrest bad guys all day long. Two important and recognizable features of traditional policing are, first, its organizational structure, and second, the crime-fighting cornerstones of policing: preventive patrol, rapid response, and investigations.

Organizational Structure of Traditional Policing. The organizational structure of traditional policing is very rigid and well defined. Policing is paramilitary, hierarchical, and shift-based. What does it mean to say the policing is paramilitary? A **paramilitary** organization is a semimilitarized force whose organizational structure, tactics, training, subculture, and function are similar to those of a professional military but which is not included as part of a state's formal armed forces. There are clearly defined lines of communication, policies, authority, and responsibilities. Similar to a military force, police use the designation of ranks—for example, captain, lieutenant, sergeant, and corporal—and wear recognizable uniforms with badge and gun. A distinct **chain-of-command**, whereby power and authority reside at the top and delegate downward, marks the organizational structure. The patrol officer is at the bottom of the hierarchy and would be expected to report only to the rank directly above them. Accordingly, it is against protocol for an officer to walk into the chief's office and complain about a shift, a fellow officer, or other matters. That officer must initially report to their sergeant, who, in turn, may then take the matter to a higher level. The hierarchy is characterized by a **unity of command** whereby an officer only has one boss, one commander. For example, a police officer assigned to a specialized unit such as SWAT (Specialized Weapons and Tactics) would not answer to a commander of the traffic division during an incident involving a SWAT response. Discussed later in the book, we will see that the strict hierarchical management model is not conducive or ideal to a community policing model, which calls for greater involvement and decision making at the police officer rank, asking officers to take risks and be innovative. The paramilitary organizational structure is an enduring aspect of policing that is unlikely to change significantly

▶ **Photo 1.1** What other aspects of policing could be considered paramilitary?

because it provides control, discipline, uniformity, accountability, loyalty, and a certain amount of predictability in outcome.

Crime-Fighting Focus of Traditional Policing. In addition to the hierarchal and rigid organizational structure as a recognizable feature of traditional policing, the second attribute is the focus on crime fighting. While crime fighting is an important part of their job, police officers under the traditional model tend to devalue other duties. The strategies police promote are based on three foundational pieces of policing called cornerstones. In this next section, we will discuss how those cornerstones reinforce traditional policing. Despite research that challenges the value and efficacy of these cornerstones, they are considered indispensable and fundamental to policing.

Cornerstones of Traditional Policing. The three cornerstones of traditional policing are preventive patrol, rapid response, and investigations. Police have three opportunities to impact crime. First, they can prevent it from happening through deterrence, usually by their mere presence in a neighborhood. Second, when they respond to a call of a crime in progress, they can intervene and stop the criminal activity, such as in a case of domestic violence. Lastly, the police can solve the crime after it has occurred through investigations. Therefore, if the police were neither effective at preventing the occurrence of a crime nor successful in its intervention, what do you think the chances are that police will be able to solve crimes at a later time, sometimes decades later?

The first cornerstone involves the ability of police to deter crime before it happens or discover criminal activity when it is happening. Preventive patrol involves walking or driving around an area with the goal of discovering and/or deterring criminal activity by increasing police presence. Preventive patrol, also known also as random patrol, occurs in geographical areas in the city or county called a beat. The **beat** is a geographic area with set boundaries such as streets or buildings. The number and/or size of beat areas in a law enforcement agency's jurisdiction are subject to change due to factors such as population growth, recession, and demographic shifts. For example, in Southern California, unrestrained population growth contributed to the housing boom of the 1990s, thus increasing the demands placed on policing services. Police officials divide the city into beat areas, determining deployment needs based on population density and crime rates. Officers, then, are assigned to beats with specific jurisdictional boundaries and authority. Officer Smith, for example, is assigned to Beat 3 during the day, and Officer Brown is assigned to Beat 3 for the swing shift (or afternoon-to-night shift). Several officers may be assigned to a specific beat and shift depending on the needs of the community. Officers will have partner officers within the beat area who respond together when the situation warrants additional units. At the start of their shift, after briefing, police officers leave the station and proceed to their beat area where they will patrol and await calls for service.

The Kansas City Preventive Patrol Study examined the effectiveness of preventive patrol. Law enforcement communities consider patrol the backbone of policing. Billions of dollars are spent on the deployment of uniformed officers in marked patrol vehicles with the objective of deterring crime. Until 1972, that assumption had never been challenged. The Kansas City Preventive Patrol Experiment was launched in October 1972 and was conducted till 1973. In this study, 15 beat areas were divided into three groups. The first group of five beat areas responded to calls but did not patrol. The second group of five was *normal* patrol with no changes. Finally, the third group of five beat areas was *proactive* where patrol was intensified by two to three times the normal rate. Data collected included: victimization surveys, reported crime rates, surveys of residents and business in the areas, arrest data, and trained observers. The research questions posed by researchers were:

- Would citizens notice changes in the level of patrol?
- Would different levels of visible patrol impact crime and victim surveys?
- Would citizens' fear of crime and change of behavior be apparent?
- Would citizens' degree of satisfaction with police change within the areas?

The findings revealed that citizens did not notice the difference in patrol level, there was no visible impact on increasing or decreasing crime, and citizen satisfaction with police did not vary in the three groups. Trained observers on ride-alongs noted that police spent a considerable amount of time waiting on calls for service rather than interacting with citizens or patrolling.[15] Interestingly, these findings did not motivate change by the police administration. The perceived positive value of patrol, despite the lack of evidence of its success to reduce crime, prevent crime, or increase citizen satisfaction, means there is little incentive to revamp or divest themselves from this practice. Surprisingly, both wide acceptance of the findings and equally fervid criticisms of its methodology can be found. The experiment has not been replicated; However, it is both valued and hotly debated since its publication in 1974.[16] It would take years before the law enforcement community would reassess its commitment to this practice, reevaluate it, and include aspects of community engagement into new and alternative patrol strategies (e.g., bicycle patrol, foot patrol, and mounted patrol).

The second cornerstone of policing is **rapid response.** An obvious symbol of traditional policing—the patrol car—revolutionized policing, allowing officers to respond quickly to the crime location. Over the years, that notion became the "quicker the better." It was believed that rapid response to 911 calls was the mark of effective policing. Research revealed, however, that less than 5% of the time, rapid response resulted in an arrest. The Kansas City Experiment in 1974 showed that there was no significant impact on crime deterrence, citizen fear of crime, community attitudes toward the police, or police response time.[17] No evidence supported the notion that rapid response either increased apprehension rates or decreased crime.[18] Findings suggest that delay between the crime and a reporting party's call to police was the problem, not when the police got to the scene. Delays in reporting a crime may result from a number of reasons. For example, the affected person may not be sure that a crime has been committed; they may call a friend instead of the police; they may be unable or unwilling to call; or they may not discover the crime at or near the time it was committed. Under the traditional policing model, the public insists on a timely response regardless of the nature of the call, and when police take too long, the public understandably is dissatisfied. In addition to dissatisfaction of citizens when response times are long, responding to every call quickly could mean that police would have little time to spend on each call; thus, the quality of service would suffer as well.

The use of rapid response coincided with the reliance on the 911 system. The connection between 911 and rapid response comes from the expectation and belief by both the public and police that 911 calls are inherently urgent. Since the first 911 call in Haleyville, Alabama, in 1968, 99% of people in the United States have access to the 911 system. People call 911 instead of calling the police on another phone line, even when the matter is not imminent or immediate, thus initiating rapid response. It is estimated that 80% of the calls to 911 are nonemergency requests for service.[19] Dispatchers must sort calls and prioritize them before giving them to police officers in the field. Most agencies prioritize calls for service by using a number system as a kind of shorthand to officers for the type of response needed. For example, Priority Three calls require that officers respond when they are able, Priority Two calls require that officers respond quickly but not urgently, and Priority One indicates great urgency, necessitating use of lights and siren by responding officers. When calls stack up, officers take high-priority calls first, pushing Priority Two and Three calls to the back of the line. In some situations, an officer may request additional help on a call. When fellow officers hear that request, they respond quickly even when the original call does not warrant the high priority and despite not being dispatched to assist. Of course, police officers do follow departmental policy; however, they will often break policy if necessary due to the unwritten code of solidarity among police officers.

Rapid response has been sold to both the police and the public as something police should do; however, the price of such a practice may be too high. A concerning downside to rapid response is officer and public safety. Are fatalities and serious injuries of police and members of the public worth the off chance of catching the bad guy? Evidence, both

▶ **Photo 1.2** Five police cars were involved in a road accident during a vehicle pursuit in Chicago, Illinois.

anecdotal and empirical, challenge that assumption. Rapid response in cases where there is imminent threat would be warranted; however, the possibility of creating additional threat or bodily harm must be carefully assessed. Stories of the horrific consequences of officers involved in traffic accidents while responding to or chasing bad guys make the news too often. Since the initiation of patrol in marked police cars, traffic-related deaths have continued to climb. Data from the Federal Bureau of Investigation (FBI) and the national nonprofit that tracks police deaths in real time show that traffic-related incidents are the leading cause of death among police officers.[20] Interestingly, most police officers who died in automobile crashes were going too fast for the conditions and nearly 60% were not driving with lights and siren.[21] In the past 10 years, more than one officer per week has been killed in crashes (2006–2016 = 64 deaths per year). Some of the causes for the fatalities included not wearing a seat belt, speeding, being distracted while using the mobile data terminal (MBT), and/or experiencing tunnel vision from increased stress.[22]

Public safety is also at risk for those same reasons. In one tragic example, a Somerset, Massachusetts, police officer killed 20-year-old Hailey Allard in a traffic collision while responding to a call of a car burglary. Despite his own injuries, the officer rendered aid to Hailey before they were both transported to the hospital where Allard succumbed to her injuries.[23] In another case, the police chief of the South Bend, Indiana, Police Department, said an officer should be fired for killing 22-year-old Erica Flores, when he drove through a red light while responding to a reckless driver call.[24]

THINK ABOUT IT: DO OFFICERS NEED TO RESPOND TO CALLS RAPIDLY?

Police officers may be authorized to respond with emergency lights and siren; however, they have an obligation to use due care for public safety. **Due care** is conduct that a reasonable person will exercise in a particular situation, in looking out for the safety of others.[25] In a case currently being adjudicated in the courts, the officer responding to a call with lights and siren had a responsibility to ensure public safety of the people at the call and in route to the scene. In this example, a police officer is suing the mother of a six-year-old boy he killed while responding to a call, which many people find offensive that the officer would add insult to injury to the family. Albuquerque police officer Jonathan McDonnell claims Antoinette Suina turned in front of him and failed to yield to an emergency vehicle. Suina's son died and her daughter was seriously injured; however, the officer claims that he too was harmed and is currently on disability. In his nine years as a police officer, McDonnell has been disciplined for six prior driving-related incidents, including one unauthorized pursuit. He was responding at 80 mph and entered the intersection where he hit Suina's vehicle, even though the call had been downgraded from a Priority One to a Priority Two call. The original call reported that a man was threatening people at a supermarket with a machete. An accident reconstructionist found that the officer was going too fast for the conditions and hit Suina's vehicle at a calculated 67 mph, in a 40-mph zone.

What do you think?

1. Did the officer have the right-of-way and the legal authorization to speed?
2. Did this officer exercise due care for public safety?
3. When would responding rapidly be worth the risks to public safety?

We will learn that community policing practices do not negate the need for rapid response; however, one major difference is the way police services are evaluated. For traditional policing, rapid response has been the mark of effective policing; however, for many reasons, the reliance on and importance of rapid response may need to be reassessed and downplayed.

The third cornerstone of tradition policing is investigations. The third opportunity for police to impact crime is by solving the crime after it occurs. All law enforcement agencies have investigative units. The effectiveness of traditional policing services are measured on how many crimes are solved. This method measures investigative effectiveness. **Clearance rates** are all the crimes solved by an arrest. For example, if 30 bicycles are stolen, and an arrest is made, that solves all those thefts, and the police will record that 30 crimes were solved. One can see the incentive to have good clearance rates, because it makes the police department look as though it is doing a great job. To be sure, we do expect that police will solve crimes. However, let's take a look at the problem with reliance on investigative clearance rates as a measure of police effectiveness overall.

The Rand Study (1973) examined investigative units, how they were organized and managed, in order to assess the contributions to overall police effectiveness. Most studies focus on police activity in the field, police officers, police practices, and policies, but no studies had examined investigative units. Among the objectives of the study, researchers wanted to assess the contributions of police investigations to crime, arrest, and clearance rates. Another objective of the study was to examine staffing and productivity of investigative units. Data were gathered from all municipal or county agencies that had more than 150 officers employed and where the jurisdiction's population was over 100,000.

Additionally, interviews and observations were conducted on more than 25 departments. Researchers looked at samples of case outcomes from the Uniform Crime Report (UCR) compiled by the FBI. The findings of the study suggested that investigative units were not effective in solving crime. Stated as one of the major findings about investigative effectiveness, differences in investigative training, staffing, workload, and procedures appear to have no appreciable effect on crime, arrest, or clearance rates (p. vi). In fact, study recommendations included a reduction of half of the investigative efforts and a reassignment to more productive uses. While an overly simplistic summary of the findings is presented here, there should be greater scrutiny on investigations and the role they play in measuring the effectiveness of police services.[26] The study and policy implications were not well received by police officials due to heavy reliance on investigative units.

The three cornerstones of traditional policing—rapid response, preventive patrol, and investigations—have been challenged; however, they persist as attributes of policing today. Traditional policing is marked by insular, siloed communication and command, meaning that police officers work independently with little or no supervision. They are call-answerers, responding to incidents, writing reports, and waiting for the next call. A single call, especially one that involves an arrest, can take an officer out of service for four hours or more. The incentive is for officers to handle the call and go "10–8" (back in service) so that they are available for the next call. In the traditional model, police are adept at assessing the situation, determining if it is a police matter, and resolving problems quickly and efficiently. A police matter is one that entails criminal activity; nothing else is police business according to traditional model thinking.

Traditional policing is about "crook-catching" and locking up "the bad guy." It is about fighting crime with limited focus on long-term problem solving. That image of catching the bad guy, good versus evil, the excitement and danger, featured in media portrayals of police, draws people to a career in police department and perpetuates traditional perceptions. The reality is that police do fight crime, and they do catch bad guys. However, police spend a lot less of their time on law enforcement–related activities than on other calls for service.[27] Moreover, when they do make arrests, they usually arrest for property crimes rather than violent crimes.[28] Therefore, the notion that traditional policing is only about catching bad guys is false because it precludes a majority of police activities. Ironically, what police do in the traditional model resembles much of what police do in the community policing model—they solve problems in the community. What the traditional model does not do, however, is to recognize and honor the full capacity of the job. What the community policing model effectively does is give more credit to the majority of duties that do not necessarily include crook-catching.

Community Policing

Now that we have a better understanding of traditional policing, we are ready to explore community policing. What is community policing? Some might argue that community policing is whatever a particular agency says it is. In a series of articles about community policing, Public Safety Director Tom Casady of the Lincoln Police Department remarked:

When an agency claims to have "implemented" community policing last week, that's a pretty good indication that it has not. Individual programs or projects that form part of this change may be implemented, but community policing is not implemented. You don't start it at the beginning of the fiscal year. It is a process that evolves, develops, takes root and grows, until it is an integral part of the formal and informal value system of both the police and the community as a whole. It is a gradual change from a style of policing which emphasizes crime control and "crook catching," to a style of policing which emphasizes citizen interaction and participation in problem solving.[29]

It may be true that some agencies do claim to have community policing but do not, while others are fully engaged in community policing. However, over the past few decades, a clearer definition of community policing has emerged. Additionally, there is greater detailed and organized protocol, including the notion that not every innovation constitutes community policing.

While some police experts focus on understanding what community policing is, others focus on dispelling the myths associated with community policing—that is, that it is not soft on crime, it is not a program, it's not a panacea, nor is it a Band-Aid for all problems.[30] Myths come about from a lack of understanding or exposure to community policing and those myths generate resistance, especially among the rank-and-file officers who are essentially the purveyors of community service ideals. One officer, when asked about community policing, remarked, "We don't have time for that crap" (conversation with officer by author). It is easy to dismiss something when it is not clearly articulated and when you do not have the buy-in of critical players, especially the police officers.

As community policing programs and concepts have evolved, each development brought excitement and discoveries, and in some cases, new directions. Greater knowledge led to more intentional and systematic integrations and implementations of community policing. In part, the community policing movement included the creation of a federal unit to support the police reform. The U.S. Department of Justice Community Oriented Policing Services (COPS) was formed in 1994 to advance community policing. The objectives of COPS include funding and resources, such as training, to assist agencies in the implementation process as well as providing evaluation feedback of existing community policing programs. In fact, in order to obtain and continue to receive funding, law enforcement departments must comply with all requirements of COPS, which include periodic progress reports. Foremost, COPS has had played a large role in tailoring the definition of community policing into operational components.

COPS Definition of Community Policing. The definition of **community policing**, according to COPS, involves a detailed overview and identification of key components. Community policing is

> ... a philosophy that promotes organizational strategies that support the systematic use of partnerships and problem-solving techniques to proactively address the immediate conditions that give rise to public safety issues such as crime, social disorder, and fear of crime.[31]

The three key components of community policing are community partnerships, organizational transformation, and problem solving. **Community partnerships** include collaboration among law enforcement agencies, individuals, and organizations they serve to develop solutions to problems and increase trust. **Organizational transformation** is the alignment of organizational management, structure, personnel, and information systems to support community partnerships and proactive problem solving. **Problem solving** is the process of engaging in the proactive and systematic examination of identified problems to develop and evaluate effective responses.[32] Let's take a closer look at these components and discover how they work in the community policing model.

Community Partnerships. Partnerships refer to consultation and cooperation between the residents and the police. It also includes collaboration with all stakeholders in the community. Partnerships include legislative bodies, prosecutors, probation and parole, public works departments, neighboring law enforcement agencies, health and human services, child support services, city or county ordinance code enforcement, local business, churches, and schools. The media is also an important partner. In a later chapter, media relations and role will be discussed in depth. Police rarely solve crimes or facilitate solutions to community problems without the help of others.

Partnerships between police and the community mean that communication must work both ways, where police share information with the public and the public share information with the police. Early iterations of community policing began as a public relations campaign to inform the public about what police wanted them to know. True community policing is about a dialogue between key stakeholders and police. Input from the community helps police respond to problems more effectively. Additionally, the residents are accountable and responsible for their role in public safety. Under the traditional model, the public were considered passive receivers of police services and were viewed as a source of information when needed. Often police officers considered citizens as obstacles to important police work. Community policing is an opportunity to mobilize community support and assistance. Partnerships are formed through face-to-face contact and through Neighborhood Watch and other programs. More than ever, police departments use social media and other media sources to engage and inform residents and businesses about problems pertinent to their neighborhoods.

Under the community policing model, police are expected to build relationships and trust with the public, especially in areas where historic animosity existed. Under the traditional model, police legitimacy was something bestowed upon the police as a legitimate arm of government. Today, police have to earn the respect of the public and are quick to lose it when there is a controversial incident like an officer-involved shooting. Now, policing has become more responsive, more sophisticated, and the organization has had to make changes. Later in the text, creative alliances with special populations are presented. For example, police are now promoting positive relations with juveniles through teen police academies, where teens engage with police officers and learn about policing. Such programs would not be possible without empowering officers to think creatively and to work closer with people in the community. In order to empower officers in the field and bring about a higher level of commitment to the communities they serve, there must be an emphasis on internal changes within the police organization.

Organizational Transformation. Any organizational change, whether large or small, is challenging, but especially when much of the values, policies, practices, and norms are institutionalized. Because community policing involves a different type of response to crime as well as a different relationship with the public, the organization must be transformed to support new objectives. COPS defines organizational transformation as the alignment of management, structure, personnel, and information systems to promote community partnerships and proactive problem solving.[33] The restructuring comprises a shift in philosophy for the entire agency, encompassing climate and cultural transformation of the leadership and personnel. Another important aspect is decentralized decision making and accountability, which allows officers to take ownership and to form relations with members of the community, and to feel empowered to be creative and take risks in finding solutions for crime and disorder. Organizational transformation requires clear strategic planning, policies that articulate the community policing values, measures of police performance that include community satisfaction, address fear of crime, and focus on quality of life of citizens, and establishing greater transparency with the public.

Organizational transformation includes geographic assignment of officers to establish strong relationships, enhanced customer service, and mutual accountability. Some of the original iterations of community policing efforts involved specialized units; however, this was found to create a competitive atmosphere as well as dual police services that were at cross purposes. For example, while community policing officers were trying to form relationships with at-risk youth, traditional officers applied aggressive tactics to gain compliance through fear. Newer implementations of community policing are department wide and not limited to specialized units.

Recruitment of community-minded individuals to serve as police officers is another objective that needed change for the organization. Decades ago, police officers were recruited and hired with traditional practices emphasizing law and order, command and control, and crime fighting. Today, as officers hired with those traditional perspectives are retiring or promoted to administration, new officers are hired in the era of community and problem-solving policing. Organizational transformation promotes the concept that officers should be selected with community policing values in mind, encouraging the hiring of individuals who come to policing with a "spirit of service" rather than a "spirit of adventure."[34]

Organizational transformation embraces new technologies to enhance problem solving. Information systems technology enhances data-driven decision making and problem solving. Today, computer-aided dispatch, MDTs in police units, reverse 911, online reports, interactive mapping, email alerts, and other new technology facilitate problem solving in the field. In recent years, police have adopted new technologies and are in use today—license plate readers, iris and face recognition, body-worn cameras, DNA, biometrics, robots, drones, thermal imaging, and artificial intelligence.[35] To address the backlash by the public regarding fatalities by police, much effort is being put into less-than-lethal weapon development. All these innovative technologies, including nonlethal and less than lethal weapons, are tools police currently use, making law enforcement safer and smarter. Along with their tools, police are also using problem-solving methods that look for long-term solutions to community problems.

Problem Solving. The third key component, problem solving, is the new way of thinking about crime and disorder, not just reacting and responding to apply a Band-Aid for the short term but also for prevention, deterrence, and long-term resolution. Problem solving is the process of engaging in the proactive and systematic examination of identified problems to develop and evaluate effective responses. The problem-solving aspect of community policing is proactive rather than reactive. The emphasis on proactive problem solving encourages agencies to develop solutions to the immediate underlying conditions contributing to problems and issues in the community. The attention to the underlying conditions is what makes the difference over time, with the goal of reducing or eliminating the threat or crime altogether. Problem-solving methodologies guide decision making and support innovative solutions. While arrest is not ruled out, it becomes one of the many tools available to police rather than a solution to a problem. A popular, widely used problem-solving tool employed by police is the SARA (scanning, analysis, response, and assessment) problem-solving model. This model, more than others, has been the one most recognized and implemented by countless agencies. SARA is discussed in detail in the Chapter 3.

The definition and description of community policing's key components give us a sense of how different this model is from traditional policing. For example, community policing encompasses the formation of community partnerships, transformative changes in organizational structure, management, personnel, information and technology, as well as the implementation of problem-solving strategies. However, the definition tells us little about what it is that police do differently and why there is a need to do things differently. In later chapters, concrete examples will be presented to showcase innovative programs and why these programs were important to the community. For now, we will examine principles of community policing that serve as the foundational underpinnings of successful programs.

PRINCIPLES OF COMMUNITY POLICING

Thus far, we have defined community policing and identified its key components. In this section, we lay out the core principles of community policing. Two of the most prominent proponents and pioneers of the community policing movement, author Robert Trojanowicz, and co-author Bonnie Bucqueroux, set forth 10 principles of community policing. Throughout the past 30 years, these principles have remained applicable despite changes in technology and information systems and kept pace with the evolution of policies and practices in policing. Below is a summary of these principles:

1. Community policing is a philosophy and organizational strategy that emphasizes new ways to solve community problems of crime, fear of crime, and disorder. Law-abiding citizens should have input into police decisions.

2. Everyone in the department, both sworn and nonsworn personnel, must be committed to operationalizing the philosophy into practice. Enlisting the help of the public in their role in policing themselves and by using creative means to solve community problems.

3. Assigning an officer to serve as a Community Policing Officer (CPO) who works directly with the public, through face-to-face contact and who is not held to the isolation of the patrol vehicle.

4. Giving the CPO more time to commit to sustained contact with citizens in the community in a cooperative way to explore creative means to solve community issues. CPOs act as the ombudsman for the community, connecting people to services in the community.

5. Community policing will imply a new contract with the public, promoting greater public participation, less apathy, and less impulse for vigilantism. It also ramps down response time to noncritical incidents, giving police more free time to find long-term solutions.

6. Community policing promotes a proactive component whereby police officers will work to prevent problems rather than just respond to them, although responding to immediate concerns will remain a priority. Community policing expands the police mandate from a hyperfocus on crime toward addressing smaller issues. The intent is to have a greater impact on the quality of life issues facing communities.

7. Community policing stresses the importance of working with vulnerable populations, for example, juveniles, people of color, older people, people living in poverty, people experiencing homelessness, and people with disabilities with the goal of enhancing outreach efforts to these groups.

8. Although technology is important, community policing demands that it not get in the way of human contact and collaboration. It promotes people working together to find new approaches to community issues.

9. Sharing information about the community and its important issues should be understood by every member of the police department. Everyone in the police department should be onboard and supporting community policing objectives.

10. Community policing is a new way of thinking about the police and what they do. The public should see the police as a resource to be used to help solve problems in the community. This new way of viewing the police should not be thought of as a tactic to be abandoned for the next new approach, but it should be incorporated into the organizational matrix of the department. Police should also be able to modify responses and strategies as needed. (pp. xiii–xv)[36]

These principles demonstrate the unique philosophy of community policing. The entire organization must be committed to the objectives to make it successful. There were problems associated with the establishment of separate units of CPOs to perform specialized tasks and to serve as liaisons between "traditional" officers and the community. This divided force resulted in officers working at cross-purposes. For example, officers were trying to establish relationships with the community, while traditional officers continued to view the members of the community as potential threats. The problems associated with two different types of officers on the same force drove home the principle that everyone in the department needed to be on board with community policing. Today, many departments do employ nonsworn CPOs to extend the services of the police; however, CPOs do not work in the same capacity as the officers themselves.

CONTRAST OF TRADITIONAL AND COMMUNITY POLICING MODELS

Traditional policing is not undone by community policing but is enhanced by allowing officers to be innovative, to use their talents, to collaborate and form relationships with the community and with key stakeholders, not feel less powerful but more supported, and ultimately, more fulfilled in their role as police officers. Three particular aspects separate traditional from community policing models. For one, under the community policing model, police have a very different relationship with the community. Second, police have a broader mission than just crime fighting. And third, community policing involves new ways of solving problems with a focus on long-term solutions.

Police and Community Relations

How does community policing succeed in areas where traditional methods have failed? Above all else, it is the formation of a productive relationship between the police and the public. Traditionally, police are not good at sharing power and authority, especially with civilians. They are trained to take control, be powerful, issue orders, and expect compliance. Collaboration between police and citizens was neither expected nor desired in traditional policing. Under the community policing model, working with citizens in identifying, prioritizing, and solving problems is necessary and expected. It was the lack of collaboration, consultation, and communication that contributed to serious dissention, distrust, and dissonance between the police and the public. In fact, it was the animus between the police and the public that led to this reform movement in the first place. Today, police recognize the importance of gathering insider knowledge of situations, people, and idiosyncrasies in the communities. They understand that the lack of information makes their jobs more difficult.

Expanding the Police Mandate. They have the expertise to handle matters that, in many circumstances, the average citizen is not prepared to solve on his or her own. Under the traditional model of policing, police preferred to circumvent noncriminal issues by referring people to civil avenues of problem solving. For police, at that time, crime fighting was their primary mandate. How does a community policing model change that approach? Community policing broadens the mandate to include handling other community problems, signs of disorder, fear of crime, quality of life issues, and crime in a partnership with the community.[37] **Fear of crime**, a fear of being a victim rather than the actual likelihood of being victimized, was not viewed as a legitimate police concern. In the past, police did not believe that fear of crime was a tangible, solvable problem. Under the traditional model of policing, officers would say to a citizen: "If something is stolen, call us. If you are afraid of it being stolen, don't call us." The Flint, Michigan, Foot Patrol Study, conducted in the early 1980s, was one of the first studies that examined the impact that fear of crime has on people and the detriment to the quality of life. This study had a huge influence on the move toward community policing and will be discussed in more detail in Chapter 2. Under the community policing model, police believe fear of crime is a legitimate concern, and they provide a sense of safety and security to the community by validating those concerns with empathy and action.

Contrasting Problem Solving, Information Systems, and New Technologies

Differences between the two models involve some aspects that are not necessarily due to different philosophies but have more to do with modernization of equipment and information systems. Police using problem-solving strategies can tap into highly sophisticated data collections and analytics as well as innovations that were not even imagined in earlier policing eras. For example, DNA, drones, body cams, voice and face recognition, and license plate readers were all developed during the community policing era. There is some concern with the possibilities of overreach, widening the net, invasion of privacy, big brother oversight, and perhaps, even social distancing—all problems community policing hoped to address with improved relations.

Improvements in Problem Solving. Community policing encourages officers to come up with creative solutions to address a vast array of problems, from disorder to murder. For example, innovative solutions for the handling of domestic violence may include follow-up visits and home checks (see Case Study 1.1). Community policing pushes a more intentional and focused approach to problem solving, which includes targeting particular crimes, social disorder, and fear of crime as well as looking for long-term solutions. Problem-solving strategies are aimed at both urgent and critical incidents, such as robberies, assaults, mass shootings, as well as persistent circumstances, such as drugs and homelessness. In future chapters, we will present some of these programs aimed to tackle endemic community problems.

Predictive Policing and Information Systems. Problem solving is not new; however, police departments are employing information systems, data collection, and analysis to make fully informed strategic decisions for long-term solutions. Predictive policing is a method of data collection and analytics to target current and future crime trends.[38] Police can work smarter. Even under the traditional model, departments collected huge amounts of data, but such data were not effectively used and applied. Even crime mapping systems have become so sophisticated, and they are able to provide instantaneous updated information for operations and deployment. Larger law enforcement agencies employ specialized units of crime analysts to assist them with data collection and analysis to inform decision making.

New Technologies and Policing. Certainly, one of the most significant differences between traditional and community policing eras has to do with technological developments and less to do with philosophical transformations. Throughout history, police have adopted new technologies, sometimes with great reluctance and even resistance from the troops. Traditional policing was marked by the innovation of motorized patrol, and ironically, police departments in the community policing era reverted to foot patrol in an attempt to renew closeness to the community. Originally, police officers resisted things like body cams and dash cams because they provided a record of their encounters with the public. They believed the recorded data could be used against them. Later, they realized that this technology also protected them against false accusations. Such technologies fit into the need for greater

CASE STUDY 1.1: CHULA VISTA POLICE DEPARTMENT RECEIVES INTERNATIONAL AWARD FOR DOMESTIC VIOLENCE REDUCTION PROGRAM

One Southern California agency, Chula Vista Police Department, won top recognition at the 28th Annual Problem-Oriented Policing Conference, where it received the 2018 Herman Goldstein Award for Excellence in Problem-Oriented Policing for its study on domestic violence prevention and enforcement strategies. The police department, in partnership with South Bay Community Services, the San Diego County Probation Department, the District Attorney, Child Welfare Services, Adult Protective Services, crime analysts, and research partners, designed and implemented proactive strategies. The undertaking was a huge task to the small but busy department. After the first year, "the results were impressive," said Police Chief Roxana Kennedy. Rates of domestic violence in the research area dropped by 25%. Additionally, victims reported greater satisfaction of police handling and 92% said they feel confident to call police in the future if they needed help. As the chief noted:

> Domestic violence is one of the most common, dangerous, and frustrating problems facing our officers. Officers often find themselves dispatched to the same addresses again and again. It can be difficult for victims to escape these types of situations. There is a sense of futility and helplessness among both victims and police.

Traditional police strategies require mandated arrest of physical abusers and a list of resources to be supplied to the victim. The victims, often overwhelmed by their circumstances and fearful of their abuser, are unlikely to take action even with the list of helpful resources. The Chula Vista project focused on the offenders, giving them written warnings and continued follow-up at the residence with the offender and victim. The follow-up, repeated warnings to the offender put the offender on notice that the police were coming back. The officers checked on the victim and left notices for the offender if not present. This strategy gave victims more support and prevented further violence in most cases. Police Chief Kennedy stated, "I am very proud of the innovative work of our officers. This is just another example of the commitment our personnel make to keep our community safe."[39]

transparency of police and citizen interactions. Innovations of DNA, facial recognition, and the like provide better accuracy in convictions.

COMMUNITY ISSUES AND COMMUNITY POLICING

The nature of policing is such that officers do have to sort through problems and help facilitate resolutions, regardless of whether the situation is serious or minor, criminal, or noncriminal. Police officers, under the community policing model, partner with members of the community to identify problems, prioritize, and resolve them. Some issues are serious and may involve strategic, long-term problem solving, such as gangs, drugs, or homelessness. Other problems may involve immediate resolution, such as those involving domestic disputes. No issue is beyond the purview of community police officers, including problems of gangs, drugs, at-risk juveniles, violence, civil disturbances, homelessness,

THINK ABOUT IT: NEW GADGETS FOR POLICE, AND THE FUTURE HAS ARRIVED

Customers wait in long lines for release of new cell phones, often willing to pay big money for all the bells and whistles with new features and capabilities. Sometimes less than a year later, another product hits the market with even better features. It is difficult to keep up with technology and police agencies are no different. Although hindered by limited operational budgets, law enforcement agencies adopt technologies that make their job smarter and safer. Some of the new gadgets police employ today were nothing but science fiction back in the day when community policing was first introduced. The future is here, and in many ways, police can work smarter, safer, and quicker with less chance of wrongful arrests, convictions, or harm to others or themselves. The following are some of the newest developments in technology.

- *Drones*: Police are using drones as first responders to enter dangerous situations, to monitor emergency scenes, for example, a mass shooter or traffic accident. The drones can be used in search-and-rescue efforts to help find missing persons. Drones are used to surveil and map out highly trafficked areas where drugs or other criminal activity occur. They can also be deployed to assess bomb threats or hazardous material spills. And finally, they provide documentation for later use in court.[40]

- *Social media*: Social media is used in many ways and especially effective in the community policing era where relationships and partnerships are key to police effectiveness. Social media increases transparency, and disseminates information widely and quickly, and provides a venue for the public to ask questions. Police can use social media to provide tips for safety, road closures, and announce upcoming community events. Social media is employed to prevent or investigate criminal activity by seeking and receiving information about suspects and their whereabouts.[41]

- *Automated license plate readers* (ALPR): This system has had international adoption, making vehicle license plate comparisons easy and efficient. They utilize high-speed cameras either attached to police patrol vehicles or mounted in strategic areas to capture all plates that come into view. The data captured includes license plates, vehicles, and in many cases, people.[42] ALPR has successfully identified stolen vehicles and has led to the identification and capture of criminals.[43]

- *Biometrics*: Fingerprints and blood typing seem rather archaic when compared to the advances that have been made in the area of biometrics. Facial and voice recognition can instantly identify subjects in the field. DNA has improved to the point where even the smallest amount can render impressive information. Everyday there are advancements in the area of biometrics.

- *Domain Awareness System (the Dashboard)*: This system, developed by Microsoft Corporation and the New York City Police Department, connects officers out in the field with real-time information and offers pictures and videos of calls in progress as well as providing instantaneous analysis. Officers can make informed decisions prior to arriving at a call.[44]

There are several new technologies currently being developed, too many to list here. For example, voice-activated systems for the siren and other devices make it easier and less dangerous for officers to perform simple tasks. Body cameras and eye wear that record police/citizen encounters are currently being used. It should be noted that some of these inventions are being challenged for privacy issues. Much of our lives are no longer private, however, as our use of computers and social media increases, and law enforcement can take advantage of information that comes through social media and other computer technologies.

Continued

(Continued)

What do you think?

1. What privacy issues do you see arising out of some of these new technologies?
2. What are some of the advantages and disadvantages of new technology and community relations?

terrorist activity, active shooter situations, domestic violence, prostitution, sex trafficking, and serial killers. Ultimately, police using the community policing proactive model are looking for long-term solutions versus the traditional reactive model where police resolve the issue over and over again.

Unfortunately, because community policing it is thought to be "soft on crime," many departments abandon it during times of serious crime, such as the 1980s crack epidemic.[45] Community policing is often viewed as a luxury and not a necessity. Serious crime calls for serious measures. Currently, the move toward the militarization of police is in direct response to incidents of terrorism, mass shootings, and other serious attacks on our communities. This is not a time for community policing, some might argue. Surprisingly, this is exactly the time for community policing. Embracing community policing does not negate military tactics but incorporates military-type weapons and tactics into the toolbox officers have available to them. Throughout this text, we will discover the reason why community policing works, why it is the ideal response to serious crime, and why it is here to stay, as well as understand the challenges and obstacles going forward.

SUMMARY

We have learned that traditional policing is the default model of policing with its rigid paramilitary, hierarchical organizational structure, highly developed chain-of-command, and a focus on reactive, call-answering policing. Community policing is proactive, with the goal of preventing crime and disorder. Under the traditional model, people are labeled as victims, witnesses, or suspects. Under the community policing model, the community members are full-fledged partners with police, and they too have responsibilities to participate in addressing issues in their own neighborhoods. The community policing model softens the organizational structure allowing greater discretion at the patrol officer level, encouraging creative and innovative thinking. It empowers police and the public to form a bond of trust and resiliency to share information and work together. Community policing is proactive, whereby police address issues before they manifest into unmanageable problems.

A little bit of irony exists here in the fact that police, even under a traditional model, do handle a multitude of situations, both criminal and noncriminal. One might argue that community policing does not broaden the role, it broadens the understanding of all that police do. Moreover, the value of police and policing to the public is increased under the community policing model. For the most part, there is little difference in what police did in prior times and what they do now. In the past, police did handle all calls for service; however, only the calls involving crime gave them the status they desired. Handling calls involving crime fulfilled the expectations of the public as well.

In the next chapter, we will elaborate on the lessons learned from the implementation of experimental programs of police–community relations (PCR), team policing, foot patrol, and application of broken windows theory. We will also show how community policing incorporates the principles of Sir Robert Peel in 1829 London by proscribing police function and accountability to the public it serves. Reflecting back to the beginning of this chapter, regarding Delmonte Johnson's death in Chicago, we wonder what could have prevented the tragedy. The consent decree in Chicago recommended the implementation and expansion of community policing. It did not recommend the deployment of the National Guard. Both community and police need to work together to resolve the rampant violence in Chicago and across America. Police did fail to prevent Johnson's death, but the community failed as well. What can community policing do differently? The overarching objective of this textbook is to examine the claim that community policing is an effective model to address and resolve most community issues using smarter and more sophisticated strategies.

KEY TERMS

Beat 8
Chain-of-command 6
Clearance rates 11
Community partnerships 14
Community policing 13

Consent decree 4
Defund 3
Due care 11
Fear of crime 18
Organizational transformation 14

Paramilitary 6
Preventive patrol 8
Problem solving 14
Rapid response 9
Unity of command 6

DISCUSSION QUESTIONS

1. Discuss the concept of defunding the police. How would it benefit or harm the community and/or the police?

2. Identify and discuss the cornerstones of traditional policing? Why are they important to policing?

3. Define community policing. What are some of the key differences of community policing and traditional policing?

4. What are the principles of community policing? What principle do you think is most important? Which, if any, principle may be problematic and why?

5. What are examples of community issues and problems? How would traditional policing and community policing respond similarly or differently to these problems?

CHAPTER 2

COMMUNITY POLICING IN THE 21ST CENTURY

Learning Objectives

1. Explain how community policing evolved.
2. Identify and describe the three eras of policing.
3. Identify and describe the historical underpinnings of community policing.
4. Explain why community policing is not a passing fad.

COVID-19: TESTING POLICE AND COMMUNITY RELATIONS

The relationship between police and the community is imperative at all times, especially during a time of crisis. The current crisis caused by the Coronavirus, also referred to Covid-19, is a crisis that touches everyone, including police officers, health care workers, and everyone else who typically takes care of the safety and health of community members. In fact, the police and health care workers are on the frontline of this crisis. Much of the effectiveness of the police officers during a crisis depends on their ability to connect to the residents and convince them to comply with certain orders. Almost all states have implemented shelter-at-home orders for their residents and prohibit them from congregating and doing many other things that would be normal, such as having parties, and attending sports events, and weddings. Police officers must be able to effectively communicate with people who are not inclined to comply with these restrictions on their freedom. If police have failed to build relationships with residents, that communication becomes a lot more difficult and police–resident encounters could result in harm.

For instance, during the Covid-19 crisis, states, counties, and cities have imposed shelter-in-home orders, quarantine orders, self-quarantine orders, mandatory face-masks orders, and social distancing measures. For instance, in most states all nonessential businesses have closed. In many states, cities also closed beaches, parks, trails, and playgrounds. These closures are meant to help enforce social distancing measures during the Covid-19 crisis. Given that states such as California have a population of 40 million, police need citizens to comply. It would be impossible to arrest or ticket millions of people. Effective relationships between the police and residents will result in higher levels of compliance and avoid violent encounters or public protests. Community policing can lay the groundwork to establish effective police–community relations.

INTRODUCTION

In this chapter, students will learn about the different eras of policing and the political and social influences that have shaped community policing. We will discuss the different operational styles of policing and how community policing can improve the relationship between police and residents to help arrest criminals, keep the peace, and assist during a time of crisis. The chapter will also talk about the challenges police officers have faced and are still facing. The chapter will also discuss why community policing is not a passing fad. Additionally, we will examine what led up the community policing movement and why the shift to community policing was deemed to be necessary. Impetus for organizational changes comes from a variety of forces, both internally and externally. For example, police organizations are impacted by environmental factors, such as population growth and demographic fluctuations. Modern technologies create innovations in crime and criminal behavior, such as cybercrime, identity theft, hacking, and sex trafficking; or conversely, new technologies can be used to solve crimes and otherwise aid police such as facial recognition, DNA, infrared cameras, and less-than-lethal weaponry. Changes can also be forced on an organization through legal and legislative means. A glance back in history will reveal that many changes came about following civil uprisings against overreaching or corrupt government actions against citizens.

Transformative change often occurs incrementally and slowly so much so that people undergoing that transformation may not recognize that it is happening. An examination of the history of policing reveals many transformative milestones, such as the introduction of the radio, motorized patrol, and record-keeping, which modernized policing but also had a powerful impact on the transformation of police–community relations. On a larger scale, there is transformation in philosophy and values of an organization. Extraordinary occurrences brought about paradigm shifts in policing, resulting in distinct policing eras. A **paradigm shift** occurs when there is one set of thoughts, ideas, beliefs, values, and practices associated with an organization (in this case policing) that are challenged or discredited, and a new set of thoughts, ideas, beliefs, values, and practices replace the old ways. Police historians recognize three major eras in policing: political, professional (reform), and community policing. Because each era is presented in the chapter, the problems associated with each will be revealed to lend understanding of what needed to be changed and how it led to the next era. Second, the historical underpinnings of community policing will highlight how community policing evolved from individual stand-alone programs to a comprehensive ubiquitous era of its own.

THREE ERAS OF POLICING

An in-depth historical account of policing from its introduction to the present day will not be presented here; however, we will examine and identify aspects of particular time periods and discuss why change was indicated. There are three commonly acknowledged eras of policing: political, reform or professional, and community policing. These eras can be separated into three distinct time periods, each with noteworthy and unique attributes. Although certain elements are common to all three, such as the paramilitary organizational structure and mission of crime fighting, there were many differences as well. Each era improves on the previous one; however, much of what is wrong with policing has been institutionalized, thus resistant to change despite internal and external forces. We will examine how each era became defined as such, why certain aspects became problematic, and how those problems were addressed.

Law enforcement throughout the United States has a somewhat chaotic and uneven history because of the lack of standardization and its decentralized nature. The concept of **decentralized policing** means that police organizations are under control by regional divisions, that is, city, county, and state government, not necessarily under federal authority, although federal forces were established earlier than city police; for example, the U.S. Marshals Service was established in 1789.[1] The fragmented nature of American law enforcement differed greatly and purposefully from Britain's structure, where all law enforcement was tied directly to a central commissioner under Parliament's Home Secretary. Following the Revolutionary War, those charged with creating the United States of America wanted to move away from centralized governance and monarchy of England. The notion of a localized sphere of control and authority appealed to them. As various law enforcement agencies popped up across America, more distinct jurisdictional boundaries were established for handling the various types of crimes rather than simply handling crimes based on geographic areas. For instance, there are special units for sex crimes and homicides. However, some jurisdictional overlap, conflict, and/or collaboration of crimes and geography are inevitable among the many law enforcement agencies. An offender who engages in drug trafficking, for example, may be tried by the state in which they were arrested but also by the federal government because drug trafficking is a federal offense. Or, an offender who commits homicides in several states creates the dilemma of determining which state will prosecute the offender. You might remember the Washington snipers, John Allan Muhammad and Lee Boyd Malvo, who killed 10 people and injured three others in the Mid-Atlantic/Washington Area in October 2002. They killed people in Maryland and Virginia and after their arrest on October 24 in Maryland, prosecutors had to decide in which state to prosecute the two snipers. In the end, they were put on trial in Virginia because Virginia has the death penalty and Maryland does not. In such cases, states have to collaborate and resolve conflicts.[2]

Describing police history in three distinct time periods helps us recognize and identify dominant philosophies and strategies of the time; however, the history of policing in America is much more complex and dynamic, beginning with a very informal type of policing to the very formal policing we know today. For example, the introduction of police to American municipalities may be better understood from a regional perspective, for example, Northeastern, Southern, and Western expansion. Cities within the regions developed policing services for different reasons. The Northeastern beginnings having the greatest ties to England, and Sir Robert Peel developed a more rigid concept of the policeman with a uniformed, paramilitary ideal. In this region, police were needed to protect property of the landowners from transitory people, especially immigrants. Early Northeastern police were developed to protect the shipping industry in Boston, for example. Southern policing had its roots in slave patrols, preventing runaway slaves, slave uprisings, and enforcing Jim Crow laws. The move West, through an unforgiving and unsettled landscape, called for a more rogue style of law enforcement and the focus was on land claims and lawlessness. The regional perspective may seem to show differences by region; however, clearly the overarching impetus for policing was to serve and protect the rights and property of property owners from transient, low socioeconomic immigrants and non–property owners.

From the introduction to present, law enforcement in America has undergone significant transformation in terms of services, philosophy, strategies, and technology. Its most important aspect is the relationship with the public. Problems with police and the public interactions and relationships were the impetus for change. The three eras of policing have distinct characteristics that differentiate them from each other; however, as we will learn, one key difference concerns the relationship between police and the public they serve. In the first era, there was a close relationship but rift with corrupting influences.

Political Era

The earliest period was known as the political era. The political era, so called because of the political influence on police services, highlighted the sphere of control of police rather than on the role of police within the community. It lasted from the 1840s until the early 1900s and was characterized by the close relationship between police and politics. During this period, various interest groups influenced the police. Some have referred to them as the "adjuncts to local political machines."[3] Politicians recruited police officers and assisted with their promotion and, in return, police helped politicians in getting reelected and even assisted in rigging the elections. From our modern perspective, this all-encompassing political control would be viewed as corrupt. It was business as usual to the players and participants at that time. Certainly, there were concerns, eventually leading to the need for change and ushering in the next era.

How did the political era become the political era? Beginning in the early 1840s, cities began to adopt the idea of a paid police force. The introduction of American police forces was haphazard, chaotic, not well-thought out, and very disorganized. Being different from the British ideals was the underlying premise; however, it led to a rather unprofessional, corrupt enterprise that begged for intervention and change in the end. In part, the political era came about because of the need to control opportunities for personal enrichment and power. Controlling the law enforcement arm of the government seemed to be a logical step in attaining power over others. Unfortunately, this simplistic view fails to illustrate the depth and breadth of corruption of the times. The corruption was at all levels of government. Several factors contributed to what was known as the **Spoils System**. The Spoils System came from the notion "to the victors, go the spoils." In other words, those who were victorious, or in power, would reap all the goods and benefits they desired.[4] After Andrew Jackson became the president in 1829, he instituted many changes, including rewarding people who campaigned for him with cabinet positions. The opposition, angered by the changes, coined the phrase as a bitter reminder that the loser gave up everything to the winner, and the winner could change things to his liking. This idea trickled down to local government whereby newly elected politicians had the power and authority to revamp city government as well as to hire and fire at will, rewarding those who voted for them.[5]

▶ **Photo 2.1** What impact did the Industrial Revolution have on the development of modern policing?

In the early part of the 19th century, during the Industrial Revolution, rural agrarian society gave way to the building of factories, changing the means of production and services, resulting in urbanization and rapid population growth. Such economic shifts in production and employment contributed to significant social changes. People flocked from farmlands to the cities with the promise of factory jobs. Freed slaves made the journey north to freedom and work. Immigration from Europe to America went unchecked, flooding the cities with masses of impoverished people, seeking employment. As the population in cities grew and changed, fear and distrust of the newcomers prompted demands for protection from these "dangerous masses."[6] Challenges and disruption to cultural norms, rather than rampant criminal behavior, sparked the notion that police were needed. Policing was introduced, moving away from the unpaid night watchmen service of earlier, simpler times, to one that was paid and more formalized.

However, departments popped up in cities and towns in a very disorganized and random fashion, with little or no planning. Leadership came from local political authority. Early police departments were loosely organized and structured, poorly managed, and lacked standards in duties, functions, and employment practices. Early policing had little, if anything, in common with today's highly sophisticated law enforcement. Duties of early policemen included feeding poor people in soup lines, removing dead animals from the street, and providing overnight lodging for wayward families. The expectation that police were there to handle the crimes and disorder of the dangerous new immigrants and migrants was somewhat ironic in that policemen were often recent immigrants themselves, one step further along than the people they policed. Policemen were unskilled or semi-skilled, but the position offered more money than most skilled labor and had the opportunity for graft. Given a badge, perhaps a uniform, their own gun, and sent out on patrol, policemen were given the freedom to police as they saw fit.[7] Could you even imagine if this was our modern practice?

Early police were politically affiliated and therefore politically controlled. When politicians were elected or defeated, entire police forces were hired or fired. Positions within the police force, such as sergeant or captain, were bought for a price. Being hired as a policeman was the reward for campaigning for and supporting a particular politician. Policemen kept their jobs by being beholden to the politician they supported. The reason the era was named the "political era" had much to do with these special arrangements. Arrangements included taking graft or monies so that certain business would thrive, while penalties and fines were assessed against those who were persona non grata. Policemen had many financially lucrative opportunities; however, they did engage in community service, which is why, in the community policing era, many reformers wanted to return to the good old days when police walked the beat in neighborhoods and everyone knew them by name. However, before a return to those early days of beneficial relationships would be relevant, an era of separation and alienation was necessary to put an end to the corruptive influences of political control. The rampant corruption in government and the overly intimate community relationships were the impetus for the reform era.[8]

Reform Era

The reform era, also known as the professional era, began during the 1920s and early 1930s and transformed policing from its chaotic, random, and disorganized state to a modern, sophisticated, well-trained, and highly structured force. The police chief of Berkeley, California, August Vollmer, rallied police executives to focus on the moral vision. Vollmer's protégé, O. W. Wilson, would become the main force behind the reformation of the American police force.[9] Wilson was not only influenced by August Vollmer but also by J. Edgar Hoover, the director of the Federal Bureau of Investigation (FBI) from 1924 until 1972. Hoover was the leading administrator transforming the FBI into a national security organization. Hoover also transformed the image of the FBI and created a reputation of integrity, professionalism, competence, and power.[10]

Wilson used Hoover's strategies as a model for reforming the urban police force. He opposed the idea that politics was the basis of the legitimacy of the police because the involvement of politicians caused the problems of corruption. Thus, the separation of politics and policing was the main goal. The reforms were not universal, however. In some states, the state government took control over the police. In other areas, civil service eliminated patronage and ward influences. For instance, in Los Angeles and Cincinnati, the chief of police was a civil service position and the police chief was subject to regular examinations. In other cities, such as Milwaukee, the chief of police was a tenured position and the chief could only be removed by the police commission. And in some cities, such as Boston, the contracts of police chiefs were made so that they would not overlap with the city mayor's tenure. The police took control over all decisions, including tactical decisions about whether to arrest people who participated in a riot. The police were guided by the law. All these different strategies aimed to separate politics from policing.[11] The police function in the reform era focused on crime prevention, apprehension of criminals, and deterrence. Police stopped responding to other community problems as those were perceived as social work functions. The police also stopped providing emergency services and these services were transferred to other organizations.[12]

In 1967, President Johnson's Crime Commission Report on Law Enforcement and Administration reconceptualized the police as part of the criminal justice system. This report also laid the groundwork for the change in how police were organized. Police organization was now based on classical theory, which was based on two assumptions. First, workers are not interested in work. Second, they only work for economic incentives. Based on these assumptions, police organization focused on division of labor and unity of control. Division of labor means that tasks should be broken into components to enable increased efficiency because workers will become highly skilled in performing this task. For instance, police officers should be trained to perform a particular task, such as foot patrol of a particular beat. They will get to know the beat and the people who live there. They will learn the problems that exist in the neighborhood, such as the sale of drugs and how to solve these problems. This also led to the development of special units, such as homicide units and domestic violence units. Unity of control means that there is a pyramid of control where a central office has the final authority over decisions. This structure limited discretion. Police officers were raised to enforce the law. Officers are closely supervised and instructions flow downward.

In addition, the relationship between the police and citizens were linked together. The police were perceived by the citizens as neutral and distant, focused on solving the crime rather than responding to the emotional needs of a victim. The citizens were perceived by police as passive recipients of police services because the police were professionals. This perception of the role of citizens was based on medical services where doctors were professionals and patients the passive recipients of medical services. Citizens were supposed to call 911 as the central communications system. Foot patrol was abandoned because it was perceived as an expensive and inefficient practice. Police cruisers became increasingly popular and police became more disassociated from citizens.[13]

During the reform era, the government also established the Uniform Crime Report (UCR) and began collecting information about crimes reported to the police. The main purpose of the UCR at the time was to assess the effectiveness of police measured by the number of arrests, response time, and the number of times a police car passes a given point on a city street. However, the reform era certainly had its own problems. First, in the 1960s crime started to increase, and police failed to meet the expectations of the public and their own. Second, fear of crime rose rapidly and did not necessarily correspond to the level of crime. In some areas, the fear of crime was much higher than the actual crime rate and in other areas, the fear of crime was low even though crime rates were high. Third, many people of color did not perceive the police as impartial but rather as being discriminatory. In addition, people of color felt mistreated by police. Fourth, the civil rights and antiwar movement increased the tensions between police and citizens. Numerous riots and student protests showed the disconnect between police and citizens. Citizens increasingly began questioning police tactics as they were able to watch police action on television and listen to it on the radio. The wide-ranging public exposure of police behaviors was the beginning of a growing scrutiny of police actions, which today has expanded to social media and body cameras. The police reacted by changing their hiring practices, training of recruits, and supervision of officers. Police departments also slowly started recruiting women and people of color. Fifth, the public had realized that the idea that police were simply enforcing the law with little or no discretion did not reflect the reality of police work. Police officers had wide-ranging discretion in most decisions, including whom to stop-and-frisk and whom to arrest. Sixth, there were also problems within the police force. One of the main issues was that patrol officers were treated as low-status workers who had to follow petty rules, including rules for off-duty behavior. In contrast, police executives had much latitude and wide discretion. This differential treatment led to resentment by patrol officers and other officers of low status toward management. Seventh, the financial support for police departments decreased substantially, which led to the downsizing of many police departments and, of course, fewer policemen on the street. Finally, private businesses and organizations began hiring private security personnel and the private security industry became a thriving business, which was perceived as competition by the police force. Later, of course, police officers would also work for private security when they were off duty.[14]

While the reform era led to important positive change during the 1940s and 1950s, the social changes that occurred in the 1960s and 1970s showed that these reforms were not sufficient any longer. The police force had to adapt to the challenges society was facing, such as increased immigration, racial tensions and discrimination, changing crime trends and a rising fear of crime, and technological innovations, which increased public exposure of criminals and police. Police needed to reconnect to the citizens if they wanted to be effective in fighting crime.

Community Policing Era

Community policing, which started in the 1970s, has had many iterations, some successful and some less so. The risks and efforts taken along the way and the lessons learned from each aspect and each program have given birth to the community policing we have today. An examination of U.S. law enforcement history reveals several paradigmatic shifts in philosophy and practice. Change is inevitable, especially with technological advances. Impetus for change can spring from internal or external forces, and it can be slow or rapid. In terms of the shift to community policing, many factors influenced the need for change, the biggest being the alienation of police from the public that epitomized policing in the 1960s. For a particular change to endure, it has to be successful and it must be enculturated into police officer behavior and practices. Additionally, what worked in the past may not be effective in today's changing climate and diverse communities.[15] Community policing, in some form or another, is in its fourth decade, which speaks to its success and endurance. More accurately, once introduced and experienced, community policing is not easily dismissed as a "passing fad." Community policing has been, and is, transformative to the police–community relationship. Let us look at what brought us to where we are today.

The history of community policing reveals a path that is complex—full of twists and turns, successes and failures—but it has endured. It is difficult to trace an exact path because many of the innovations have been overlapping, or discarded, and then reinvented and given new names. Early ideals of police accountability to the community came from Sir Robert Peel as he established the Metropolitan Police Force of London in 1829. However, the barrage of innovations, experiments, and research has come about in the past 50 years. These include police community relations (PCR), team policing, foot patrol, broken windows policing, and community policing. Each of these ideas, concepts, and programs are incorporated into what we know as community policing.

Sir Robert Peel's Nine Principles

Historical underpinnings of community policing proscribing the way police should provide services to the public date back to Sir Robert Peel's Nine Principles (Peelian Principles) formulated in 1829 by the two commissioners of London's Metropolitan Police Department under Peel's leadership. Sir Robert Peel was known as the "father of modern policing." The principles are summarized below:

1. The basic mission for which the police exist is to prevent crime and disorder.
2. The ability of the police to perform their duties is dependent upon public approval of police actions.
3. Police must secure the willing cooperation of the public in voluntary observance of the law to be able to secure and maintain the respect of the public.
4. The degree of cooperation of the public that can be secured diminishes proportionately to the necessity of the use of physical force.
5. Police seek and preserve public favor not by pandering to public opinion but by constantly demonstrating absolute impartial service to the law.
6. Police use physical force to the extent necessary to secure observance of the law or to restore order only when the exercise of persuasion, advice, and warning is found to be insufficient.
7. Police, at all times, should maintain a relationship with the public that gives reality to the historic tradition that the police are the public and the public are the police, the police being only members of the public who are paid to give full-time attention to duties which are incumbent on every citizen in the interests of community welfare and existence.
8. Police should always direct their action strictly toward their functions and never appear to usurp the powers of the judiciary.
9. The test of police efficiency is the absence of crime and disorder, not the visible evidence of police action in dealing with it.[16]

Peelian Principles, outlined almost 200 years ago, are still as relevant today as they were back in 1829. They were unique in that that they were the first in history to proscribe police legitimacy, that is, police could only do their job if they had the public's consent, trust, and cooperation.[17] Perhaps, somewhere through the years, police prioritized their role differently and many turbulent times were in the future for police.

Police Community Relations

Another historical underpinning of community policing had to do with the relations between police and the public in order to provide the kind of services that the public deserved and desired. Police community relations (PCR) was the first attempt at altering and improving the police and community relationship. Two major reports, the Kerner

Report (1968) and the Report of the Presidents' Crime Commission (1968), called for immediate and major changes in policing in America. These reports exposed the weaknesses of traditional policing. Community relations was the first attempt to address police relations with the community, especially people of color. The PCR movement had several objectives, including bolstering relations with the public, especially racial-ethnic minority citizens, involving citizens in crime prevention, creating youth programs, and providing education and training.[18] PCR focused on changing the image of policing. Special PCR officers were assigned to build bridges with the community. PCR officers met with community leaders with the intent to give citizens an opportunity to voice their concerns and complaints about police services.

Some aspects of PCR were successful in that it forced police to realize that there was a need for change. Unfortunately, PCR leaned toward being a public relations tool for police and less about relating with the community. Issues arose with the split force structure. Patrol officers continued to fight crime and PCR officers fielded the complaints. In many cases, the PCR unit was staffed with civilians who had little clout with either the community or police. PCR failed miserably in its mission to bring police and community together; however, PCR became the underpinnings of the community policing evolution.[19]

Team Policing

PCR was an important foundation for what would become community policing in the future. Similar to PCR in terms of the focus on the importance of community relations, team policing emerged as a means to not only increase the presence of police in a neighborhood but to also have a team respond to community problems. In 1967, The President's Commission on Law Enforcement and the Administration of Justice specifically recommended team policing: Police departments should commence experimentation with a team policing concept that envisions those with patrol and investigative duties combining under unified command with flexible assignments to deal with the crime problems in a defined sector (p. 118).[20] Many departments across the nation did just that—implementation of team policing experiments. Unfortunately, although immediately popular, police administrators had little idea what to do and how to do it. Every city had their own version of team policing. Three basic principles were part of most team policing efforts. First, geographic stability of patrol, whereby officers were permanently assigned to specific neighborhoods. The idea was to encourage ownership by a team of officers. The second principle was maximum interaction between all team members in that area, including all shifts within a 24-hour period. A team leader insured team collaboration and communications. Lastly, the third principle proscribed maximum communication among team members and the community. Meetings between community and police were established to ensure flow of information and building of relationships and trust.[21] Success of implementations of team policing varied due to size of city, understanding of the elements by police administrators and participants, commitment to the experiment, and other factors.

President Johnson's Crime Commission Report in 1967 indicated the theory behind team policing was to decentralize and soften the rigid quasi-military structure, which many believed accounted for the dissonance between police and the community and contributed to the angst of the 1960s civil unrest. Unfortunately, that quasi-military organizational structure was instrumental in the failure of team policing. Officers on the teams were given more discretionary authority of decision making in the field, with less input from higher levels of the police organization. Power at the bottom of the structure is a threat to the hierarchical nature of traditional policing as discussed earlier. Although team policing was considered a failure and abandoned quite readily, it did contribute to ideals of community policing. Some of those concepts—such as decentralized authority and greater latitude of innovation and decision making at the lowest levels, ironically those that contributed to the failure of team policing—are critical to successful community policing.

Efforts to bring police and community together are central to PCR and team policing innovations. Additionally, police administrators thought that returning to foot patrol might bring police and residents in close contact with one another, hopefully fostering positive relationships and building trust. Foot patrol embodied the bygone days of early policing, before the patrol car, when simpler times prevailed and officers knew people by names. Foot patrol forced police officers out of their vehicles and back into the neighborhood. In theory, this was going to reverse the damage the patrol vehicle had done to the trust of the community with their police.

Foot Patrol

During the civil unrest of the 1960s, relations between police and citizens, especially racial-ethnic minority citizens, deteriorated to the point of open hostility. Conflict and confrontations resulting from arrests were common, leading to further unrest and violence. The distance between citizens and police increased and to many people, the police seemed to be an alien occupying force. Officers insulated themselves in the isolation of the patrol vehicle. Researchers, social scientists, and police officials wondered what could be done to remedy the dismal relations between the citizenry and police. The value of motorized patrol was in question and soon to be tested.[22]

One idea to bring citizens back together was foot patrol. Before motorized patrol, policemen walked a beat, interacting with the public throughout their shift. It seemed logical to bring back that kind of public contact where police officers were easily accessible in and around the neighborhoods on foot, not whizzing by in a patrol car. A few police agencies ventured forth to embrace the long-gone method of policing and to test its viability to improve PCR. Well-known foot patrol experiments included The Newark Foot Patrol Experiment (1973) and Flint Foot Patrol Study (1979). As with most police innovations, grant money allowed departments to engage in experimental tactics. Incentives to improve relations of police and communities began in earnest in the 1970s following a publication from The President's Commission on Law Enforcement and Administration of Justice (1967).[23] The final report called for detailed reorganization plans and a range of reforms for police departments. This report launched the community policing movement, although not in the form it is today; however, it was one of the pushes toward innovation and change. Foot patrol, while not the innovation it promised, signaled a return to times when police were considered part of the community and not an alien force.

Social scientists and police officials worked together to evaluate the advantages of foot patrol; however, not everyone was excited about foot patrol. The first problem came in the resistance of officers to leave the sanctity and safety of their patrol vehicles.[24] For example, one of the first foot patrol studies placed officers right out of the academy to foot patrol. Some implementations recruited officers by using incentives such as flex days off, allowing officers to choose their own schedule, and greater opportunity for individual innovative actions. Most officers view the patrol unit as their office on wheels with easy access to a computer, high-powered weapons, riot gear, reference material such as vehicle codes and penal code books, and other accoutrements of the job. It is easy to see why officers were reluctant to give up the safety and mobility of motorized patrol for something as archaic as walking a beat. Another problem was created because foot patrol and motorized patrol were separate units within the department, sometimes working at cross-purposes. Foot patrol officers were charged with the duty of being the face of the police department, to get to know the citizens, and to engage in relationship building. Motor patrol officers continued to be call answerers. Studies revealed animosity between these units.

The Newark Foot Patrol Experiment had eight foot patrol beats and utilized crime data and citizen perceptions of crime. Included in the study were citizen attitudes toward foot patrol officers. Objectives of the study, specific to the areas where foot patrol was implemented, included:

- Improve police–citizen relationships.
- Increase citizens' perception of safety.
- Decrease crime.
- Increase willingness of citizens to report crime.
- Increase arrests.
- Increase job satisfaction of foot patrol officers through formation of positive relations.
- Lessen citizens' fear of victimization.[25]

While most of the findings did confirm that crime had not decreased significantly, there was a definite improvement in the perception of safety by the residents in the foot-patrolled beat areas. Additionally, despite initial reluctance of police officers to be assigned to foot patrol, foot patrol officers demonstrated increased morale and greater satisfaction with their jobs.

The study in Flint, Michigan, in 1979 implemented foot patrol in 14 neighborhoods and involved 22 officers. The study was designed to address three areas:

- The lack of neighborhood organizations and services.
- Apathy of citizens toward crime prevention.
- The lack of positive relations between police and citizens.[26]

The Flint Foot Patrol Experiment was considered an abysmal failure. When the grant money ran out, foot patrol was abandoned. The crack epidemic was also to blame, because police believed the violence and deaths from crack meant a return to traditional policing tactics and removing police from the danger of being out on foot. Lessons learned from both Newark and Flint foot patrol experiments were not all negative, however.

Interestingly, two positive outcomes included that the public were more satisfied with police and police were more satisfied with their jobs. Foot patrol has not been totally abandoned. Most recently, police administrators are seeing the benefits of using many alternative forms of patrol such as horses, bicycles, Segways, skates, and other unique methods; not only to increase police presence but to ensure that the public had ready access to police, to increase opportunity for communication—all critical to fostering positive relationships and trust. Case Study 2.1 illustrates a modern foot patrol.

CASE STUDY 2.1: MODERN-DAY FOOT PATROL IN CAMDEN, NEW JERSEY

"We are knocking on doors, introducing ourselves—letting them know that we're here to serve them," Officer Matias said.[27]

Foot patrol was one of the earliest strategies to bring police and the community in closer contact, with the hopes of forging a positive, productive relationship. Among many other changes, the police force in Camden, New Jersey, changed their name to Camden County Police Department, and rolled out foot patrol in a dramatic way, requiring every officer to get out of their vehicles and walk a beat. The city of Camden had been plagued with violent crime and homicides. In 2012, Camden was fifth nationwide in homicides, with 87 murders per 100,000 residents.[28]

▶ **Photo 2.2** Camden County police officers Anson Simmons and Bianca Rivera check in with a customer at a small store as they patrol in Camden, NJ. **Mel Evans/Associated Press/AP Images.**

In the summer of 2012, it had 21 homicides; and after the new force took over, the number of homicides decreased to six. Camden eliminated the entire force,[29] which reinvented itself with community policing as the foundation. The new force has been given its walking orders, literally. Camden's police chief believes that human interactions between officer and residents enhances relationships and sets the legitimacy and trust of the police force. Chief J. Scott Thomson wants to change the culture of policing by addressing citizen distrust. "We're not going to do this by militarizing streets," Chief Thomson said. Instead, he sent officers to knock on doors and ask residents their concerns.[30] Although everything is not rosy in Camden, it appears that things have gotten better, and they have gotten better quickly. Now, police officers often play a spontaneous ball game with the youth who ran from them in previous years. One resident remarked that her seven-year-old son used to be afraid of the police but now wants to be one.

The officers themselves are happier about the job they are doing. Officer Virginia Matias reads to kindergartners each week, hoping to establish a connection that will extend beyond the classroom. She patrols the streets and interacts with people along the way, even knocking on doors and letting people know that the police are here to serve them.[31]

Some critics say that this cannot work in the long haul and show that statistics are not necessarily showing that improvements can be measured in lower crime rates. "The statistics are one thing, but how the people in my city measure public safety is not on a piece of paper," Chief Thomson said. "It's by what they sense when they open their front door. And that's where the change in the city is absolutely visceral."[32]

Discussion Questions:

1. What are the benefits of foot patrol?
2. What might be some of the downsides of foot patrol?
3. Should every officer be required to patrol on foot?

Broken Windows Theory

One of the lessons learned from the foot patrol studies was that citizens were equally concerned about noncriminal matters as they were about criminal ones. If citizens call police, they want and expect help immediately. While crime is a problem, people are more likely to complain about the disorder in a neighborhood. Signs of disorder, public vagrancy, graffiti, prostitution, boarded up buildings, and broken windows contribute to a fear of crime.[33] What disorder represents is the consequences it has on community stability, undercutting informal social control, discouraging investment, and stimulating the fear of crime.[34] People have little confidence in police who seem unable to do anything, thus increasing fear of crime. Addressing fear of crime was new to police. They could not wrap their collective heads around the notion that fear of crime was a legitimate police concern. Fear of crime as a topic has been studied in many ways. Some would argue that a person's perception of being victimized is a psychological response; others might argue that environmental factors contribute to a person's fear level. Criminologists Wilson and Kelling (1982) developed the **broken windows theory**[35] based on the notion that if a broken window goes unrepaired, it seems as if no one cares and more windows would be broken, creating an environment ripe for crime. This theory was based on an earlier study by Stanford psychologist Philip Zimbardo in 1969. In that study, Zimbardo set up two abandoned cars with their hoods up in two separate neighborhoods. The car left in the poorer neighborhood where disorder was common was completely stripped within 24 hours. The car left in an affluent neighborhood was untouched.[36] The broken windows theory was further popularized in the 1990s when New York City police commissioner William Bratton and Mayor Rudy Giuliani applied the theory to policing in New York City. Central to this policing strategy is reducing disorder, reducing fear of crime, and increasing quality of life for residents. Did it work?

The original application of broken windows policing that focused on both petty and serious crime showed a tangible drop in serious crime through the 1990s and 2000s. This drop, some critics argue, could be attributed to a number of coincidental events that took place around the same time. Other cities tried their hand at this type of policing, and to some extent, similar success resulted. Variations of implementation and changes in the original application, however, did result in mixed outcomes. Because of varied results, there has been much criticism of the broken windows theory. Some critics of the theory say that it fails to address true violent crime and that broken windows have little do with that. Other studies that refute the broken windows theory argue that there is no evidence that attention to lower-level crime impacts other types of crime. Some argued that broken windows policing focused on and criminalized communities of color by defining what constitutes disorder.[37] In many cases, getting rid of disorder brought on by undesired gentrification of neighborhoods, which led to higher-priced properties and upscale businesses, forced poor people, mostly people of color, to leave. Also, the targeted heavy-handed enforcement on lower-level crime, such as vandalism, graffiti, gambling, and drinking in public, was disproportionately targeting people of color. Fear of crime, some would argue, is media generated and has little to do with the environment because many people live in these conditions and it is normal to them. Fear of crime has more to do with the fear of victimization, an emotional response from past victimization or media portrayals depicting victims that are similar to a person who then becomes fearful, or from new that focuses on violent crime. Overall, no one can argue against the important role the theory has had on the community policing evolution. Many early community policing programs, for example, foot patrol studies, did show that public satisfaction of police did increase when police paid more attention to signs of disorder and lower-level crime complaints. Peer-reviewed studies that make cases both for and against the broken windows theory continue to be made into the 21st century, as more and more cities adopt the approach.[38] Case Study 2.2 illustrates the strategies applied for reducing fear of crime.

CASE STUDY 2.2: REDUCING FEAR OF CRIME IN HOUSTON AND NEWARK

The cities of Houston and Newark were very concerned about the high levels of fear of crime among their citizens. Fear of crime was perceived as a threat to organized society due to the negative side effects. People become suspicious about one another, which erodes their sense of community and creates tensions. People also start to distrust that the government can protect them from criminals, making them dissatisfied and more willing to take matters into their own hands. Strategies that focus on increasing arrests and solving crimes have proven ineffective in decreasing fear of crime among citizens. Similarly, simply telling people that their fear does not match the reality of crime rates does not reduce their fear of crime. Some departments also attempted to lower the fear of crime by minimizing the number of crimes reported to police. Not surprisingly, this was not a successful strategy either. When one quarter of American households experience incidents of crime every year either in their own family or their circle of friends, their own experiences weigh much heavier than official reports or statements.

The cities of Houston and Newark believed that community policing strategies, such as publishing newsletters, creating community police stations, contacting citizens about their problems, and stimulating the formation of neighborhood organizations, might prove successful in reducing fear of crime. Thus, Houston and Newark implemented these strategies. Both Houston and Newark published monthly newsletters with and without crime statistics to educate the population about the actual level of crime. Houston also implemented a Victim Recontact Program, which was meant to determine how victims were doing, how police could assist them, and to show that the police cared about them. In addition, Houston created a Police Community Station where citizens could speak to the officers about their problems and concerns. By reducing the physical distance between police and citizens, it was hoped that citizens would be more willing to work together with the police in reducing neighborhood problems. The officers also participated in community programs, such as monthly neighborhood meetings and school programs to reduce truancy. A Citizen Contact Patrol program was implemented to better understand the perceptions of the citizens. For that purpose, police would proactively approach people in the community to talk to them about the problems and what police could do to solve these problems. Finally, Houston developed a Community Organizing Response Team (CORT), which operated from October 1983 until May 1984. CORT was meant to build relationships between the members of the community and build a group of citizens who would define and solve neighborhood problems.

Newark implemented the newsletter but also two other programs. The Signs of Crime Program was meant to reduce the signs of public disorder. They implemented order maintenance programs and random enforcement. The Directed Patrol Task Force consisted of 24 officers who completed a three-day training program and then engaged in foot patrol, radar checks, bus checks, enforcement of the state disorderly conduct laws, and road checks.

The cities of Houston and Newark then assessed the impact of the different programs on the fear of crime. The newsletters were not very effective in Houston or Newark, mainly because very few people read them, especially residents with less than a high school education. In Houston, the Community Organizing Response Team, the Citizen Contact Patrol Team, and the Police Community Station Program substantially reduced the fear of crime. However, the Recontacting Victims Program had the opposite effect in that victims who were recontacted believed that there was more crime. In Newark, the Signs of Crime program was not effective in reducing fear of crime. However, the CORT program did have positive outcomes in that people perceived less disorder in the community, less fear of property crime, and less fear of personal victimization. In addition, satisfaction with police services and satisfaction with the area overall increased. Thus, overall, the combined effects of the programs led to significant reductions in the fear of crime.[39]

Discussion Questions:

1. Why do you think some of the programs implemented in Houston and Newark were more successful than others?

2. Have you experienced a community policing program in your community? If so, describe the program and explain what you think were positive outcomes.

SUMMARY

We have learned that many components of community policing were experiments and implemented as individual stand-alone programs within the organization. Programs such as PCR, team policing, foot patrol, and broken windows theory generated high expectations but resulted in marginally successful and disappointing outcomes. Many administrators and police personnel were not supportive of the move to get closer to the community and hoped that their careers survived to return to the normalcy of traditional policing—many aspects could not be undone. Aspects included the importance of public trust and police legitimacy as enumerated by the Peelian Principles so long ago. There would be no return to times where police could be unresponsive to community concerns and whiz by in their patrol vehicles like an occupying army. Neither the public nor the police would stand for apathy. Lessons learned included new expectations of accountability and new demands. The lessons learned in the piecemeal application of community policing resulted in identifying the core tenets of community policing, features that needed to be included in all efforts going forward.

In this chapter, you learned about three eras of policing: political era, reform era, and community policing era. During the political era, there was a close relationship between the political machinery and the police. The politicians hired the police officers who in turn assured that these politicians had no competition during elections and would surely be reelected. The political influence on policing led to much corruption and distrust between citizens and police. The growing tensions led to the reform movement and the beginning of the reform era. During the reform era, also called the professional era, policing was professionalized. Politics and policing were divorced, and the police implemented a new structure and line of command. The main function of police was crime control, to be achieved via preventive patrol and rapid response to calls for service. At this time, police services were centralized, and the 911 emergency system was implemented. Police officers were supposed to be impartial and professional. However, the distant relationship between police and citizens, civil rights and other social movements, and dissatisfaction within the police force led to the demise of the reform era and its replacement with community policing. Community policing focused on building better relationships between the police and citizens and building community support. The organizational design is decentralized, using foot patrol and problem solving to prevent and control crime and solve problems in the community that citizens are concerned about. The quality of life and citizen satisfaction are important goals for community policing programs.

KEY TERMS

Broken windows theory 37

Decentralized policing 26
Paradigm shift 25

Spoils System 27

DISCUSSION QUESTIONS

1. Identify the three eras of policing and what was unique to each era.

2. Why is policing in the United States considered to be decentralized?

3. What were the objectives of the foot patrol studies?

4. Why do you think the broken windows theory is important to the community policing movement?

CHAPTER 3

IDENTIFYING AND RESPONDING TO PROBLEMS WITHIN THE COMMUNITY: THREE CASE STUDIES

Learning Objectives

1. Describe the challenges police departments and officers are facing in the community.

2. Describe the SARA model and how it guides community policing.

3. Explain what practices work in addressing alcohol-related crimes.

4. Discuss the issues police face and possible solutions in responding to people with mental disorders.

5. Discuss the dangers associated with domestic violence calls for service and how police can effectively reduce domestic violence incidence.

On February 2, 2017, 19-year-old Timothy J. Piazza participated in a hazing ritual at the Beta Theta Pi fraternity at Pennsylvania State University. A surveillance video showed that Piazza had 18 alcoholic drinks in one hour and 18 minutes. Piazza never obtained his own alcoholic drinks. Every drink was provided to him by the fraternity brothers. Piazza, looking intoxicated, fell down a staircase. Later, Piazza was helped onto a couch. The next morning, fraternity members found Piazza unconscious. They carried him upstairs but did not call an ambulance until 40 minutes later. He died at the hospital on February 4, 2017. The medical exam showed that Piazza had suffered a broken skull, irreversible brain damage, and a fractured spleen, which resulted in internal bleeding and hemorrhagic shock. Sixteen fraternity members are facing charges from lying to police, tampering with evidence by deleting videos from the night in question, to manslaughter.[1]

INTRODUCTION

Police officers face a diverse set of problems in their community, including alcohol-related offenses, people with mental disorders, and domestic violence. There are certainly many other problems and one of the important tasks of law enforcement is to identify the main problems in the community and develop strategies on how to approach them. This chapter will begin by discussing strategic planning for identifying specific problems. Following that, we will describe the process of analyzing the problems and developing community policing approaches. Finally, we will demonstrate these processes via three case studies that span diverse problems officers face on a regular basis.

SARA MODEL

Problem-oriented policing can focus on crime hot spots, repeat offenders, repeat victimization, repeat times, or any other nongeographic concentration of crime. In order for problem-oriented policing to be effective, the police department must identify a specific problem and determine different targeted responses to the problem. The police department must also find partners in the community who will use their resources to assist the police. Finally, the police department must collect and analyze data pertaining to the problem prior to and following the intervention to determine the effectiveness of their response.[2]

A popular, widely used problem-solving tool employed by police to identify problems and develop strategies on how to approach these problems is the **SARA model**. It consists of four phases: (1) scanning; (2) analysis, (3) response, and (4) assessment. This model, more than others, has been the one most recognized and implemented by law enforcement agencies.

The SARA model builds on Herman Goldstein's concept of problem-oriented policing and was developed and coined by John Eck and William Spelman (1987).[3] The first phase of problem-oriented policing using the SARA model is "scanning." Scanning involves the identification of potential problems within the community. This step often involves identifying the consequences of these problems and finding out how often and why the problem occurs. This enables law enforcement agencies to prioritize the problem and select problems for closer examination.

The second step is referred to as "analysis," which focuses on learning more about the problem by collecting and analyzing data. The main question the law enforcement agency is trying to answer is why the problem is occurring. If possible, the agency will narrow the scope of the problem and determine the most appropriate response to solve or minimize the problem.

Following the analysis, the third phase is the "response." The law enforcement agency decides what is the best response or intervention for a specific problem. Agencies will research what other agencies and communities with similar problems have done and brainstorm new and creative interventions. During the response phase, the agency develops specific objectives to be achieved through the intervention. It also outlines the responsibilities of the law enforcement agency and its various partners. The response or intervention is then implemented.

Finally, the agency has to do an "assessment" of the effectiveness of the response or intervention. The agency must determine whether the intervention was implemented in a way that is consistent with the response plan. The agency also collects and analyzes new data about the occurrence of the problem to find out whether the intervention was effective and whether the objectives set during the analysis phase were attained. Agencies should also continuously assess whether the problem has recurred and whether a different intervention is needed.[4]

There are numerous agencies that have implemented problem-oriented policing using the SARA model. The following part of this chapter will describe three case studies that focused on different problems very commonly encountered by police: (1) alcohol-related crime and victimization; (2) offenders with mental disorders, and (3) domestic violence calls for service.

CASE STUDIES

There are many varieties of problems police officers face on a daily basis. Three of the most common problems are alcohol-related crimes and victimization, people with mental disorders, and domestic violence calls for service. All of these involve offenders and victims who can pose a serious threat to police officers and others as the offenders may act in ways unpredictable to the officer. People who are under the influence of alcohol or are suffering from acute mental distress do not behave rationally, which creates unique problems for police officers. Similarly, domestic violence calls for service also pose serious risk of bodily harm to officers and victims. Officers may not be able to distinguish immediately who the aggressor is if both people involved have visible injuries. In some cases, officers also get attacked by the victim if they fear that the offender will get arrested. The following case studies will lay out the special circumstances of these situations and solutions police departments have created in response.

Addressing Alcohol-Related Crimes and Victimization

The National Household Survey on Drug Use and Health (NSDUH) reports that 86.4% of people aged 18 and older have had alcoholic drinks during their life, 70.1% reported having had alcoholic drinks in the past year, and 56% reported having had alcoholic drinks in the past month. The survey also asks questions about binge drinking and heavy alcohol use. **Binge drinking** is defined as drinking alcohol until the drug alcohol level is above 0.08 g/dL. For the average woman, that would be about four drinks and the average man five drinks. **Heavy drinking** is defined as binge drinking on five or more days during the past month. In 2015, 26.9% of people reported binge drinking and 7% took to heavy drinking in the past month. Alcohol contributes greatly to premature deaths among the general population. Every year, approximately 62,000 men and 26,000 women die as a consequence of alcohol use. This makes it the third leading preventable cause of death in the United States. About 31% of deaths are caused by driving under the influence of alcohol. The misuse of alcohol also creates a heavy economic burden due to loss of productivity, profit, and competitiveness with a total of $249 billion each year.[5]

Part of the economic burden includes the cost of police encounters because of the close relationship between alcohol and crime. Alcohol impacts a person's judgment, actions, and aggression levels. Research shows that alcohol is a factor in 40% of all violent crimes. With a total of 3 million violent crimes committed each year, that is a staggering number. Of these violent crimes, alcohol was a factor in 37% of rapes and sexual assaults, 15% of robberies, 27% of aggravated assaults, and 25% of simple assaults.[6]

The Problem of Alcohol-Related Incidents in College Towns

The NSDUH shows that alcohol use is common among college students. In fact, drinking alcohol in general, binge drinking, and heavy drinking are more prevalent among college students than the general population. In 2015, 58% of students between the ages of 18 and 22 reported having had alcoholic drinks in the past month. This is almost 10% more than persons in the same age group in the general population. About 37.9% reported binge drinking in the past month, which is about 5% more than persons

▶ **Photo 3.1** What is the relationship between alcohol and crime?

in the same age group in the general population. And 12.5% reported heavy drinking in the past month, which is 4% more than persons in the same age group in the general population.[7]

The consequences are very serious for the students, their family, and the community. Every year, 1,825 college students between the ages of 18 and 24 die as a consequence of drinking alcohol. This also includes motor-vehicle accidents. In addition, 97,000 students between the ages of 18 and 24 reported that they were raped or sexually assaulted by another student who had been drinking. And one in four students experienced negative effects of their drinking on academic progress.[8]

Not surprisingly, police working in communities with large numbers of university students often have to respond to 911 calls for alcohol-related offenses and public disorder offenses. Studies have shown that neighborhoods with a high alcohol outlet density (e.g., liquor stores, bars, and so on) also experience substantially more violent crime than other neighborhoods.[9] As mentioned above, there is much evidence that alcohol consumption increases the potential for aggression, which can lead to violent crime. In fact, alcohol is the number one drug associated with aggression and violence, and, as a result, violent crime. A large-scale study in 11 countries found that 62% of all violent crimes involved offenders who were under the influence of alcohol.[10]

Before we can talk about the causes of the alcohol–violent crime connection, we need to define the terms *aggression* and *violence*. The term **aggression** is defined as "any form of behavior directed toward the goal of harming or injuring another living being who is motivated to avoid such treatment." The terms *aggression* and *violence* are mostly interchangeable, but the term *aggression* is used in the empirical literature and the term **violence** is generally used as a forensic term for investigations of interpersonal aggression or aggression between intimate partners or family members, also referred to as "domestic violence." Violence is also often used to describe individuals who have come into contact with law enforcement due to their behaviors after alcohol consumption.[11]

The alcohol–crime connection is closely related to the pharmacological effects of alcohol on the brain, and as a result, individuals' behaviors under the influence of alcohol.

Pharmacological Effects of Alcohol

1. Psychomotor Stimulant Effects

   Alcohol has rewarding properties similar to cocaine and amphetamines, increasing sensation seeking and impulsivity. These effects may increase the likelihood of confrontational and provocative behavior, resulting in violent altercations.

2. Interrupted Threat Detection

   Alcohol may alleviate subjective feelings of stress. People who are intoxicated and who engage in confrontational and provocative behavior may not appropriately appraise the situation and its potential for a harmful result. Stated differently, alcohol has a disinhibiting effect coupled with diminished fear of injury or death, therefore increasing the likelihood of aggressive behavior.

3. Alterations of the Pain System

   Alcohol can also reduce pain sensitivity and in the 19th century it was used as a surgical anesthetic. If this is true, then alcohol may contribute to an escalation of aggressive behavior because the intoxicated person does not feel pain as much as they would without alcohol. However, there is also research that shows that alcohol can increase pain sensitivity depending on individual levels of absorption and elimination of alcohol. Under this theory, aggression may escalate because the intoxicated person has a heightened reaction to pain and the significance of the provocation.

4. Cognitive Interference

   Alcohol may interfere with cognitive functioning, especially the functioning of the prefrontal cortex. The prefrontal cortex is responsible for the executive function, such as planning, inhibition, and active monitoring. Diminished executive functioning of the brain can lead to inappropriate responses in certain situations. For instance, an intoxicated person may perceive the words and tone used by another person as offensive. Had they been sober, they may not have perceived the words and tone as offensive.[12]

The alcohol–aggression relationship is complicated and varies for each individual. What remains to be a consistent result across studies is that alcohol is a main contributing factor to aggression and violent crime. See Case Study 3.1 to understand how making students aware addressed the problem of alcohol among them.

CASE STUDY 3.1: OREGON STATE UNIVERSITY

Corvallis is a town of about 57,000, of which about 20,000 are students at Oregon State University. The Community Livability Unit of the Corvallis police department encountered a great number of calls for service for alcohol and drug use, loud noise and littering, vandalism, sexual assault, fights, and violations of the liquor law. These offenses negatively impacted the neighborhoods and the residents. Police implemented different approaches and the number of incidents did decrease, but the residents did not experience much progress because there was still much noise and problem behaviors. Thus, the Community Livability Unit developed its own approach—focusing on alcohol use and abuse—because it believed that alcohol was the main contributing factor to several offenses. Data from Oregon State University showed that 50% of vehicle fatalities, 60% of suicides, 90% of campus rapes, and 95% of violent crimes on campus involved prior alcohol consumption by the victim, the offender, or both. The data also demonstrated that about 45% of students had engaged in high-risk alcohol use, defined as five drinks or more a day.

In response, the Community Livability Unit of Corvallis created a comprehensive educational outreach program in cooperation with Oregon State University Health Services, Benton County Health Services, and the City of Corvallis. It used money from a grant that supported the development of strategies to curtail underage drinking to hire a marketing company. The university provided all printing services. A marketing company designed the Red Cup Campaign "Know Your Limits, Know the Laws" that focused on four questions:

1. What's the open container law?
2. Does the open container law apply on game day?
3. I've been drinking, but I don't have any alcohol with me. Can I get in trouble just for being in public?
4. If my underage friends bring alcohol to my party, can I get into trouble?

The campaign was a great success. Students gravitated toward the Red Cups. A pre- and post-survey demonstrated that 73% of students had seen the Red Cups, 35% reported increased awareness, 42% reported increased knowledge, 47% reported increased knowledge of limits, and 35% reported modifying their behavior. The data from calls for service and crime data also support the success from the Red Cup Campaign. Calls for service for alcohol-related cases dropped from 189 to 138 per year, calls for service for livability-related offenses decreased from 1,462 to 1,183, and Oregon State University student conduct cases related to alcohol decreased from 52.5% to 49.9%.[13]

Discussion Questions:

1. Why do you think the Red Cup campaign was more successful than handing out information in the form of flyers?
2. What other strategies would you employ to reduce binge drinking and ensure that residents are not disturbed by late-night partying and noise?

Alcohol Outlet Density and Violent Crime

We know that alcohol consumption has a significant impact on violent crime. Thus, it is not surprising that studies suggest that there is also a significant relationship between alcohol outlet density, also referred to as bunching, and violent crime. Neighborhoods with a large number of bars, liquor stores, and other alcohol outlets have a higher violent crime rate than neighborhoods with few or no alcohol outlets.[14]

The relationship between alcohol outlet density and violent crime is not only due to the actual consumption of alcohol but also because a large number of alcohol outlets suggest that there is physical decay and disorder. Broken windows theory suggests that the physical characteristics of a neighborhood can explain neighborhood crime rates. Neighborhoods with evidence of physical decay and disorder have higher violent crime rates. The theory explains that people believe that in neighborhoods with physical decay and disorder, no one cares about what is happening, including crime. People may also believe that no one will do anything about crime in the neighborhoods. If people don't do anything about cleaning up garbage, repairing broken windows, abandoned cars, and graffiti, then why would they intervene if someone commits a crime?[15] Alcohol outlet density may also result in lower alcohol prices due to greater competition between the outlets. Lower prices may increase alcohol consumption and therefore violent crime and premature death.[16]

Reducing Alcohol-Related Violence

Over the past centuries, there have been several attempts to reduce alcohol-related violence. For instance, taxation was used to curb alcohol consumption by increasing the prices, and alcohol was banned during Prohibition. Policymakers believed that violence would be greatly reduced if alcohol was not legally available. Reduction of alcohol-related violence has also focused on the individual using strategies such as enhanced punishments for crimes committed while under the influence of alcohol (e.g., driving under the influence). These strategies omit the impact of the broader social and cultural settings, however. More recently, intervention strategies have shifted toward addressing alcohol-related violence within the context in which it occurs. For instance, violence may occur in social settings, such as fraternity parties or at football games, in physical settings, such as certain neighborhoods, or in relationship settings, such as domestic violence. Some researchers suggest that by reducing the alcohol outlet density within cities and neighborhoods, they may be able to decrease alcohol-related violence.[17]

This type of intervention is closely related to urban planning and zoning. For instance, liquor licensing agencies could reduce the density by only licensing a certain number of alcohol outlets within a certain geographic area. Licensing agencies could also issue licenses based on distance, that is, there has to be a certain distance (i.e., a certain number of miles) between alcohol outlets to reduce bunching.[18]

Another proposed solution is based on deterrence, which includes policies and laws that have criminalized carrying a firearm while intoxicated. For instance, in Ohio, it is a first-degree misdemeanor to carry a firearm while intoxicated, which is punishable with fines of no more than $1,000 and up to six months in jail.[19] By 2008, 46 laws in 31 states prohibited the possession or use of a firearm while intoxicated. Of these 31 states, 18 prohibit firearm ownership or use on the basis of "habitual alcohol use." There is also a wide range of laws prohibiting driving under the influence of alcohol and enhancing punishments for injuries and deaths caused by intoxicated drivers.[20] For instance, in California, the punishment for vehicular homicide—a felony, usually punishable by 10 years in prison—can be more severe if the driver was intoxicated.[21] Case Study 3.2 looks at the reduction of alcohol outlet density.

CASE STUDY 3.2: REDUCING ALCOHOL OUTLET DENSITY IN VALLEJO, CALIFORNIA

The city of Vallejo, California, has a population of 11,000 and about 200 on- and off-premise alcohol outlets. An outside evaluation, which assessed the problems associated with high alcohol outlet density, found that geographic areas with high alcohol outlet density had a higher number of calls for service to police for several offenses, including sexual assault, fights, drinking and driving, loitering, and other nuisances. The Vallejo Alcohol Policy Coalition (VPAC) collaborated with the Vallejo Police Department and the California Alcohol Beverage Control Department to develop a Responsible Beverage Service curriculum for the city. In 1994, training was developed and required for all alcohol retail businesses to obtain the Conditional Use Permit. The training had an immediate effect, reducing calls for service by 6.5% within the first year of the training, thereby reducing the number of hours police spent responding to calls by 20%. In addition, the city passed a "Deemed Approved" ordinance with the help of the VPAC, and the police vigorously enforced the ordinance. The result was a 53% reduction in overall calls for service between the last 10 months in 1998 and the last 10 months in 1999. Calls for service for battery and assault decreased by 25% and for other crimes linked to alcohol by 22%. VPAC also worked to increase compliance of alcohol merchants with regards to sales to minors by implementing undercover operations in cooperation with the police department. Between 1997 and 1999, the compliance rate increased from 74% to 98%.[22]

Discussion Questions:

1. Why do you think that requiring alcohol outlet business owners to complete training would reduce alcohol-related crime and police calls for service?

2. What other community policing strategies can you think of that would reduce calls for service for alcohol-related crimes?

POLICE RESPONSE TO PEOPLE WITH MENTAL DISORDERS: A GRAY ZONE

It is estimated that a total of 8.3 million individuals suffer from severe mental disorders, such as schizophrenia, bipolar disorder, and severe depression. About half of them receive no treatment.[23] When individuals with mental disorders are in a crisis situation, it is typically police officers who are called to deal with the situation. Police have long been functioning as interventionists for people with mental disorders. Between 7% and 10% of all police encounters involve people with mental disorders. On average, each police officer responds to people with a mental health crisis six times per month. Research suggests that the majority of these encounters do not involve violence or major crimes. They also do not typically require emergency apprehension. Instead, police have to use their problem-solving skills in their encounters with people with mental disorders. Research has identified three core features in their work with this population: (1) solutions to chronic vulnerability; (2) using local knowledge to guide decision-making; and (3) negotiating with complainants and people with mental disorders.[24]

People with mental disorders have several vulnerabilities, including a higher risk of being assaulted, being homeless, being arrested, and being injured or killed during police encounters.

Police have to make decisions on the spot about whether they can resolve the problem, whether to arrest the person, or to contact other agencies, such as social services and health services. These encounters are complicated by the fact that many people with mental disorders also often have substance abuse issues and other problems, such as homelessness. The deinstitutionalization of people with mental disorders has shifted the burden of caring for this population to police and jails and prisons.[25]

The deinstitutionalization movement was part of the civil rights movement and it was driven by three beliefs. First, many people believed that the psychiatric hospitals were cruel and inhumane and that people were simply kept captive. Second, many people also believed that psychotropic drugs would evolve as a cure for mental disorders. Finally, the costs associated with caring for people with mental disorders were too high. Unfortunately, deinstitutionalization did not improve the lives of people with mental disorders as many of them became homeless, are locked up in jails and prisons, and lack basic support services. Medications have not proven to be the cure for mental disorders as they have significant negative side effects, people forget to take them, or they don't want to take them. Individuals with mental disorders also stop taking medication that has worked well for them because they believe they are cured. State hospitals have either been closed, have a greatly reduced capacity to care for people, or can hold individuals with severe mental disorders only for a few days until they are not in an immediate crisis any longer. In many states, state hospitals do not admit individuals with severe mental disorders because they are covered by Medicaid and should be treated by private facilities. But private facilities often do not have space, resources, or the manpower to care for all individuals referred to them. Thus, individuals with severe mental illness are left without adequate services, which often results in them getting arrested for minor offenses, such as loitering, indecent exposure because their bathroom is the public park, property damage, drug use, or open bottle or similar offenses.[26]

As a result, the incarceration rate for individuals with severe mental illnesses has greatly increased. In 2016, about 20% of the jail population and 15% of the prison population have been diagnosed with severe mental illnesses. Speaking in total numbers, approximately 383,000 individuals with severe mental illnesses are behind bars, which is 10 times the number of individuals currently housed in state mental health hospitals.[27]

It also means that police have a greater exposure to individuals with severe mental disorders and have to recognize their conduct as symptoms of a mental disorder. This can be difficult during chaotic encounters. In addition, police also have to decide whether the individual creates a public safety risk and how to best serve the needs of the individual and ensure the safety of people in the community. The National Alliance on Mental Illness reports that every year about two million individuals with mental illnesses are arrested and held in jails. Most of them will serve a short sentence for a small crime and then be released to the community, often without any support services. They rarely receive adequate treatment while in jail and even if they do receive treatment, the treatment ends once they are released. There are far too few community service providers to care for them. The result is what is called the **"revolving door of incarceration"** where individuals are returning to jail over and over throughout their lives. This creates a heavy burden for the individuals and their families but also for law enforcement and the corrections system, and, of course, communities that have to deal with homelessness and minor crimes on a daily basis.[28]

Solutions for Chronic Vulnerability

A research study by Woods examined police encounters with individuals with mental health disorders. The researchers conducted ride-alongs with police and recorded their experiences. Police do consider arrest a solution to an immediate crisis situation, but they recognize that many of the arrestees have little to lose by getting arrested as they are often homeless, in poor health, and with little family support. Thus, police often try to find other solutions, especially if the individual is not dangerous, although they may be engaging in disturbing behaviors that are bothersome to members of the community. They also found that officers often relied on their knowledge about these individuals from prior encounters. Police had their "regulars" who engaged in certain disturbing behaviors on a regular basis and often in the same vicinity or neighborhood. One solution by police is to escort individuals away from a business or place where they are causing the disturbance. For instance, a woman regularly caused disturbances at a McDonalds by knocking things over and bothering the customers. The officers learned over time that if they came with two or three officers, she was cooperative and would let them lead her away from the McDonalds. They also knew that they would be called again sooner or later. They did not arrest her as they did not perceive her as a threat to anyone.[29]

Using Local Knowledge for Guided Decision-Making

Officers have a certain "area knowledge" about the areas they are called to. They know which areas are more dangerous than others and they know a great deal about many of the residents. They use their knowledge during calls for service. They know which individuals could be a threat and they may arrest them. They also know which individuals they can calm down by taking a walk with them. In some instances, officers may take the individual to a hospital in hopes that they will receive treatment. But if the individual does not consent, they cannot be coerced into treatment. Many people do not want to be admitted to a hospital or receive treatment. This is also true when officers take an individual to a housing or community care center. The officers may still take the individual to the treatment facility because there is nothing else they can do except arrest them.[30]

Negotiating Peace With Complainants and Call Subjects

Health or social service providers may call the police when they have an individual who is aggressive or refuses to leave. These individuals are also referred to as **call subjects**. The typical approach of police in such situations is to talk to the individual, understand their circumstances, and offer advice until they feel that everyone is safe and they can reach a state of normalcy. If the individual has not committed a crime, officers typically listen to what the person wants to do, escort them out of the vicinity, and let them go on their way.

Police are also regularly called by family members of individuals with severe mental disorders. Family members often lack the resources to respond to a crisis. But family members also sometimes call police when the individual is simply a nuisance and they want the police to take the individual away. Officers then have to explain to the family why they cannot arrest the individual or take them to a hospital. Being a nuisance to someone is not a crime. Officers also stated that at times family members may make up things to convince the officers to take the individual to a hospital and get a break from caring for them. Many officers do not feel comfortable completing a petition for involuntary emergency psychiatric evaluation based on information they cannot confirm. Thus, the individual is back home within a few hours even after being taken to a hospital.

In sum, the reality of community police work is that during encounters with individuals with severe mental disorders, officers have to rely on temporary solutions they know are insufficient to the individual and the community. They rely on their "area knowledge" to guide their decisions and restore the peace. These decisions also depend on whether the individual actually engaged in criminal behavior or appears to be a threat to another person. During their encounters, they have to negotiate with the complainants and the individual who caused the disturbance and find the best solution.[31]

Crisis Intervention Teams as a Best Practice

Crisis intervention teams (CITs) were established in an effort to reduce the criminalization of conduct by individuals with severe mental disorders. In this capacity, officers serve more like "guardians" and less like "enforcers." You have already learned in the prior section that police use their communication skills and knowledge during encounters with individuals with severe mental disorders to negotiate peace and accomplish an outcome that is satisfactory to complainants and the call subjects.

Crisis intervention teams are specialized police-based programs based on a systematic response intervention model. The purpose of this program is to increase the safety for officers, complainants, and the call subjects during the encounters. In addition, the program is intended to improve the interactions between police and individuals with severe mental disorders. The standard CIT course is a 40-hour course that focuses on the disease process and symptoms of various mental disorders. It teaches officers the skills to assess whether an individual likely has a mental illness. It also teaches communication and de-escalation techniques for encounters with call subjects, communication with mental health providers, and completing emergency evaluation petitions.

The first city to develop and implement the CIT training was Memphis, Tennessee, in 1988. The city's police department worked together with mental health professionals, local advocates, and the National Alliance on Mental Illness (NAMI). CIT has eight anchoring pillars:

1. "partnerships between law enforcement and mental health advocacy;
2. community ownership through dedicated planning, implementing, and networking;
3. law enforcement policies and procedures;
4. recognitions and honors of CIT officers' accomplishments;
5. availability of mental health facilities;
6. basic and advanced training for officers and dispatchers;

7. evaluation and research; and

8. outreach to other communities."[32]

One of the main goals of CIT training is to improve the knowledge of police officers about mental illness and improve the attitudes toward individuals with mental disorders. This is imperative to achieve positive outcomes during police encounters with these individuals.

One of the counties that has successfully implemented CIT is Miami-Dade in Florida. With a population of 2.7 million, this county has the highest concentration of people with mental disorders. A total of 192,000 adults and 55,000 children, or 9.1% of the population, have a serious mental illness. The police department found that every month, one individual with severe mental illness died during police encounters; the department had made 20,000 arrests and incarcerations from this population group alone, putting great strain on the county's their jails, budget, and resources. The City of Miami also paid out millions of dollars in wrongful death lawsuits. In 2010, the county implemented the CIT training. It trained 54,000 officers representing all 36 police departments. The program not only trained officers on how to deal with individuals with severe mental disorders, it also provided opportunities for officers to seek help for their own problems, which include stress from work and post-traumatic stress disorder (PTSD) after witnessing traumatic events or after being injured. Officers who experience PTSD may respond inappropriately during encounters with individuals with severe mental disorders. Thus, offering mental health advice and treatment to police officers is imperative to avoid fatal shootings and injuries. About 150 officers reach out to the program coordinators and seek help every month. This is quite remarkable given that acknowledging stress and fear can be perceived as a weakness in the police culture. The training not only changed officers' attitudes toward individuals with severe mental disorders but also toward themselves.

Since 2010, the two largest police departments, the City of Miami and Miami-Dade County, has handled 71,628 mental health-related calls for service and made only 138 arrests. They were able to close a jail and have saved taxpayers $12 million per year overall. The number of fatal shootings has decreased by 90%. The success of the CIT in Miami was also recognized by the state legislature, which appropriated funds to build a new treatment center for individuals with acute mental illnesses. These individuals need treatment and that is difficult to find around the country.[33] Case Study 3.3 explains how ineffective management led to adverse outcomes in a mental health facility.

DOMESTIC VIOLENCE

On November 19, 2018, Juan Lopez met with his ex-fiancé, Dr. Tamara O'Neal outside Chicago's Mercy Hospital and Medical Center. O'Neal, an emergency physician, had broken off her engagement to Lopez, so Lopez had requested that she return the engagement ring he had bought for her. O'Neal called the police and told them that Lopez was upset, possibly armed, and that she was concerned about her safety. When a friend of O'Neal tried to intervene, Lopez pulled out a handgun and shot the 38-year-old O'Neal to death. Lopez then shot at the police car before running into the hospital where he killed first-year pharmacist resident Dr. Dayna Less. Two police officers also ran into the hospital and during the ensuing

CASE STUDY 3.3: THE CROFT UNIT—WHO IS BEHAVING BADLY?

The Durham Constabulary has a total population of about 31,000 people. After 5 p.m., there are only two or three police officers available to respond to calls for service. Most of the time, these officers were not available to calls for service because they were at the Croft Unit. The Croft Unit was a short-term residence (6, 12 months) for 20 adults with mental health issues. It was run by a for-profit organization. The Croft Unit was very challenging for the staff. One problem was that after 5 p.m. only two staff and one supervisor were present at the residence. A second problem was the limited mental health training of the staff. The staff members were recruited from all different types of backgrounds with no requirement of mental health training. There was a high staff turnover, low pay, and a lack of experience working with this population. Finally, the residents were allowed to consume alcohol until 11 p.m., which meant that several of them were regularly intoxicated. In addition, many of the residents were taking strong medications, which coupled with alcohol, led to difficult-to-control behaviors.

The Croft Unit made 164 calls for service within an 18-month period. Of these, 81% were made by the staff and 71% were made after 5 p.m. when the Durham Constabulary was short on officers. A total of 42 crimes were reported and officers made 29 arrests. Also, 73% of the victims were living or working at the Croft Unit. The vast majority of crimes were minor incidents, such as property damage and simple assault. For instance, the residents would stomp on a picture frame, scratch staff who were trying to put them in restraint, or hit the staff with a bottle when the staff attempted to take the bottle away.

On average, police were called to the Croft Unit every three days and they typically responded with two officers and two police cars. The perception of the community was that the residents were difficult people and the police were ineffective in solving the problems because they were there so regularly. In order to get a better understanding of the causes for the repeat calls for service, the police met with the staff of the Croft Unit, members of the community, and the Care Quality Commission. The Care Quality Commission oversees the providers of care facilities. During its assessment, it found out that the supervisor had no mental health training, the residence was designed for 16 residents, not 20, medications were not handled safely, and adequate care plans were lacking. It had rated the Croft Unit as "inadequate" several months in a row. Every time the management of the Croft Unit received the rating, it was told what the problems were and how to address them. That did not happen, however, and, as a result, the Care Quality Commission issued a fixed penalty notice. The for-profit organization closed the facility because it could not make a profit anymore.

The lesson the Durham Constabulary and the community learned was that not everything is as it seems. The problems were not caused by the residents but by an ineffective management driven by the desire to make a profit.[34]

Discussion Questions:

1. If you were the chief of the police unit responsible for a community with a halfway house where people from prison are living temporarily, how would you help them reintegrate into society?

2. You have a high number of calls for service from a halfway house. What strategies would you use to determine what causes the high call volume, and how would you attempt to reduce the number of calls for service?

gunfire, Lopez killed 28-year-old police officer Samuel Jimenez and injured another officer. Lopez, who had a history of domestic violence, was fatally shot during the incident. Four years earlier, his ex-wife had received a restraining order after Lopez had repeatedly threatened her. Lopez also had a history of bullying colleagues at the Chicago Fire Department and was dismissed from the department in 2014 for failing to show up for work after one of the bullying incidents.[35]

Police departments and officers face a variety of challenges in the community, especially in cases involving domestic violence. Domestic violence is typically defined as abusive behaviors within an intimate partner relationship, including physical, psychological, and sexual abuse. The U.S. Department of Justice estimates that there are 1.3 million nonfatal domestic violence victimizations per year. The National Crime Victimization Survey collects data from individuals via self-report surveys from which the above figures are obtained. Only 56% of these victimizations were actually reported to the police and in only 48% of these reported cases were the victims willing to sign the criminal complaint against the offender. The rate of arrest for the reported cases was 39%. Victims of serious violence, such as rape and aggravated assault, were no more likely to report the violence to the police than the victims of a simple assault. Thus, the severity of the injury did not predict reporting of the crime to the police. However, police were more likely to arrest the offender when the victim had serious injuries.[36]

The National Center for Victims of Crime estimates that in 2015 approximately 5.4 out of 1,000 women and 0.5 out of 1,000 men experienced domestic abuse. However, the true number of victims is substantially higher because domestic violence is highly underreported due to the personal nature of the crime.[37] Research has identified four main reasons for the failure to report domestic violence: (1) personal privacy (32%), (2) protecting the offender (21%), (3) the crime was minor (20%), and (4) fear of reprisal (19%).[38]

A substantial number of calls for service are made for domestic disputes and many police departments that serve more than 250,000 people have a full-time specialized unit for domestic violence. These specialized units include police officers, counselors, and social workers who work together to provide a number of services. These services include investigating the crime; interacting with service and treatment agencies; assisting the victims; training officers; victims, and community members; and acting as a liaison for officers.[39]

Domestic Violence and Intimate Partner Homicide

Domestic violence has numerous emotional and physical consequences and may also result in intimate partner homicide. Between 40% and 50% of all women murdered are killed by their intimate partner and the vast majority of these homicides (70–80%) occur after the woman had experienced domestic abuse. Thus, the most effective way to prevent these homicides is to identify women who are at high risk and intervene promptly. There are four main risk factors: (1) past violent incidents, (2) timing of past violence, (3) type and severity of violence, and (4) alcohol and drug abuse by the offender.[40]

Jacquelyn C. Campbell developed the Danger Assessment Tool, which lists risk factors associated with intimate partner homicide. The assessment tool measures the four main risk factors mentioned above. Listed below are some of the assessment questions:

1. Partner used or threatened with a weapon?
2. Partner threatened to kill the victim?
3. Gun in the house?

4. Physical violence increased in frequency?

5. Partner controls all or most of the women's daily activities?

6. Partner uses illicit drugs?

7. Partner is drunk every day or almost every day?

8. Partner is violent outside the home?

The assessment tool is meant to be used by the victim and by victim service providers to determine whether the victim is at a high risk of being killed.[41]

Domestic Violence and Officer Safety

Incidents of domestic violence cannot only have dire consequences for the victims but also for officers responding to the calls for service. Domestic dispute–related calls for service are a great concern for police officers for several reasons. Victims of domestic violence typically do not call the first time the violence occurs. When they call police, there have been repeated assaults and the officer is coming in during a volatile situation. In domestic disputes, it may also not be immediately clear who was the primary aggressor if several parties have injuries. Officers have very little time to assess the overall situation and to determine whether to approach the residence or wait for backup. Also, backup may not always be available.

Domestic dispute calls for service have the highest officer fatality rate of all calls of service. Between 2010 and 2014, a total of 91 officers died while responding to a call for service. Of these, 20 officers died responding to domestic dispute calls. All but one officer were killed with a firearm. Officers who respond to these calls are well aware of the danger, but they also know that the victim may be in imminent danger. This is especially true when officers know that the offender has made threats to kill the victim or others, when the offender was known to be armed, or when the offender had a history of violence. Thus, it seems advisable to have two or more officers respond to these calls to decrease the number of fatalities by providing immediate support, including life-saving measures. However, the reality of many police departments is that there is a shortage of funding and officers, making it impossible to send more than one officer to calls for service. In addition, research suggests that even when two officers responded to a domestic dispute call, the death toll was still high. In 35% of the fatalities, officers were alone at the scene. In 40% of the cases, two officers responded to the scene, and in 25% of the cases, three or more officers responded to the call for service. Several officers were shot when they approached the residence. Other officers were killed while investigating the case. Having several officers at the residence can save lives as they can restrain people present at the scene and prevent attacks on the officer who is investigating the incident. Also, additional officers are often necessary to separate the parties if the dispute is ongoing.

Similarly important to increasing the safety of officers is the access to information about the offender and any prior domestic disputes. The sharing of information and discussing the approach to take at the residence help officers act in a coordinated way. Knowledge of the potential for violence against officers can save lives. For instance, in one situation, two officers were killed because they were unaware that the offender was armed. In order to avoid such incidents, many police departments have implemented intelligence-led policing methods, which include technology and social media, such as instant access to call details and criminal databases.[42]

There are some other measures officers can take to lower the risk of harm to themselves. First, waiting for backup is an important strategy. Supervisors should stress the importance of waiting for the backup officer when two or more officers have been dispatched. Second, dispatchers should be made aware of the dangers of domestic dispute calls and instructed not to send officers alone. This also applies to calls for service for "child custody disputes," "assistance with clothing," and "assistance in serving a protection order" as they all relate to domestic disputes and may be highly charged situations. Finally, dispatchers should be diligent to obtain as much and as accurate information as possible about the offender, prior calls, and weapons charges and share the information with the officers dispatched. Officers are at a great disadvantage if they don't know who the offender is and whether they are armed.[43]

The Problem of Officer-Involved Domestic Violence

Although much research has been done on domestic violence rates in general, there is very little research on the rates and consequences of domestic violence where the officer is the offender, which is also referred to as **officer-involved domestic violence (OIDV)**. Estimates of OIDV vary widely, from 10% of all officers to more than 40%. Victims of domestic violence by police officers face unique challenges. First, the victim knows that the abuser has access to lethal weapons and is well trained in using these weapons. Second, the abuser knows the location of women's shelters and other places where the victim could seek help, which means that these places are not safe for the victim. Third, the abuser is part of the criminal justice system and knows how the system works, which gives them the opportunity to manipulate the situation and put the blame on the victim. Other police officers in the department are typically more likely to believe the person they work with. Even victim advocates tell the victim that it will be difficult to get help from the criminal justice system if they file a report against the officer. Fourth, some victims have great concerns that their partner's job would be at stake and that they would not be allowed to carry a firearm any longer. Fifth, the so-called Code of Silence among police officers creates a great barrier for victims and protection for the offender. The Code of Silence, which emphasizes loyalty among officers, impedes the willingness of other police officers to report the domestic violence perpetrated by a colleague and provide assistance to the victim even if they know that the officer is abusive. In addition, officers are much less likely to arrest a colleague even if they know the colleague has assaulted their intimate partner repeatedly. Officer solidarity is imperative in a job where officers are exposed daily to dangerous situations, violence, and traumatic situations. All of these factors reduce the reporting by victims of OIDV.[44]

Research has identified three main risk factors related to the job that may induce domestic violence by police officers. First, the police officer training on subduing and interrogating suspects can have a spillover effect on their relationship. Officers may use their training to assert power and control over their intimate partner. Second, the police culture emphasizes hypermasculinity and authoritarian behavior, which may also spillover into their private life. Finally, the daily stress at work and burnout may contribute to domestic violence as officers take out their frustrations on their intimate partner.[45]

The National Center for Women and Policing outlines the failures of police department policies and how they contribute to the problem. Most departments handle OIDV cases informally, which contradict most departments and legislatively required responses to other domestic violence cases. The most common intervention by departments was counseling for the officer. Several studies document the exceedingly lenient treatment of officers who assault their intimate partners. For instance, in San Diego, the rate of prosecution of domestic violence cases is 92% in general, but only 42% when officers are the perpetrators. Between 1990 and 1997, the Los Angeles Police Department investigated 227 cases of domestic violence against officers. In 91 cases, the investigation found evidence of domestic violence. But only four officers were convicted on criminal charges and the punishment was very lenient. One officer was suspended from his job for only 15 days. A second officer had his conviction expunged. But the lenient treatment not only applies to criminal prosecutions, it also applies to officers' personnel files. For instance, at the Los Angeles Police Department, a sustained allegation of domestic violence was not mentioned in the officers' performance report in 75% of all cases and 26 of the 91 officers were promoted within the next two years. Instead of providing assistance to victims, department policies serve as a deterrent for the victim to report the abuse.[46]

A study by Oehme et al. (2016) shows that officer training focused on responding to OIDV calls can improve the response by officers and their willingness to arrest a colleague and provide help to victims in the form of filing a report, informing them of their rights, and connecting them to victim services. Some police departments have responded with increased screening of applicants, increased training, and adopted departmental policies that address OIDV cases specifically. However, this is only true for a few departments and departmental changes are only one part of the problem. In order to have a substantial positive impact, the response by other critical players in the criminal justice system, including prosecutors, must also change.[47]

The Challenges of Community Policing and Domestic Violence

Community policing requires police to develop working relationships with stakeholders in the community and change the way they have traditionally conducted police work. The reality of community policing often does not include actual inclusion of community members in the police organization and police work. However, domestic violence has been one of the more successful examples of community policing.[48]

Community policing did not focus much on domestic violence until the mid-1990s when funding for domestic violence projects increased. The increased focus on domestic violence was mainly spurred by social movements, such as the battered women's movement, and public protest. Police departments were heavily criticized for their responses to domestic violence. These social movements resulted in an understanding that domestic violence responses and community policing practices can be successfully integrated. There are three domestic violence organizing principles. The first principle is maximizing the safety for battered women. This principle aligns with the community engagement/partnership principle of community policing. The goal is to strengthen the community

▶ Photo 3.2 What are some special considerations officers need to take into account during domestic violence calls?

support for the women by providing a coordinated community response. The second principle is holding batterers accountable for their violence. This principle aligns with the practice of solving problems within the community. It is imperative to provide individualized solutions by determining what combination of arrest, court sanctions, education, and community sanctions are the most effective for a particular offender. The third principle is challenging the cultural underpinnings of battery. This principle aligns with the practice of organizational change, including changing the broad public acceptance of violence toward women and girls.[49]

Promising Practices in Community Policing of Domestic Violence

The Battered Women Criminal Justice Center (BWJP) conducted a field study in four communities: Chicago, Illinois; Marin County, California; Duluth, Minnesota; and London, Kentucky.[50] They looked at five promising practices and how these practices fared in different cities. The five promising practices are (1) coordinated partnerships between the police, domestic violence service providers, prosecutors, and the community; (2) neighborhood-based organizing and problem solving; (3) Community Policing Action Team; (4) access and collaboration, and (5) the safety and accountability audit as a community problem-solving tool.

Coordinated Partnerships. One promising practice is to assign designated Domestic Violence Liaison Officers (DVLOs) who are working in a particular police district. The

DVLOs connect police officers in that particular district to community residents and domestic abuse service providers. They are actively engaged in the community with the task of building a community/police response to domestic violence. These DLVOs conduct a variety of activities, including reviewing incident reports, flagging repeat offenders, checking on the victims and connecting them to service providers, attending meetings, and presenting information about domestic violence to various organizations and schools.

Neighborhood-Based Organizing and Problem Solving. A second promising practice is to establish a Mayor's Office on Domestic Violence (MODV) as implemented in Chicago. The MODV connects the police department, domestic violence service providers, and other agencies and assists their efforts by planning, monitoring, and evaluating services. The MODV also developed and implemented a public education campaign called "There Is No Room for Domestic Violence in This Neighborhood." The educational materials included posters, radio spots, and safety plan brochures printed in nine different languages and were distributed to the entire city and to groups who would likely be contacted by victims of domestic violence, including family and friends, faith-based groups, health care providers, employers, and the gay and lesbian communities. They also met with people in different communities and developed several recommendations for community outreach:

1. Encourage clubs to meet and talk about domestic violence.
2. Develop culturally sensitive domestic violence messages.
3. Increase access for all population groups by providing bilingual speakers and translation services.
4. Organize speak-outs.
5. Identify safe places for women in cases of an emergency.

In addition to these outreach services, the city also recommended the development of affordable housing, holding police accountable for their response to domestic violence incidents, and the creation of a special court for domestic violence cases.

Community Policing Action Team

Marin County created a Community Policing Action Team (CPAT) that consisted of officers of each police department within the county working under the lead of a full-time deputy specifically assigned to coordinate domestic violence responses. The deputy works with the Marin Abused Women's Services and is positioned at their office. The goal was to transform the community using three main strategies.

1. Transforming individuals by increasing knowledge and changing their attitudes, behaviors, and beliefs about domestic violence.
2. Giving the community the opportunity to take ownership of the issue and increase participation in prevention.

3. Changing the social norms that contribute to domestic violence via public policies and procedures.

Community members are encouraged to develop campaigns and strategies that hold the offenders accountable. CPAT also developed training for police officers as they recognized existing domestic violence among police officers. In order to change the social acceptability of domestic violence, it is imperative to ensure that police and community members have a good understanding of domestic violence, the consequences thereof, and effective responses.

Access and Collaboration. The Duluth Police Department developed its own response to domestic violence incidents. It was the first police department to implement a mandatory arrest policy if police officers had probable cause to believe that a misdemeanor assault or a greater crime had occurred. Duluth also gave community-based advocates access to training within the police department. Duluth founded the Domestic Abuse Intervention Program with the goal of developing collaboration between community groups and the police department. Another purpose was to revise law enforcement policies and reassess its community response by working with community groups. The revised policy gives officers the power to determine who was the predominant aggressor when there was aggressive behavior by multiple parties. The revised policy also addresses how to respond to children, assist victims, and be reviewed by supervisors.

The Safety and Accountability Audit as a Community Problem-Solving Tool. The Duluth Police Department also examined its community response to domestic violence, including protection orders and the policies used by police and prosecutors. This decision to conduct the audit was made after four battered women were murdered in the area. The two main goals in domestic violence cases are ensuring the safety of the victim and maintaining the accountability of the offender. When battered women are murdered by their batterer, those two goals are not met. The audit revealed that in the 100 cases initiated by law enforcement, there were 55 incidents of compromised security and officer accountability. Thirteen different agencies came together to address the problems and ensure the safety of domestic violence victims. One of the main findings was the time lapse between time of arrest and disposition, which gave the offenders much time to continue to injure the victim.[51]

SUMMARY

Law enforcement agencies face a range of challenges within their communities. Crime greatly impacts the life of community residents, and agencies spend significant resources on identifying and responding to crime. The SARA model is widely used across agencies to reduce crime and make neighborhoods safer. The chapter also discusses three types of offenses and offenders that have proven to present unique challenges. Police officers face unique issues with regard to domestic violence cases, calls for service for individuals with mental health issues, and alcohol-related crime. Alcohol is a main contributing factor to violent behavior and injuries due to drunk driving. This is especially a problem in college towns with a high alcohol outlet density. Alcohol and the effects thereof can have substantial negative effects on residents, such as loud noise late at night, destructive and aggressive behavior, littering, and creating an environment where residents feel generally unsafe. Reducing the alcohol outlet density and education of students and other constituents about the law and the negative effects of excessive alcohol consumption have shown some success.

Another common problem for police officers is calls for service involving individuals with mental disorders. The deinstitutionalization movement is the main contributing factor for the great increase in people with mental disorders who are homeless or warehoused in jails and prisons. Many officers have developed informal strategies in working with this population group to avoid arrests. The crisis intervention team is one of the best practices that has been successful in reducing arrests and providing services to this population group.

Finally, domestic abuse calls for service are a great strain on police departments and constitute a unique problem for officers and victims. Many victims do not report the crime, of if they do call for service, they often do not want to file a police report. Officers respond to the same couples on a regular basis and may witness escalating violence. Officers are also at great risk of injury or death during these encounters, and when officers are the offenders, the victims have very little recourse. Many departments have developed specialized units and have implemented unique practices to reduce the number of calls for service and increase officer safety. But domestic abuse remains one of the main challenges for the police.

KEY TERMS

Aggression 45
Binge drinking 44
Call subjects 51
Crisis intervention teams 52

Heavy drinking 44
Officer-involved domestic violence (OIDV) 57
Revolving door of incarceration 50

SARA model 43
Violence 46

DISCUSSION QUESTIONS

1. Explain the SARA model and how it helps law enforcement agencies identify and address crime and disorder in their communities.

2. Discuss the alcohol–aggression–violent crime relationship. Discuss the four pharmacological effects of alcohol. Which effect do you think is the most salient with regard to explaining why alcohol is a main contributing factor to violent crime?

3. Discuss the problems faced by police in neighborhoods with a high alcohol outlet density? What can police do to effectively address these problems?

4. Which factors contribute to the high number of individuals with mental health issues in jails and prisons? What strategies have proven successful in reducing arrests?

5. Discuss the problems domestic violence cases pose for victims and police officers. Which strategies have been effective in reducing calls for service and increasing the safety of officers?

6. Discuss the unique issues domestic violence victims face who are being abused by a partner who is a police officer. What should police departments do when dealing with these cases?

PART II

RESPONSES TO SPECIFIC CRIME TYPES

CHAPTER 4

COMMUNITY POLICING AND TERRORISM

Learning Objectives

1. Describe the different types of terrorism.

2. Explain the role of community intelligence in countering terrorism.

3. Discuss why it is important to build partnerships between police and Muslim and Arab American communities.

4. Describe the tensions that exist between the police and Muslim and Arab communities when they are working in a partnership and how these tensions may be overcome to build trust.

5. Explain why it is important to build resilience to violent extremism from right- and left-wing extremist groups within communities.

On December 2, 2015, Rizwan Farook and his wife, Tashfeen Malik, took their automatic weapons—the Colt ArmaLite Rifle 15 or AR-15—to the San Bernardino Inland Regional Center (IRC) and killed 14 civilians and injured 22 civilians.[1,2] The incident lasted several hours until police eventually killed the attackers. During the shootout with police, Farook and Malik also injured two police officers. The investigation revealed that the IRC incident was a terrorist attack that had been planned and prepared for some time.

The day at the IRC started at 8 a.m. as a training day for 80 employees of the San Bernardino County Environmental Health Department. Rizwan Farook was one of the employees, working as an environmental health inspector. The meeting room had been decorated for a Christmas party later that day. At around 10:30 a.m., Farook left the meeting shortly after reading a text message on his phone. He returned 30 minutes later, together with Malik, armed with automatic rifles. The couple fired more than 100 rounds outside and inside the meeting room. Some employees were able to escape through the doors and others took shelter underneath the tables. Witnesses stated that the shooters walked through the room and shot everyone who moved or made any sounds. The shooters abruptly exited the room after two to three minutes, got into their black SUV, and drove away.

Four police officers arrived at the IRC within three minutes of the attack and started an immediate search for the shooters inside the building. After the Columbine school shooting in 1999, during which two high school students killed 12 students and one teacher, officers had been trained to search first for the shooters and engage them to prevent further shootings. But as they were working their way through the smoke-filled room, it became very difficult to keep going. One officer stated, "It was the worst thing imaginable—some people were quiet, hiding, others were screaming or dying, grabbing at our legs because they wanted us to get them out, but our job at the moment was to keep going. That was the hardest part, stepping over them."³

In the months after the terrorist attack at the IRC, the Community-Oriented Policing Services (COPS) conducted an investigation of the incidence response and made recommendations for the training of police officers. They also made recommendations for communication and collaboration with the community. The recommendations emphasized the importance of building trust and long-term relationships with local leaders and community faith groups and identifying a liaison to work with in case of a terrorist attack. The report recognized that good community relations and trust by its residents are imperative to prevent future terrorist attacks.

Over the past decades, community police officers have built bridges with immigrant communities. However, the 9/11 terrorist attacks and their fallout have made it more difficult to build partnerships and trust in immigrant communities of Muslim and Arab origin due to a variety of factors, including racial profiling, federal laws, and an increase in hate crimes. At the same time, community policing has helped reduce crime and violence across the United States. The same principles can also be used to combat terrorism and violent extremism. Community policing emphasizes police–community partnerships, the building of trust in the police, and interacting with community residents and community leaders. These relationships are important to prevent violent extremist and terrorist attacks because police need information about suspicious activities and individuals. Without relationships and trust by the community residents and leaders, this type of information would not be communicated to police and police could not prevent violent attacks or arrest individuals following an attack. In addition, community–police partnerships are imperative to solicit the help of community members in identifying, preventing, and eliminating terrorist ideologies before they manifest themselves.⁴

INTRODUCTION

This chapter will discuss the difference between domestic and international terrorism. Students will learn about the role of community intelligence in countering terrorism and the importance of community policing in preventing violent extremism and terrorist attacks. Students will also learn about the importance of community policing for intelligence gathering, the tensions that exist between the police and Muslim and Arab communities, and how to overcome these tensions as well as the importance of building resilience within communities toward extremism. Finally, the chapter will explain community policing approaches to domestic terrorism and political extremists from both the left and the right.

TYPES OF TERRORISM

Defining Terrorism

Federal law defines **terrorism** as any violent or dangerous crimes that "appear to be intended" to either (1) intimidate or coerce a civilian population, (2) influence government policy by intimidation or coercion, or (3) affect government conduct by mass destruction, assassination, or kidnapping.[5] The first requirement is that the crime is actually listed as a violent crime in federal or state laws. For instance, murder is a violent crime. Thus, murder would fulfill that requirement. Second, the act must intimidate or coerce a population, influence government policy, or affect government conduct. The intimidation could be targeted at the civilian population. The courts have not clearly defined what the term *civilian population* exactly means. For instance, some people may classify a school shooting with numerous fatalities as terrorism because they define *civilian population* as the population of students in the school where the shooting occurred. Other people may disagree with this classification and argue that *civilian population* refers to the entire population within the United States. Thus, this second requirement is a matter of interpretation. The act may could also influence government policy by intimidation or coercion. For instance, antiabortion activists may attack a clinic where doctors perform abortions. One can argue that these are acts of terrorism because they are trying to influence abortion policies by intimidation and coercion. Finally, the violent crime could affect government conduct. For instance, the Unabomber Ted Kaczynski mail bomb attacks between 1978 and 1995 may have aimed to influence government conduct. In 1995, Kaczynski published his manifesto, which was later printed in the *New York Times* and the *Washington Post*. In his manifesto he criticized the industrial revolution and technology. Kaczynski might have attempted to change the conduct of government and reverse the industrial revolution and technology.[6]

International and Domestic Terrorism

The Federal Bureau of Investigation (FBI) distinguishes between domestic and international terrorism. **International terrorism** is defined as "violent, criminal acts committed by individuals and/or groups who are inspired by, or associated with, designated foreign terrorist organizations or nations (state-sponsored)."[7] Even though international terrorist groups have suffered setbacks and defeats, they are still a substantial threat to the homeland. In particular, the FBI highlights the continued threats emerging from Al-Qaeda and its affiliates, such as the Al-Qaeda in the Arabian Peninsula and Syria. In addition, the Department of Homeland Security is paying close attention to the Islamic State of Iraq and the Levant (ISIS), which still has significant resources and global support of its branches. The United States is especially concerned about soft targets that could be attacked by ISIS supporters from their many locations. ISIS has been very successful using social media to post propaganda and recruit followers. For instance, in March, the FBI arrested a man from Maryland. He was planning to use his vehicle for a ramming attack with the goal to kill as many people as possible. In August, the FBI arrested a person in Queens who was planning to stab people in the name of ISIS. The FBI is also paying close attention to the Islamic Revolutionary Guard Corps-Qods Force and its partner Hizballah. In 2019, the FBI arrested two men who had been surveilling Jewish and Israeli facilities and dissidents from Iran. The two men pled guilty to the charges. The FBI is not only concerned about these international terrorist groups, but they are even more concerned about homegrown terrorists.[8]

Domestic terrorism is defined as "violent, criminal acts committed by individuals and/or groups to further ideological goals stemming from domestic influences, such as those of a political, religious, social, racial, or environmental nature."[9] All of the recent terrorist attacks were classified as domestic terrorism.

With regard to domestic terrorism, the FBI distinguishes between **terrorism committed by large groups**, such as Al-Qaeda, and terrorist acts committed by lone offenders, such as the Unabomber. Lone offenders are classified into two categories. First, there are **homegrown violent extremists** who are inspired by international terrorist groups. The FBI found that about one third of all identified homeland attackers during 2018 were juveniles inspired by terrorist groups such as ISIS and Al-Qaeda.[10] For instance, on June 12, 2016, Omar Mateen shot and killed 49 people in a nightclub in Orlando. He was an American citizen who had pledged his allegiance to ISIS. This was one of the deadliest mass shootings since the 9/11 terrorist attacks. Mateen had been on the FBI radar and was interviewed by the FBI in 2013 and 2014. The FBI did not believe that he was a credible threat. After the attack, ISIS supporters praised the attack.[11]

Second, there are **domestic violent extremists**, who are inspired by domestic ideologies that are often spread via social media and the Internet.[12] In 2020 the FBI made racially motivated violent extremism its top priority. FBI director Christopher Wray stated that the threats of violence stemming from these groups are as high as the threats arising from international terrorist groups. Investigation of domestic terrorism motivated by racial or religious hatred makes up a large part of the current FBI investigations. According to Wray, most of the threats and terrorist activities are fueled by beliefs of white supremacy. Hate crimes, which are a close cousin to racially motivated violent extremism, are also a top priority. These terrorists typically attack soft targets, such as houses of worships, shopping centers, and public events where large crowds gather.[13]

For instance, on August 3, 2019, Patrick Crusius shot and killed 22 people outside a Walmart in El Paso and wounded 24 others. The FBI classified the shooting as a case of domestic terrorism, and Crusius was charged with capital murder, which could result in the death penalty. Crusius stated that "his targets were Mexicans," he opposes "race mixing," and he wanted to stop the "Hispanic invasion." The FBI believes that Crusius posted a document full of hatred toward Mexican immigrants and Latinos on an online messaging board about 20 minutes prior to the attack. This has been characterized as an example of homegrown right-wing extremism.[14]

In 2019 the FBI established the Domestic Terrorism—Hate Crimes Fusion Cells, which are tasked with combat terrorism and investigation of hate crime. The main purpose of the fusion is sharing information and improving investigative resources to bring justice to victims and prevent future crimes.[15]

There are three other main categories of domestic terrorism: (1) antigovernment/antiauthority extremism; (2) animal rights/environmental extremism, and (3) abortion extremism.[16]

Antigovernment/antiauthority extremists are also called *anarchist extremists*. They are loosely organized groups with no central leadership who oppose any type of authority, such as a central government, laws, and police. Their goal is to destroy the symbols of government and capitalism, including police agencies, government agencies, and corporations using violence. They destroy property, use firebombs, and attack government officials.[17] During the 1970s and 1980s, a group called Weather Underground detonated a bomb at the headquarters of the U.S. State Department in Washington, DC. No one was killed or injured, but there was substantial damage to 20 offices spread across three floors. The group claimed credit for many more bombings over the years. The group stated that "Our intention is to disrupt the empire . . . to incapacitate it, to put pressure on the cracks."

Even though the group ceased its activities in the mid-1980s, many of its members were not arrested until several decades later.[18]

Animal rights/environmental extremism is classified as special interest extremist groups. This includes groups such as the Animal Liberation Front (ALF) and the Earth Liberation Front (ELF). These groups are different from traditional right-wing and left-wing terrorist groups because they are targeting specific issues, including releasing lab animals and resisting the killing of animals for consumption. For instance, the ALF and ELF have cut break lines on trucks transporting frozen fish or meat, set fire to meat processing factories, and placed firebombs in private homes and other places. They do not intend to kill human beings during their attacks and there have been no casualties resulting from their attacks. They are mainly recruiting young people between the ages of 18 and 25 and have engaged in thousands of terrorist acts resulting in more than $100 million in damages.[19]

Between 1995 and 2010, ALF and ELF combined were responsible for 239 arsons and bombings. Their main targets are private homes, meat processing plants, automobile and truck dealerships, universities, and fur and leather companies. A relatively small number of people were responsible for the vast majority of attacks. It is very difficult to identify the extremists, however, because they often do not engage in legal protests or movements.[20]

Abortion extremists use violence against abortion providers, including murder, bombings, arson, assault, kidnapping, and vandalism. They also regularly make death threats against doctors, nurses, and staff of clinics where abortions are performed.[21] Between 1973 and 2003, there were more than 300 attacks on abortion providers by anti-abortion extremists. The vast majority were arsons, bombings, and butyric acid attacks. Most of the attacks resulted in property damage, but there were also people killed during the attacks.[22] One of the most publicized attacks occurred in Kansas. In May 2009, Dr. George Tiller, the sole abortion provider in Wichita, Kansas, was killed by Scott Roeder, who is now serving a life sentence. The clinic where Dr. Tiller had been working stayed closed for four years until it reopened in 2013.[23]

One of the main concerns over the past few decades has been the increasing use of the Internet and social media by domestic and international extremists for the purpose of spreading their ideology, recruiting new members, and planning and carrying out attacks. The FBI is especially concerned with their ability to recruit American citizens and carry out attacks within the United States. The FBI is partnering with law enforcement agencies to identify extremists living in the United States and disrupt terrorist attacks before they occur. The Nationwide Suspicious Activity Reporting (SAR) Initiative is an example of collaboration between the FBI and other agencies such as Homeland Security and local law enforcement. The main goal is to gather, document, process, analyze, and share information about extremist activity and potential terrorist attacks. They are also asking citizens to report suspicious activity to their local law enforcement.[24]

THE ROLE OF COMMUNITY INTELLIGENCE IN COUNTERING TERRORISM

Since the 9/11 terrorist attacks, the United States has greatly improved its antiterrorism efforts. Federal, state, and local law enforcement have built stronger relationships and improved their sharing of intelligence. But information sharing between law enforcement agencies is not sufficient to counter terrorist attacks. Law enforcement must gather intelligence from community members to further its investigations.

The community is a very important part in preventing violent extremist and terrorist attacks, such as the attack in San Bernardino, because the residents have valuable information about people who may have become radicalized and who may pose a future threat. In addition, the community often needs the help of police and other services because the residents are negatively impacted in several ways by the attack, and especially if the attacker was part of the community. The community loses some of its members and others suffer from anxiety and fear of future attacks. The community may also become the target of revenge and hate crimes against the residents as they may be assumed to be potential terrorists also. But instead of help from police, communities often come under increased surveillance and the target of racial profiling, which creates significant tensions. This is very problematic because residents who feel discriminated are less likely to help police when needed. Information from the community is critical for law enforcement during a terrorist attack as it may help police apprehend the suspects. For instance, information from the community helped police apprehend the two brothers who had carried out the Boston Marathon Bombings in 2013 (Case Study 4.1).

The example of the Boston Marathon Bombings exemplifies why good relations between the police and the community are important to stop violent extremism. One of the most effective ways in which the Boston community helped the police was through social media. The Boston Police Department used its Twitter and Facebook accounts to distribute information, keep the public informed, and correct misinformation. Even media outlets waited for the police tweets and Facebook posts before reporting on the events as they unfolded.[25]

Much of the success of preventive policing depends on the ability of police departments to generate intelligence from the community and incorporate such information into police priority setting. **Intelligence** is defined by the Bureau of Justice Assistance as

> the collection of critical information related to the targeted criminality that provides substantive insight into crime threats and identifies individuals for whom there is a reasonable suspicion of relationship to a crime. Information collection is a constant process, along with ongoing information verification and analysis.[26]

Without intelligence, police cannot prevent the commission of crimes and terrorist acts. Community policing is imperative to the goal of generating intelligence. If community police officers are able to establish relationships with the residents, they can identify community concerns about crime, suspicious behaviors, and extremist behaviors. These relationships also enable joint action and problem solving between the police and the community.[27]

The executive assistant director for the Office of Intelligence at the FBI in 2005, Maureen Baginsky, stated that

> Intelligence is not something that is only collected by covert agents attempting to subvert another government or organization, or even prevent

CASE STUDY 4.1: COMMUNITY INTELLIGENCE FAILURE AND THE BOSTON MARATHON BOMBING

On April 15, 2013, three people were killed and more than 264 people were injured by two pressure cooker bombs that detonated within a few seconds of one another near the finish line at the Boston Marathon at 2:49 p.m. Eastern Time.[28] The bombs were hidden in backpacks. They were self-made and included explosives, nails, shards of metal, and ball bearings. Following the bombing, police desperately searched for the perpetrators. Police recorded witness statements and statements by the surviving victims. The police released video footage and pictures of two suspected perpetrators, suspect 1 and suspect 2. Three days later, on April 18 at 5 p.m., the police asked the community to help identify the suspects. A few hours later, at 10:30 p.m., police officer Sean Collier was killed by the suspects while sitting in his police cruiser. The suspects had attempted to take his gun.

The suspects then carjacked a car and took the owner "Danny" hostage. They pulled up behind "Danny's Mercedes," pulled a gun on "Danny," and stated that they are the Boston Marathon bombers. They started loading heavy equipment into the trunk of the Mercedes and then drove away. They used Danny's ATM card to withdraw money. The next morning on April 19 at 12:15 a.m., while stopping to get gas, Danny managed to flee and call the police. Via Danny's iPhone and the car's satellite system, police were able to trace the car. They eventually found the abandoned car in Watertown. The suspects had taken possession of a Honda after a gun battle with police. Police canvassed the neighborhood and talked to the residents who identified one of the suspects as Tamerlan Tsarnaev. Police later found Tamerlan and he was shot during the encounter. He died at the hospital a few hours later.

The second suspect was still on the run and police ordered the residents of Watertown to shelter in place. Police went door to door, trying to find suspect 2. One of the residents checked his boat because the cover was hanging loose. He saw blood and a body inside his boat and called the police. At 7:30 p.m. on April 19, police and a Specialized-Weapons and Tactics (SWAT) team surrounded the boat and started firing. The second suspect, Dzhokar Tsarnaev, who was hiding inside the boat eventually surrendered to the police and was taken into custody.[29]

The community was an integral part of the police investigation and the capture of suspects. Without the videos, pictures, and other information, such as tweets by the brothers, it would have taken longer to identify and capture the brothers and more people may have gotten killed and injured.

Discussion Questions:

1. What role does the social media play in helping police catch terrorists?

2. What may be the dangers of using social media to pursue terrorists?

attacks on our own government. This view is antiquated and no longer valid. The truth is, the collection and analysis of intelligence is no longer limited to government agencies.[30]

Types of Intelligence

Loyka (2013) identified three types of intelligence: (1) strategic intelligence; (2) operational intelligence, and (3) tactical intelligence. First, strategic intelligence

▸ **Photo 4.1** What role can state and local police play in counterterrorism efforts?

includes information about criminal activity, criminal groups, and threats. **Strategic intelligence** allows the police department to plan and allocate resources and understand specific intelligence targets, including the target's philosophy, structure, motivation, and characteristics. For instance, police may try to collect intelligence on a specific terrorist group, such as the ISIS, and people in the community who support ISIS and may engage in terrorist acts. It is very difficult to collect this type of intelligence and it is often lacking in police departments, making it difficult to prevent crimes. For instance, on October 31, 2017, Sayfullo Saipov rented a Home Depot truck and intentionally drove into a bike path full of people in Downtown Manhattan. The police later found a handwritten note saying that Saipov was acting on behalf of ISIS. Unless police know which individuals have a violent extremist ideology and are plotting an attack, it is impossible to prevent such attacks. This type of information is hard to come by, of course, because these individuals typically stay hidden.

Second, **operational intelligence** refers to imperative operational decisions, such as the surveillance of an individual or a group. The main purpose of gathering operational intelligence is to maintain public safety. Operational intelligence provides information about people who should be monitored because they may commit a terrorist attack or other criminal act. The Patriot Act and Executive Order 123333 ("United States Intelligence Activities") provide the legal authority for the surveillance of people. These two authorities have greatly expanded the ability of law enforcement to collect data. After the leaking of information by Edward Snowden about the collection of information from millions of people in the United States and other countries by the National Security Agency (NSA) in 2013, there was much concern about privacy and individual rights. Following the Snowden exposé, the NSA ended mass surveillance, but some people have warned that effective counterterrorism strategies rely on operational intelligence and without comprehensive surveillance, terrorists may not get caught until they have carried out an attack. The goal of the mass surveillance was preventing terrorist attacks and the revelations by Edward Snowden and the ending of the NSA mass surveillance have substantially impaired law enforcement.

Finally, **tactical intelligence** is typically used for "raid planning" or in crisis situations. Tactical intelligence is imperative for the formulation of criminal investigations as it helps build a case against a suspect. During a crisis situation, tactical intelligence is important to mitigate the threat and manage a response to the threat. For instance, in the event of a terrorist attack, tactical intelligence enables law enforcement to act decisively and either prevent the attack or mitigate the damages caused by the attack.[31]

Intelligence-Led Policing

Intelligence-led policing became a main focus of law enforcement practices after the 9/11 terrorist attacks and the 9/11 Commission Report, which detailed the failure to collect and share intelligence that could have potentially prevented the attacks. As a result of the report, the role of intelligence in policing has greatly increased across federal, state, local, and tribal levels of law enforcement.

The term *intelligence* is often used synonymously with the word information. Intelligence is different from information, however. Intelligence refers to "information plus analysis." Thus, intelligence is the product of collecting information and evaluating and analyzing the information. Intelligence is a critical component for law enforcement decision-making, planning, targeting, and crime prevention. Once a police department has received information and analyzed it, it has to decide what to do with the intelligence it has gained. The purpose of **intelligence-led policing** is to prevent crime and terrorist attacks, more efficiently allocate resources, and develop counterterrorism strategies.[32]

Intelligence-led policing is based on community policing and problem-oriented policing as police–community partnerships directly support intelligence-led policing by gathering information from the public, improving communication between police and citizens, reducing fears, and taking a scientific approach to policing. Community police officers know and understand their communities, enabling them to develop effective responses to specific problems in these communities. As we have discussed in Chapter 1, community policing is focused on the SARA model based on the four steps of scanning, analyzing, responding, and assessing problems. Officers focus on the problem rather than the incident. They systematically collect and analyze data about the problem and then tailor their response to the problem and the community. Case Study 4.2 details how intelligence-led policing helped to arrest youth trying to flee the country to join ISIS.

CASE STUDY 4.2: CATCHING TERRORISTS IN MINNESOTA

Minnesota has become a main recruiting ground for terrorists. In 2015, law enforcement arrested six men for attempting to travel to Syria and join the terrorist group Islamic State (ISIS) of Syria and Iraq. The men had tried to fly to countries near Syria from San Diego and New York. Zacharia Yusuf Abdurahman, 19, Adnan Farah, 19, Hanad Mustafe Musse, 19, and Guled Ali Omar, 20, were arrested in Minneapolis and Abdirahman Yasin Daud, 21, and Mohamed Abdihamid Farah, 21, were arrested in California. The police had received a tip from a man who had been part of the group and who had also planned to go to Syria and join ISIS.

Rick Thornton, director of Minnesota's FBI office, stated that the Somali community, especially the juveniles, has been the target of terrorist recruitment efforts. There is a high threat of radicalization among these Somali juveniles because they feel alienated from the broader society.[33, 34]

Discussion Questions:

1. If you were a police chief in Minnesota, how would you approach the problem of terrorist recruitment in the Somali community, especially juveniles?

2. How would law enforcement be able to collect intelligence about people who plan to join ISIS?

THE IMPORTANCE OF BUILDING PARTNERSHIPS BETWEEN POLICE AND MUSLIM AND ARAB AMERICAN COMMUNITIES

In the previous section, we discussed what intelligence is and the importance of intelligence for effective counterterrorism strategies, such as the apprehension of the Boston Marathon bombers and the attackers of the IRC in San Bernardino. In order to get such intelligence, police departments must develop partnerships with community leaders.

Since the 9/11 terrorist attacks, one of the main concerns has been the radicalization of Muslims and Arab Americans, especially young people. These young Muslims and Arab Americans who are at high risk of becoming radicalized often live in immigrant communities of Muslim and Arab descent. Muslim Arab Americans are at higher risk of radicalization because they are often marginalized and disenfranchised from mainstream society as a result of racism and Islamophobia. They may be invisible to law enforcement, but other members of the community may take notice of changing behaviors, especially radicalization and the potential for violent extremism. Thus, it is imperative to build partnerships with community leaders in Muslim and Arab American communities to get information about at-risk youth and adult members who could potentially plan terrorist attacks.

There is no greater source to information about community members than other members of the same community. Lyons (2002) stated:

> Until we learn to police in ways that build trusting relationships with those communities where criminals or terrorists can more easily live lives insulated from

observation—no amount of additional funding or legal authority, consistent with living in a free society, will increase the capacity of our police forces to gather the crime and terror-related information we desperately need.[35]

One of the groups at the highest risk of radicalization are Muslims and Arab Americans who have been incarcerated. This is not only true for the United States but also for European countries. Great Britain, which has a significant Muslim population and has experienced numerous terrorist attacks, has developed a pilot study where mosque groups create outreach programs to incarcerated Muslims in an attempt to de-radicalize them and reintegrate them into the community after their release. These groups are sometimes referred to as "theological fire brigades." Individuals at risk of radicalization are identified prior to their release and contacted by the mosque group. The mosque group conducts a theological intervention with the goal to prevent these individuals from joining terrorist groups or to convince them to leave terrorist groups. The mosque groups also provide education classes in prison with the goal of preventing radicalization.[36]

Office of Community-Oriented Policing Services

In the United States, the Office of Community-Oriented Policing Services (COPS), an agency under the Department of Justice, is tasked with advancing community policing practices within state, local, territorial, and tribal law enforcement agencies. The mission statement of the COPS office recognizes the importance of building partnerships with communities based on trust and mutual respect.[37]

Community policing begins with a commitment to building trust and mutual respect between police and communities. It is critical to public safety, ensuring that all stakeholders work together to address our nation's crime challenges. When police and communities collaborate, they more effectively address underlying issues, change negative behavioral patterns, and allocate resources.[38]

COPS has defined **trusting relationships** as relationships

where citizens voluntarily approach police officers with information or problems because they trust that law enforcement represents their best interests. At a minimum, they are willing to guide police activity through structured forums or meetings, and information they pass along will often pertain to substantive crime issues in their communities. Likewise, police officers trust that community members wish to aid them in their efforts to promote security and safety in the neighborhood. Police officers can also be confident that information they acquire from community members will be useful in accomplishing this task. Most important, when this level of collaboration occurs, the community will share responsibility for the effectiveness of the strategies implemented.[39]

In order to improve the ability of law enforcement to build partnerships and enhance community policing, COPS and the Police Executive Research Forum (PERF) convene executive sessions for law enforcement chief executives, policing professionals, and government policymakers to share their knowledge and discuss strategies on combating terrorism while advancing community policing. After the executive sessions, a white paper is distributed to law enforcement agencies and decision-makers at all levels of government.[40]

One of the first executive sessions in 2003 focused on engagement and partnerships of communities with a diverse population, and especially Muslim, Arab, and Sikh populations. The main task of the session was to discuss problems within these communities that made building partnerships and identifying strategies to overcome these problems a difficult task. The white paper details the concerns these communities have, including concerns about laws, law enforcement practices, and hate crimes against members of their community. The white paper also outlines federal law enforcement issues that impact the ability of local law enforcement to work with diverse communities. Finally, the white paper outlines the challenges local law enforcement agencies face in building partnerships in these diverse communities and preventing terrorist acts.[41]

Based on their experiences, COPS has proposed five key principles for law enforcement: (1) fostering and enhancing trusting partnerships with the community; (2) engaging all residents to address public safety matters; (3) leveraging public and private stakeholders; (4) utilizing all partnerships to counter violent extremism; and (5) training all members of the department.[42] Case Study 4.3 explains the importance of building relationship with the community.

TENSIONS BETWEEN POLICE AND MUSLIM COMMUNITIES, OVERCOMING TENSIONS, AND BUILDING TRUST

As discussed previously, effective police counterterrorism efforts greatly depend on their ability to partner with Muslim and Arab American communities. Establishing these types of partnerships is challenging due to inherent differences in perceptions and beliefs between police and residents. The difficulty already begins with the definition and causes of terrorism, the meaning of the term radicalization, and the seriousness of the problem.[43] In addition, there are other problems. Cherney and Hartley (2017) identified four main obstacles to building partnerships between police and leaders in Muslim and Arab American communities: (1) distrust between Arab American communities and the police; (2) balancing the priorities of intelligence gathering, community engagement, and trust building; (3) determining who to partner with in the community; and (4) levels of consultation and community input.[44]

Distrust Between Muslim and Arab American Communities and the Police

First, fruitful community partnerships are built on trust and mutual respect. Trust requires that people in the community believe that the police have the best interest of the residents in mind. Unfortunately, many actions by police have left the impression among Muslim and Arab American people that they are the target of police investigation rather than being protected by police. For instance, police use of stop-and-search powers, blanket surveillance, travel bans, and other tactics have left many Muslims and Arab Americans feeling under siege. But, it is not only the surveillance and racial profiling of Muslims and Arab Americans; there is also another dimension of discrimination that is less obvious and less talked about that has great negative effects on the development of partnerships: the exclusion of Muslims and Arab Americans from decision-making processes or the perception of being excluded. Some of these perceptions are due to poor communication. For instance, a law enforcement agency consulted with members of the community with regard to some issues that had to be addressed but then failed to communicate the outcome to the community and, as a result, the community felt that its input had no real influence on the outcome. It left a sour feeling

CASE STUDY 4.3: UNITED KINGDOM—RESPONSE TO THE 7/7 BOMBING

On July 7, 2005, three bombs went off in different parts of the underground train system and one on a double decker bus. Four suicide bombers, who were associated with the terrorist organization Al-Qaeda, killed 52 people and injured hundreds across the United Kingdom. At the time, the United Kingdom had a significant deficit on intelligence regarding the threat posed by Al-Qaeda and the main question was how to get the intelligence necessary to prevent future attacks and arrest people who were planning an attack. For instance, the United Kingdom had little information about the four suicide bombers. Traditional intelligence methods (intelligence collected by national security agencies, such as MI5) were not effective, and they were especially ineffective in Muslim communities, which was a great concern for the United Kingdom's counterterrorism efforts. It became clear that police needed to build relationships and trust within communities to gather better intelligence.[45]

Some police units developed Special Community Engagement Units with the goal to build "strategic contacts" with community leaders and opinion formers—that is, people who had close contacts and influence on groups that were important for police and their counterterrorism efforts. These community leaders and opinion formers would gather intelligence on people and groups of interest in these communities, which would then help the police arrest people who were planning a terrorist attack. The community leaders and opinion formers were also used to feed information to the community, typically with the goal of countering rumors and misinformation. There was no secrecy involved. Police engaged openly with these community leaders and opinion formers. The main idea was that even though many residents in these communities distrusted the police as an institution, they might trust certain individuals. These trusted individuals might be able to build personal relationships with the community representatives and thus be able to gather important intelligence.[46]

Discussion Questions:

1. Think about what you know about the U.S. response to the 9/11 terrorist attacks with regard to its community policing strategies.

2. What are similarities and differences you can think of between the United Kingdom and the United States?

3. What strategy would you recommend for responding to terrorist attacks? How can we engage communities to work with police to prevent terrorist attacks and arrest terrorists?

of having been consulted only for the police to be able to say that they care about the needs and concerns of the community. The residents felt betrayed, hampering future partnerships. Police must ensure that the input received from community members receives significant weight in their decision-making processes and that it must then be communicated effectively to the community.

Other problems arise from poor coordination among police. A certain police unit may have conducted a police action without consulting or involving the community police officers or the community partners. For instance, the FBI may conduct a raid in the neighborhood without informing the community police officer. The community leaders may not understand why they should work together with the police and provide information if there is no communication from the police in return. This type of situation undermines trust and cooperation and it may take substantial effort on behalf of the community police officer to reestablish a functioning partnership with community residents again. It is important that different police agencies, such as the FBI and other federal agencies, communicate with the local police to avoid undermining the efforts of the local police.

In addition, members of a community may not understand how law enforcement agencies conduct their business and they may misinterpret police actions and perceive it as illegitimate. Police react to a wide range of complex situations every day and these situations typically involve disputes, crimes, and violent events. Police actions often involve arrests and searches of residents. And even though police do their best to handle all police actions effectively and competently, sometimes things go wrong. It is in these situations that the community feels that its members have been wronged and that the police may be "out to get them." A study by the Vera Institute of Justice (2006) demonstrates the tensions that have ensued since the 9/11 terrorist attacks.[47] On September 11, 2001, four simultaneous terrorist attacks carried out by Al-Qaeda killed almost 3,000 people. Nineteen terrorists had kidnapped four airplanes. Two of the planes were crashed into the New York Twin Towers of the World Trade Center, killing 2,606 people, including 343 firefighters who were trying to save the lives of people inside the Twin Towers. A third airplane hit the Pentagon outside Washington, DC, killing 125 people, and the fourth airplane crashed in a field in Pennsylvania.[48] After the attacks, police strategies changed to an increased focus on homeland security and policing Muslim and Arab American neighborhoods. Federal and local law enforcement agencies began collecting information about people and organizations suspected to have ties to terrorist organizations, especially Al-Qaeda. This new focus on homeland security was backed by legislation, mainly the Patriot Act. The heightened surveillance of Muslim and Arab American communities had serious negative consequences, including an increased sense of being victimized, harassed, and curtailed in their civil liberties. They also felt a greater suspicion toward the U.S. government and police and anxiety about their place in society.[49] All of these different issues result in substantial distrust toward police by the residents of these communities.[50]

This distrust runs both ways, however. Law enforcement also feels distrust toward the residents in the Muslim and Arab American communities, adding to the tensions. One of the most common complaints by police and politicians is that Muslim leaders don't speak out enough against violent extremism.

Credit Line: iStock.com/400tmax

▶ **Photo 4.2** In 2016 the New York Police Department began allowing Sikh officers to wear traditional turbans and have beards up to a half-inch long.

There is a powerful conventional wisdom in the law enforcement circles that I live in: That these communities are at heart uncaring, complicit in neighborhood crime, corrupt, destroyed. Nobody cares about the crime, the law enforcement narrative goes, or they raise their kids right, get them to finish school, have them work entry-level jobs—like I did, like my kids do—instead of selling drugs on the street. They don't care about the violence: nobody will even tell us who the shooters are . . . Nobody cares about the drugs because everybody's living off drug money. . .[51] (pp. 17–18).

Such attitudes can undermine the goodwill of leaders in Muslim communities, making it more difficult for police to build partnerships to combat the rise of extremism.

Social capital are relationships built on trust. In crisis situations or situations where police make a mistake, they can draw on the social capital that they have built. In such crisis situations, police can contact the community leaders with whom they have a relationship of trust and inform them about what happened and discuss how such situations can be prevented in the future. The community leaders can then communicate with the residents.[52] For instance, in 1999, a Florida Specialized Weapons and Tactics (SWAT) team conducted a late-night drug raid. They broke into the home of Catherine and Edwin Bernhardt, threw Catherine to the ground, and held her at gunpoint. Her husband, Edwin, came downstairs in the nude when he heard the noise. The police handcuffed him and forced him to put on a piece of his wife's underwear. Police then took the couple to the police station and held them for several hours. The police eventually discovered that they had raided the wrong address.[53] These types of mistakes happen and sometimes people even get killed by mistake. If police have social capital within the community where the incident occurred, they can ask the community to assist them in reviewing the incident and providing the input needed for procedural improvements. This type of involvement during a crisis can strengthen the relationship. Conversely, a lack of communication and collaboration can weaken the relationship and destroy social capital.[54]

Community policing strategies can result in trust and respect between police and communities. Indeed, some research suggests that trust in police may be growing due to increased community policing strategies and the building of partnerships. Sun and Wu (2015) found that the majority of Arab Americans in Detroit neighborhoods have confidence in police and that trust between neighbors is closely related to confidence in the police. Thus, helping build a cohesive neighborhood with mutual respect and trust will often translate into better partnerships between police and community leaders and help identify terrorists hiding in these communities.[55]

However, many neighborhoods lack police–community partnerships and there remains a substantial amount of defensiveness and also hostility among Muslim and Arab American communities toward police, which feeds into the hands of the terrorist groups and recruiters. One of the key goals of militants and recruiters is to polarize Muslims against *infidels* and motivate them to participate in a worldwide Jihad. However, just as there are people who are susceptible to this message, there are many Muslims who feel alienated by these militants and who are concerned that the militants are hijacking their religion. Police need to identify Muslim leaders who have these concerns and develop community strategies to counter the rise of extremism.[56]

Balancing the Priorities of Intelligence Gathering, Community Engagement, and Trust Building

Second, balancing the priorities of intelligence gathering, community engagement, and trust building creates significant problems to building successful partnerships between police and Muslim and Arab American community leaders. Community-based intelligence gathering is fundamentally different from traditional intelligence gathering through surveillance and informants. Community-based intelligence gathering is driven substantially by the community concern for everyone living in the community and how providing intelligence to police will impact the life of others. In addition, residents in the communities are concerned that police officers are only interested in intelligence gathering rather than honest community engagement. Motivation is a key factor of gaining community trust. Police must be able to demonstrate to community residents and leaders that they truly want to engage in the community as a partner, not just as beneficiary.[57]

One such example of effectively engaging local community leaders, in this case the imam, occurred in Berlin, Germany. In February 2008, a museum in Berlin Moabit (a neighborhood mostly comprising immigrants) experienced an outrage by some members of the Muslim community as a reaction to an exhibition organized by a Danish artist around the theme of political and religious extremism. One of the exhibitions showed an image of the Kaaba, a sacred building inside the Great Mosque of Mecca, with the title "Stupid Stone." Fifteen young Muslims men threatened to use violence if this piece of the exhibition was not removed. The museum temporarily closed out of fear of a violent attack. Local politicians strongly protested the closing of the exhibition and took a firm stance on the freedom of speech. They also worked with the local imam in addressing the problem. The imam visited each of the young Muslim men and convinced them that such conflicts cannot be resolved with violence. The museum did not experience any further problems after reopening.[58]

As you can tell from the Berlin example, it is very important to build relationships and trust with community leaders, such as imams, in communities that are the least likely to assist the police and where extremists can easily hide to avoid violent attacks. The 15 young Muslims would likely not have listened to German law enforcement, but they did listen to their imam because in their community the imam is the authority figure rather than the German police.

Determining Who to Partner With in the Community

Third, determining whom to partner with in the community is a challenging task. Police often engage with a small number of community leaders who often represent the mainstream residents. The actual target of police, however, are extremists and especially people who are leaning toward violent extremism because they are the most likely to being recruited by terrorist groups and commit terrorist acts. Thus some researchers suggest that police must engage leaders of the orthodox and nonviolent radical extremist groups to receive information about people who may be planning a terrorist attack.[59]

Engaging with these orthodox groups is important because the members of these groups can provide insights about the radicalization process and how people get engaged and recruited by terrorist organizations. Also, these orthodox groups carry a greater voice among the population at the highest risk of becoming radicalized, which makes them good partners in counteracting radicalization.[60] Partnering with these orthodox groups also carries significant risks for police, however. Police are concerned about becoming entrapped with these extremist groups and the political consequences of these partnerships. Nonextremist Muslims may feel alienated when police partner with the extremist groups in their community, and vice versa. Muslim communities have voiced their concerns that such police–extremist partnerships can be divisive and foster new sectarian bodies that have relationships with U.S. neoconservative organizations. Neoconservatives generally believe that the United States must use whatever military strategy is necessary to preserve its status as the greatest global power. They believe that any threat to the United States must be eliminated with preemptive military action. They strongly supported the preemptive strike against Saddam Hussein and the Iraq war and they typically show unwavering support for Israel. This ideology, of course, causes great concerns to the Muslim communities who feel targeted by neoconservatism.[61]

In addition, when law enforcement partners with Muslim groups, and especially extremist groups, the police lose their ability to criticize these groups because relationships are built on trust, which would disappear if police would use information gained from the groups to criticize them. There is some evidence that partnerships with Muslim communities have reinforced the belief among law enforcement that Muslim communities are working to curb terrorism. However, these partnerships have done little to build trust toward the police in Muslim communities, making it less likely that police will receive critical information about future terrorist attacks.[62]

Levels of Consultation and Community Input

Fourth, consultation and community input must be real, that is, community engagement must include the opportunity for the community to give input and have an impact on outcomes. Consultation without any influence on the outcome can cause more damage than good because it suggests that police are not sincere about the partnership. Stated differently, a partnership cannot be a one-way street. All partners must receive benefits from the partnerships or there will be increased distrust and resentment by the group that have no influence or benefit.[63] Some researchers suggest that in partnerships between police and Muslim and Arab American communities, the community should drive the agenda and have the opportunity to make its voice and grievances about counterterrorism tactics heard. Counterterrorism tactics, including counterterrorism laws (e.g., Patriot Act), surveillance, racial profiling, random stops and searches, security checks at airports, and a lack of response to Muslim victimization and hate crimes, have had great negative effects and create an environment where Muslims feel unjustly targeted. This can become the driver of radicalization, especially among young Muslims who grow up feeling targeted and discriminated against. Terrorist groups seek them out and offer them a home where they can "get even." Breaking the cycle of distrust and resentment is a difficult task and can only be accomplished via an ongoing dialogue and sincere engagement of Muslim communities.[64]

Communities have assets that may be used to prevent extremism and terrorist attacks. Unfortunately, there are communities where police rarely receive assistance from the residents in countering crime and terrorism. It is especially important to target these communities and build relationships and trust as these may be communities where crime rates are high and where terrorists may hide and go unnoticed. Case Study 4.4 provides a feature on homegrown terrorism.

THE IMPORTANCE OF BUILDING RESILIENCE TO VIOLENT EXTREMISM WITHIN COMMUNITIES TO RIGHT- AND LEFT-WING TERRORISM

The vast majority of attacks has been carried out by members of the Muslim community, particularly young Muslims living in the country where the attack occurred.[65] For instance, the attacks in San Bernardino were carried out by two members of the Muslim community, Tashfeen Malik and Syed Rizwan Farook, who were supporters of the terrorist group Islamic State (ISIS).[66] However, there are also many other attacks that have been carried out by white supremacists or left-wing extremists as described at the beginning of the chapter. Since 9/11, white supremacist and other right-wing extremist groups committed the majority of terrorist attacks in the United States. According to a report by the Anti-Defamation League, between 2009 and 2018, 73% of fatalities caused by terrorist

CASE STUDY 4.4: HOMEGROWN TERRORISM IN EUROPE

The United States is not alone in its battle against the rise of violent extremism. Europe also struggles to address the challenges posed by what some researchers call "homegrown terrorism." Europe has a significant immigrant population, and *homegrown* in Europe refers to marginalized immigrant neighborhoods with mostly Muslim background. There have been numerous incidents across Europe where people have been attacked and killed by violent extremists. For instance, on November 2, 2004, Dutch filmmaker Theo van Gogh was first shot and then stabbed to death in Amsterdam by Mohammed Bouyeri. The knife used to stab van Gogh also pinned a letter to his chest. Bouyeri, a young extremist with Moroccan decent, wanted to set an example of what would happen to people who, in his opinion, insulted Islam. The letter contained death threats to several other people, including another filmmaker, Hirsi Ali. Van Gogh and Ali had made a 10-minute short movie titled *Submission*. The main purpose of the movie was exposing the Islam's treatment of women. The movie is about four women who detail different types of abuse they experienced, including being beaten. All four characters were played by a single actress wearing a transparent black veil. The movie stirred up much controversy due to its content, and mainly because of the nudity displayed. Many Muslims felt that the movie was an insult to Islam and were enraged that it had been shown on television. Bouyeri was eventually captured and sentenced to life in prison without parole. He had made a death list of other people who he believed had insulted Islam and deserved to die.[67]

Discussion Question:

1. How could police and community leaders work together to curtail homegrown extremism?

attacks were attributed to right-wing extremists.[68] According to the FBI, the main drivers of domestic terrorism are anti-Semitism, racism, and sociopolitical conditions. In 2019 domestic terrorist attacks resulted in the highest number of fatalities since the Oklahoma City bombing by Timothy McVeigh in 1995. The vast majority of these deadly attacks were carried out of lone actors, also referred to as lone wolfs. **Lone actors** are unlikely to collaborate with other people and they are typically attacking soft targets. Lone actors are motivated by a mix of personal, sociopolitical, and ideological grudges against their targets. Soft targets include shopping centers, houses of worship, and large public gatherings. For instance, there have been numerous attacks on Jewish communities and immigrant communities, such as in Poway, California, in April 2019, where the shooter killed one person inside a synagogue, and in El Paso, Texas, on August 3, 2019, where the attacker killed several people outside a Walmart store.[69]

Communities are critical to preventing violent extremism and the recruitment by terrorist groups. The challenge for communities is to provide an inclusive environment where people feel that they are part of the community. Research suggests that community resilience toward violent extremism can be increased by strengthening ties through community programs and reducing inequities and discrimination. Another important part is increase the civic engagement of residents in the community. The main goal is to establish law-abiding norms and expectations and the motivation to adhere to these norms and expectations.[70] This is also referred to as collective efficacy. **Collective efficacy** is the "social cohesion combined with a willingness to take action on behalf of the broader community."[71] Collective efficacy has been shown to be an effective protective factor in preventing community violence.[72] Social connections are at the heart of collective efficacy. Ellis and Abdi (2017) describe three types of social connections and how they relate to community resilience to violent extremism: (1) bonds; (2) bridges, and (3) linking.

Bonds

Bonds contribute to community resilience by providing a "sense of belonging and connection with others who are similar"[73] (p. 290). One of the main contributing factors to extremism and the joining of extremist groups—especially by young adults—is a weak social identity, a lack of belonging, and the search for a meaningful identity. Communities can enhance resilience by promoting ethnic-based organizations that provide self-help.

Bridges

Bridges contribute to community resilience by promoting a "sense of belonging and connection with people who are dissimilar in important ways"[74] (p. 290). Youth and other people who join extremist groups often feel marginalized and lack identification with the community and their country. Bridges can help overcome these feelings of a lack of belonging and identification by providing programs that foster social relationships. For instance, school programs that focus on antibullying can enhance a youth's feeling of belonging and identity with their school. Also, mentoring programs that focus on integration and the building of a social network reduce marginalization. It is important to create a sense of belonging not only to a specific organization, however.

Linking

Linking contributes to community resilience by promoting "connections and equal partnership across vertical power differentials, e.g., government and communities"[75] (p. 290). Some of the most salient factors in the formation of extremist beliefs are a lack of trust and collaboration as well as unequal access to resources. Linking can help overcome these issues by establishing advisory boards that include members from the community. Another example of linking would be multidisciplinary teams made up of members of the community who engage marginalized members or groups and involve them in the community in a meaningful way.

The key goal of bonds, bridges, and linking is to involve all residents in the community and build a sense of belonging to the community. Terrorist groups target people who are isolated and are looking for a group or community they can be part of. Humans are social beings by nature, and almost all of us need social relationships and a community where we feel welcomed and appreciated. Community resilience to violent extremism very much depends on how effective the community is in building an inclusive environment.

SUMMARY

In this chapter, we talked about the importance of gathering intelligence about individuals who are radicalizing and may be planning to join a terrorist group or commit a terrorist attack. Most individuals who are becoming more radical, joining a terrorist group, or planning an attack conduct preparatory acts close to the site of the planned attack.[76] They may also post comments on social media sites or make comments to other people. Community policing is imperative to build the trust and partnerships with community residents and leaders who may interact with people at risk of radicalization. Police must engage community leaders and continue to build these partnerships, especially in Muslim and Arab American communities, as they are a main target of terrorist recruiters. This can be very difficult due to the tensions that exist between police and residents in Muslim and Arab American communities. Community police officers must work to overcome these tensions by communicating honestly and effectively with community leaders and share information important to the community.

Despite the successes and the call for increased community policing and the development of partnerships between police and Arab American communities, there has been a lack of funding for community policing efforts. Since the 9/11 attacks, the funding for Homeland Security has increased substantially by more than 50% to $44.1 billion in 2018[77], whereas funding for community policing (COPS program) has decreased substantially to $218 million.[78] The COPS budget has consistently decreased for several years. Prior to the 9/11 attacks, the COPS budget was about $800 million, and by 2004, it had decreased to $400 million. Currently, it is only about half of the 2004 budget.[79]

KEY TERMS

Abortion extremists 68
Animal rights/environmental extremism 68
Antigovernment/antiauthority extremists 67
Bonds 83
Collective efficacy 82
Domestic terrorism 66
Domestic violent extremists 67
Homegrown violent extremists 67
Intelligence 69
Intelligence-led policing 72
International terrorism 66
Lone actors 82
Operational intelligence 71
Social capital 78
Strategic intelligence 71
Tactical intelligence 72
Terrorism 65
Terrorism committed by large groups 67
Trusting relationships 74

DISCUSSION QUESTIONS

1. Discuss the difference between international and domestic terrorism. Explain how international and domestic terrorism are related.

2. Discuss what *intelligence* is and why it is imperative for counterterrorism.

3. Discuss strategies that help police build partnerships with Muslim and Arab American communities. Why is it important to build these partnerships?

4. Discuss the tensions that exist between police and Muslim and Arab American communities.

5. Discuss how communities can become more resilient toward violent extremism.

CHAPTER 5

COMMUNITY POLICING AND GANGS

On Valentine's Day 2017, Iraheta and several other MS-13 gang members abducted 15-year-old Damaris Alexandra Reyes Divas near her home in Gaithersburg, Maryland. During the police investigation, Iraheta told the story of the killing of Reyes Divas. "I asked her if at any time she had something to do with Christian," "She said 'yes.' I said 'I'm not going to forgive you'... I told her, 'I warned you not to mess with me. I told you not to mess with Christian. I told you to stay away from him or you would see what would happen. You don't play with me.' So I hit her. I kept hitting her. Until [the other gang members] stopped me." After the beating, they forced Reyes Divas to walk barefoot and without a shirt through the snow into a wooded area. Iraheta then sliced a tattoo off Reyes Divas's hand that Iraheta's boyfriend, Christian, had given to her. She then stabbed Reyes Divas several times in the stomach, chest, and neck. During the interrogation, Iraheta described the stabbing: "I left the knife in her neck," Iraheta told police. "She eventually took it out and gave it to someone else." Reyes Divas was stabbed 19 times and left to die near Interstate 495 in a shallow puddle. She was covered with railroad tires. The police later found a video with a recording of the killing of Reyes Divas. The gang members had intended to send the video to El Salvador to be promoted within the ranks of the gang.[1,2]

Learning Objectives

1. Describe the characteristics and prevalence of gangs.
2. Describe the different types of gangs and crimes committed by these gangs.
3. Explain why juveniles join gangs.
4. Discuss the relationship between drugs and gangs.
5. Discuss how community policing can reduce gang activity and recruitment.

INTRODUCTION

This chapter introduces students to the characteristics of gangs and gang activity. It distinguishes between different types of gangs and the dangers they pose to the communities in which the operate. Further, the chapter discusses what makes gangs attractive to juveniles. Gangs are also closely related to drug use and drug markets, which often result in violence and fear among community residents. The chapter then focuses on community policing strategies that are effective in curtailing gang activity. Community police officers and community leaders play a detrimental role in this process.

WHAT IS A GANG?

There is no agreed upon definition of the term **gang**. For the purpose of this chapter, we will use the definition from the Department of Justice and the Department of Homeland Security's Immigration and Customs Enforcement (ICE).

A. "An association of three or more individuals;

B. Whose members collectively identify themselves by adopting a group identity, which they use to create an atmosphere of fear or intimidation, frequently by employing one or more of the following: a common name, slogan, identifying sign, symbol, tattoo or other physical marking, style or color of clothing, hairstyle, hand sign or graffiti;

C. Whose purpose in part is to engage in criminal activity and which uses violence or intimidation to further its criminal objectives.

D. Whose members engage in criminal activity or acts of juvenile delinquency that if committed by an adult would be crimes with the intent to enhance or preserve the association's power, reputation or economic resources.

E. The association may also possess some of the following characteristics:

 1. The members may employ rules for joining and operating within the association.

 2. The members may meet on a recurring basis.

 3. The association may provide physical protection of its members from others.

 4. The association may seek to exercise control over a particular geographic location or region, or it may simply defend its perceived interests against rivals.

 5. The association may have an identifiable structure."[3]

This definition does not include traditional organized crime groups, such as the Russian Mafia or La Cosa Nostra. Rather this definition represents the characteristics of the three types of gangs in the United States: street gangs, prison gangs, and motorcycle gangs.

CLASSIFICATION AND CHARACTERISTICS OF GANGS

The National Gang Threat Assessment has found that 80% of crimes in many communities are attributable to criminal gangs. The most common gang-related crimes are alien smuggling, armed robbery, assault, auto theft, drug trafficking, extortion, fraud, home invasions, identity theft, murder, and weapons trafficking.[4]

The U.S. Department of Justice classifies gangs into three types: (1) street gangs, (2) prison gangs, and (3) motorcycle gangs.

Street Gangs

Street gangs are widespread across the United States and vary greatly by size and racial and ethnic composition. The greatest threat stems from large national street gangs that are dealing large quantities of illegal drugs and are very violent in nature. There are also many local street gangs that are smaller in size, which distribute illegal drugs to specific local markets. These local gangs may imitate the practices of the large national gangs to gain respect. The excessive use of violence is one of the greatest problems for cities and residents who may get into the crossfire of rivaling gang activity. In the future, it is likely that street gangs will become more active in trafficking drugs into the United States from other countries and build closer relationships with drug trafficking organizations and other organized crime groups. Street gangs are in constant need of money and thus one of the main goals of street gangs is to gain access to the international drug trade.[5]

One of the most notorious street gangs is Barrio 18, with an estimated 30,000 to 50,000 members across the United States and Canada. Barrio 18, founded in Los Angeles, California, in the 1960s, is also known as the 18th Street gang. Its main criminal activities are drug dealing, burglary, assault, extortion, prostitution, human trafficking, and homicide. It heavily recruits elementary and middle school students from immigrant backgrounds. Its also very actively recruits behind bars. In addition, members who were deported by the United States have spread the influence of Barrio 18 to El Salvador, Guatemala, and Honduras. It is associated with the Mexican Mafia and most of its members are of Mexican descent. Its colors, black and blue, represent its own original color (black) and its association with the Mexican Mafia (blue). One of its main enemies is MS-13, with whom its engages in violent gang fights across several countries. It has been very challenging for law enforcement to target the gang leaders because Barrio 18 is organized in "cliques" or cells and it does not have a "godfather" style leader. Thus, the antiracketeering laws that have brought down other Mafia groups have not been successful in suppressing street gang activity.[6]

Prison Gangs

Prison gangs are defined as "criminal organizations that originated within the penal system and they have continued to operate within correctional facilities throughout the United States."[7] They have a strict code of conduct, a clear hierarchy, and consist of a select group of inmates. Prison gangs are typically structured along racial and ethnic lines. They are substantially more powerful in state penitentiaries as compared to federal penitentiaries. Prison gangs also operate outside of prison and are involved in illegal drug trafficking. They are also connected to street gangs and outlaw motorcycle gangs (OMGs) and broker the transfer of drugs.[8]

One of the most widespread prison gangs is the Aryan Brotherhood, also referred to as The Brand, Alice Baker, and AB. Founded in 1964 at San Quentin State Prison in California, the Aryan Brotherhood is a highly structured White supremacist prison gang.[9] Even though it has a White supremacist ideology, for all practical purposes, it is a criminal

▶ **Photo 5.1** Tattoos as Identifiers of Gang Members.

organization engaged in a wide range of violent and drug-related crimes, including murder for hire, drug trafficking, counterfeiting, and identity theft. It works together with Latino and other gangs for criminal business purposes. Prison gangs create substantial problems inside prisons due to their willingness to use violence against the inmates of rival gangs, such as the Black Guerilla Family but also against prison guards. On October 22, 1983, Aryan Brotherhood member Tommy Silverstein killed corrections officer Merle E. Clutts in the Marion, Ohio, federal penitentiary. A few hours later, another Aryan Brotherhood member, Clayton Fountain, stabbed corrections officer Robert Hoffman and assaulted two other officers. He wanted to have a higher body count than Silverstein. In response to the incidents, federal prisons moved known AB members into supermax units, which usually means solitary confinement. This has not stopped AB members from continuing their criminal enterprises inside and outside of prison however, and AB continues to be a significant threat to inmates and officers. It has been estimated that even though only about 1% of the prison inmate population are AB members, they are responsible for about 18% of all prison murders.[10]

Motorcycle Gangs

Outlaw motorcycle gangs (OMG) are defined as "ongoing organizations, associations or groups of three or more persons with a common interest or activity characterized by the commission of, or involvement in, a pattern of criminal conduct. Members must possess and be able to operate a motorcycle to achieve and maintain membership within the group."[11] Outlaw motorcycle gangs refer to themselves as "one-percenter" motorcycle clubs. The term goes back to 1947 when the American Motorcycle Association stated in its magazine that 99% of the motorcycling public are law-abiding citizens who enjoy riding motorcycles and only 1% are troublemakers. This statement was in response to a motorcycle rally in 1947 that had turned violent.[12]

The National Gang Threat Assessment estimates that there are between 280 and 580 OMGs operating across the United States. OMGs are a serious threat to communities due to the widespread violence and the wide variety of crimes they engage in. They are sophisticated criminals who have expanded their criminal operations to the global market. OMGs are typically highly structured with different types of clubs that adhere to clear hierarchies.[13]

OMGs are commonly divided into four categories:

1. Support clubs are mainly responsible for the support of the larger club for the purpose of bolstering the reputation of the larger club.

2. Satellite clubs serve as a recruiting ground and as helpers for criminal acts.

3. Regional clubs are limited in their membership and territory and have some contacts to the larger club.

4. The larger club, such as the Hells Angels or the Sons of Silence, is at the top of the hierarchy and determines how the club operates and which activities the club engages in.[14]

One of the most notorious motorcycle gangs is the Hells Angels. They have become a global phenomenon causing great problems for law enforcement due to their frequent assaults of rival gangs but also intragang aggressions. They recruit new members from street gangs, prison gangs, and rival OMGs. The Hells Angels have been expanding by creating new chapters across other geographic areas, which has also increased violent encounters due to turf wars.[15] On November 21, 2017, a grand jury indicted 11 members of the Hells Angels on charges of murder, assault, maiming, racketeering conspiracy, and witness intimidation.[16] It has been very difficult to convict members of OMGs, such as the Hells Angels, because witnesses often do not come forward or are not willing to testify out of fear of retaliation.[17]

PREVALENCE OF GANGS AND GANG VIOLENCE

What exactly is the gang problem, or stated differently, how much crime is committed by gangs and how many people are involved in gangs? These are two of the most critical questions for law enforcement purposes. The National Gang Intelligence Center collects survey data from federal, state, and local law enforcement agencies across the United States. The results demonstrate that gangs are a widespread problem for communities

▶ **Photo 5.2** Hells Angels Motorcycle Gang.

across the nation. About half of the survey respondents (i.e., law enforcement agencies) report that gang membership and crimes committed by gangs have increased.[18]

According to the survey, between 2012 and 2014, street gang membership increased in 49% of the jurisdictions, stayed the same in 43% of the jurisdictions, and dropped in 8% of the jurisdictions. Gang-related crime increased in 50% of the jurisdictions, stayed the same in 36% of the jurisdictions, and dropped in 14% of the jurisdictions. Most gang-related crime relates to violent crimes, drug-related crimes, weapons trafficking, and sex trafficking. A large number of the crimes included intimidation and threats, often reinforced by violent attacks. For instance, in October 2014, 12 gang members of the United Blood Nation murdered a couple in Charlottesville, North Carolina, because the husband had agreed to testify against gang members who had robbed his mattress store. In 2017, two of the gang members were sentenced to life without parole in federal prison. Several other members have pleaded guilty to charges of murder, racketeering, firearms violations, and assault.[19, 20] These types of violent acts are meant to deter others from reporting gang crimes and testifying against gang members.

Gang Violence Against Law Enforcement

Threats of violence by gangs against law enforcement officers have a long history and officers are regularly attacked by gang members. In the past years, gang members

CASE STUDY 5.1: MS-13

One of the most highly publicized gangs in the United States and internationally is the so-called MS-13, which stands for Mara Salvatrucha. It has about 8,000–10,000 members in the United States and about 700,000 worldwide. Within the United States, it has members in 33 states. MS-13 has gained much publicity for several reasons. First, it is one of the fastest growing gangs. Second, it is one of the most violent gangs, well-known for using machetes to kill its victims. Third, many of the members are undocumented youth who came to the United States without their parents or any other connections in the United States, making them very vulnerable and a target of gangs. Many of them come from countries where civil wars have destroyed their families and which have hardened them for the gang life.[21]

On May 23, 2018, Donald Trump stated, "I am calling on the Congress to finally close the deadly loopholes that have allowed MS-13, and other criminals, to break into our country." President Trump has taken a number of actions in an effort to reduce the violence caused by MS-13 members. In 2017, the U.S. Immigration and Customs Enforcement agency (ICE) arrested 796 MS-13 members and associates.

▶ **Photo 5.3** MS-13 Marks Gang Territory.
Credit Line: Walking the Tracks via Wikimedia Commons, CC BY-SA 2.0.

Continued

CASE STUDY 5.1: MS-13 (CONTINUED)

This was an 83% increase as compared to 2016. In addition, in 2017, the U.S. Justice Department worked with agencies in Central America to secure charges against more than 4,000 MS-13 members. Finally, U.S. Border Patrol agents arrested 228 undocumented immigrants connected to MS-13. These numbers demonstrate the success of the new law enforcement strategies called for by President Trump. In addition, he has asked for the implementation of measures to stop the immigration of undocumented individuals associated with MS-13. These measures would include a border wall and a quick deportation of undocumented gang members.[22]

There is, however, also a substantial amount of criticism of these strategies that target MS-13 members. Law enforcement has been giving new powers by the Trump administration and these new powers have come under much scrutiny. Hundreds of suspected gang members have been arrested, many of them juveniles with no family connections in the United States who came illegally from El Salvador. Critics contend that many juveniles are being detained who have no gang affiliation simply because of their background. Some of the juveniles were deported and others held in detention for months without a hearing or any evidence that they had committed a crime. In 2017, the American Civil Liberties Union filed a lawsuit against the government, which forced the government to bring the minors in front of a court for a hearing. Paige Austin, one of the immigration lawyers, stated: "They were disappearing into the immigration detention system, and it often took parents days or weeks to even figure out where they were, much less to get them released and brought back home. And then, the second thing that was very disturbing was the lack of evidence to support the allegations that the government was making. Many of the symbols or the items that the government claims are signs of gang affiliation are, in fact, religious symbols, or they're signs of cultural pride. If they were suspected of committing a crime, the police would arrest them, and they would be in criminal custody. The fact that they're in immigration custody means that local authorities were looking for some other way to detain them in the absence of any evidence of wrongdoing."[23]

Discussion Questions:

1. What do you think about how law enforcement should identify juveniles who are MS-13 members?

2. What would you do to decrease the violence caused by MS-13 gang members?

---

have become more bold and aggressive toward law enforcement. Even though the overall number of attacks has remained stable, one third of jurisdictions reported an increase in threats against officers. Several attacks have caught the attention of the media and public. For instance, in response to a racketeering case in Arizona against the East Side Los Guada Bloods, an officer was murdered by five members of the gang. The East Side Los Guada Bloods originated in the early 1990s on the Salt River Pima-Maricopa Indian Reservation, outside of Phoenix. In 2011, several of its members had been indicted on charges of conspiracy to commit murder, attempted murder, racketeering, assault with serious bodily injury, threatening and intimidating witnesses, and firearms trafficking.[24]

In 2014 in North Carolina, the father of an assistant district attorney was kidnapped by gang members of the United Blood Nation with the intent to kill him. The assistant district attorney had prosecuted the leader of the gang and the jury had imposed a life sentence plus 84 months for kidnapping and other charges. The leader of the gang organized the kidnapping from his prison cell using phones that had been smuggled into his cell, where he spent 23 hours each day. State officials later acknowledged that the gang member must have had help from one of the prison employees. One of the messages from the gang leader stated: "Gag him real tight. Put something in his mouth. Put something over his head." The gang members kept the victim in an apartment for four days, constantly communicating with their leader in prison. The gang leader instructed them how to kill the victim, but the investigators rescued the victim after tracking text messages to the apartment. They also arrested the kidnappers who later received prison sentences ranging from 20 to 50 years.[25]

These types of retaliatory attacks are not uncommon. Threats are often delivered in person or by phone call. But there are also other ways in which threats are delivered to law enforcement officers and judiciary officials, including judges and prosecutors. Social media, such as Twitter, Facebook, and Instagram, have become a popular means of distributing threatening messages. For instance, in 2014 the police chief of Detroit, James Craig, was openly threatened by gang members using social media. Investigators believed that gang members of a narcotics network were behind the death threat. Following the threat, Craig stated that he was determined to dismantle criminal enterprises. Under Craig, drug raids had greatly increased, which has cut into the gangs' profits. Craig warned the gang members: "No way will we back off," he said. "You don't get to threaten a police chief. You threaten a police chief, you threaten every member of that police department. You threaten a police chief, and you threaten every member of our community. That's not acceptable. This reiterates that we're doing the right thing.[26]"

Another popular method to distribute threats is graffiti. In 2014, Arkansas law enforcement officers of the Rogers Police Department were threatened by MS-13 via a graffiti sprayed on the wall of a gas station bathroom door. The message threatened an attack against an officer on New Year's Day. The department had received similar threats a few days earlier.[27]

WHY DO JUVENILES JOIN GANGS?

Especially street gangs have a large number of juvenile members. One of the main issues for communities and community police officers is how to prevent juveniles from becoming involved in violent gangs. In order to develop strategies, it is important to understand why juveniles join violent gangs or how gangs recruit members.

One of the main recruiting strategies are social media platforms. The most frequently used social media platforms are Facebook, YouTube, Instagram, Twitter, Snapchat, Google+, Flickr, WhatsApp, and kik app. Especially girls are targeted via rap videos and promises of a luxury lifestyle. Once these girls join the gang, they are often forced into prostitution.[28] For instance, a member of a neighborhood-based gang in the Bronx, New York, posted rap videos on YouTube, espousing violence and the gang lifestyle. In response, he received text messages containing requests to join the gang. For example, he received a text messaging stating, "I'm from Queens but I watch all ya videos. Imma trying be down with the WTG Move." The rapper responded, "You can be WTG under me and b official for $125."[29] Even though gangs are still male dominated, the proportion of girls and women in gangs has increased substantially.[30]

Gangs also recruit directly within communities, such as in schools. MS-13 is well-known for pressuring immigrant youth from South America to join the gang. Gang members actively recruit at middle and high schools. MS-13 is especially active in schools from Northern Virginia to Long Island to Boston. Over the past few years, more than 200,000 youth have traveled to the United States from Central America to escape the violence of gangs and civil war. Most of these youth go to school and work hard to achieve a better life. But some of them become trapped in gangs, especially MS-13. They may have been MS-13 members in their home country and hoped to escape the gang by coming to the United States. Unfortunately, some youth get drawn back into MS-13, either by free will or by force from gang members. MS-13 members bully students and the only recourse for students may be to drop out of school or transfer to another school. Teachers report not only students shouting and flashing MS-13 signs but also MS-13 members coming into classrooms and attacking students while the class is in session. Administrators often don't do enough to stop the bullying and recruitment. During "recruitment season" in the spring, gang-related altercations tend to intensify.[31]

Risk Factors and Community Programs

Youth who join gangs often lack important socialization skills, self-esteem, self-control, and refusal skills. They also often perform poorly in school. Research has shown several risk factors that play a role in a juvenile's decision to join a gang and engage in delinquent acts. These risk factors are:

- "Aggression
- Substance use (especially marijuana and alcohol)
- Antisocial or delinquent beliefs
- Family poverty
- Broken home
- Low achievement in elementary school
- Identified as learning-disabled
- Association with delinquent/aggressive peers
- High-crime neighborhood"[32]

The Story of Josephina Ramirez, 20

In 2013, Josephina Ramirez, 16 years old, was sentenced to 15 years in prison for the killing of 13-year-old Julio Marquez. Josephina and two other male gang members, brothers Ezequiel and Samuel Vasquez, had lured Julio to an alleyway where the two brothers shot him several times. Julio did not die immediately however, and the brothers bludgeoned him to death with the butt of the rifle. The two brothers pleaded guilty to first-degree manslaughter and other criminal charges. Ezequiel, who was 15 at the time of the killing, was sentenced to 20 years in prison and Samuel, who was 17 at the time of the killing, was sentenced to 22 years. Josephina pleaded guilty to attempted murder and was sentenced to 15 years. The case made national headlines and Josephina was portrayed as a cold-hearted killer with no empathy for the victim.[33]

Josephina's story very clearly reflects the impact of the risk factors listed above. Josephina grew up in Portland Oregon, with a single mom, at times being homeless and at times living with friends. When they did have an apartment, they often had their water or electricity shut off because her mom could not pay the bills. Gangs, mainly the Crips and Bloods, were a normal part of life in the neighborhoods where they lived. When she was walking to school, gang members would stab and shoot each other without regard for the children walking by. Josephina had three sisters and three brothers. Her older brother was a gang member and as a result Josephina was also involved in the gang. Her broken home also impacted her performance and behavior at school. Josephina got into trouble regularly despite the efforts of her fourth-grade teacher. She said:

> I started getting into trouble around fifth grade. I started having troubles in school. Up until sixth grade I got A's and B's. Maybe the fact that I didn't have a father figure, my dad died before I was born, and my brothers, they were the only male figures that were there, and having them abuse me, physically, mentally, emotionally, like I think that was my biggest thing when I was younger.

Becoming a gang member was not a rational decision that Josephina made but rather an almost inevitable part of her life:

> When I was 12, my cousin—she was the leader of this gang—and I hung out with them, but I wasn't really in one. But a couple of months before I got locked up, I actually got jumped into my cousin's rival gang, which was kind of like a dis on her because I didn't really like her. The motivation was my co-defendant (her boyfriend), he told me, "We need a ride somewhere," and the leader of that gang was like, "I'm not going to give you a ride unless Tiny (her nickname) gets jumped in." Because they had heard about me, my reputation and stuff, and they wanted me in their clique, and they knew my cousin was the leader of their rival gang. I didn't want to, but we needed a ride, so I did what I did, which I guess is kind of a stupid reason for getting jumped into a gang.[34]

Josephina became more active in the gang, mainly to get respect from other gang members. She was also using drugs and alcohol since she was 10 years old. At age 13, she started using methamphetamine. She was sent to a drug rehabilitation program for four months after she got caught tagging her school with gang signs. For a few months after her release she was sober, but in December, she was admitted to the emergency room with alcohol poisoning. Her mom told her she could not take care of her, and Josephina was put into a program by the Department of Human Services, which now had custody over her. She ran away and got back into the gang life. Two years later, following a request of her boyfriend, she lured 13-year-old gang member Julio Marquez to his death. She was sentenced to 15 years at Oak Creek Youth Facility in Albany.

In an interview at the facility a few years later she stated:

> He was telling me, "Set him up, set him up," and I told him no, because he was my friend, and I told him I didn't want to, and he's like, "If you love me, you would do it" and that's where he got me, because I did love him and I would do anything for him, and I said "OK."[35]

The gang conflict resolution coordinator for Oregon's youth correctional facilities, Christina Puentes, commented that these are not unusual circumstances. Puentes stated that all the girls she knew at the correctional facility were incarcerated because they had committed a crime under the influence of their boyfriends: "It wasn't something that they did on their own. At least the ones I've worked with. I can't think of one girl who had gang issues who did it on her own."[36]

In order to prevent girls such as Josephina from becoming involved in gangs, it would be necessary to address the risk factors she was exposed to, including the broken home, substance abuse, family poverty, exposure to delinquent and aggressive others, and living in a high-crime neighborhood. Community programs that successfully address these risk factors must have certain characteristics. First, the program must focus on developing the skills that youth lack to resist violence and gang membership. Second, the program must be dedicated to fidelity and evidence-based practices. Third, the program must be run long term. Finally, the program must be consistently taught, supported, and reinforced by instructors.[37]

Motivations by Juveniles to Join Gangs

The Los Angeles Police Department has identified five main motivating factors for juveniles and adults to join a gang.

1. "Identity or Recognition—Being part of a gang allows the gang member to achieve a level of status he/she feels impossible outside the gang culture.

2. Protection—many members join because they live in the gang area and are, therefore, subject to violence by rival gangs. Joining guarantees support in case of attack and retaliation for transgressions.

3. Fellowship and Brotherhood—To the majority of gang members, the gang functions as an extension of the family and may provide companionship lacking in the gang member's home environment. Many older brothers and relatives belong to, or have belonged to, the gang.

4. Intimidation—Some members are forced to join if their membership will contribute to the gang's criminal activity. Some join to intimidate others in the community not involved in gang activity.

5. Criminal Activity—Some join a gang to engage in narcotics activity and benefit from the group's profits and protection."[38]

The Story of Trei Hernandez, 21

Trei is a member of the Gangster Disciples. Trei did not have a stable life growing up with his mom. His father died when he was five. Trei and his mom moved several times between Oregon, California, and Texas, barely surviving. They were living with other family members in Oregon and even though his mom had a manager position at Jack in the Box, she didn't make enough money to feed everyone. Trei had several siblings from his father's side and they lived together in one house with other cousins and aunts. At 14, Trei was the man in the house, making money, and paying the bills. His troubles started early in school. He never had a connection to his teachers. He said:

> I never gave them that satisfaction. Never cared for school. Why be in school when I can be making money? Even as a very little kid, mama had to pick me up from the principal's office. I was seeking attention. I was a bad kid.

Gangs became part of his life at the age of 13:

> One of my best friends, his dad, he was like that dude, that dude that showed me the lifestyle. We was driving down the street one day and he just passed me one of them thangs (a revolver). It was like, "You see that dude over there?" And I looked at him and I said, "Which one?" He was like, "All of them." He was like, "Shoot," and ever since then, it's just been off the hip. "I had to. It was them or me. We all called him pops. He was an older homie."

When he was 15, Trei's girlfriend had a baby. His mom told him that she would not be able to help him and so he started selling drugs and robbing people to provide for his family. When he found out that he had two arrest warrants for robbery, he took the next greyhound bus to Oregon. He was 16 at the time. When he ran out of money, he got back with the gang in Oregon and after a day of drinking, he and another gang member robbed two stores. Trei cut one person in the hand and stabbed another one in the chest, deflating his right lung. Trei was convicted and incarcerated at McLaren Youth Correction Facility. While he was at McLaren, Robert Carson became Trei's first male role model. Carson worked with Trei and after his release he enrolled in college. This could be a happy end story if it wasn't for the gang life that never leaves a gang member behind. Trei got kicked out of college and after he was re-enrolled, he got into another gang fight. At the end of the interview, Trei summed up his outlook on life:

> I don't know if I can (walk away from his gang), but as far as being in that environment all the time, that can definitely change. I would never be around the homies unless they're trying to better themselves. I am a changed man. I have strong beliefs about it—not so much the negative things, but the positive things—the barbecues, the extra pair of shoes, "Here's a new outfit for you," "Here ya go, "a lady." They've done a lot of good things for me. I would never leave it, but I would not surround myself with the negative.

At the time of the interview, Trei was still at McLaren Youth Correctional Facility.[39]

WHITE STREET GANGS

Most of the research in academia and law enforcement has focused on racial-ethnic minority-led gangs.[40] This is caused in part by the focus of the FBI's arrest statistics on racial-ethnic minority-led gangs in urban areas, especially big cities, such as Chicago and New York. The arrest data gives the impression that the vast majority of gang members are people of color, and especially African Americans and Latinos. That is not the case, however.[41] Researchers estimate that about 40% of gang members within the United States are White. In contrast, law enforcement data paints a very different picture. According to arrest data, African Americans and Latinos are 15 times more often classified as gang members than non-Hispanic White people. Specifically, law enforcement estimates for all ages show that 49% are Latino, 37% are Black, and only 8% are White. The National Youth Survey found that for youth between the ages of 12 and 16, 42% were White, 27% were Black, and 24% were Latino. Similarly, for youth gang members between the ages of 13 to 15, about 46% were White, 22% were Black, and 25% were Hispanic.[42] Greene and Pranis (2007) suggest that these discrepancies between law enforcement data and the youth survey data are caused by prejudice of police officers who have the discretion to classify a certain conduct as gang activity or individual criminal behavior.[43] White gang members are generally undercounted by police because police classify groups of young White people as individuals who have committed individual crimes, whereas young people of color are classified as street gangs and are held criminally liable for their individual crimes and the crimes of other gang members. This results in an undercount of White gang members and an overcount of racial-ethnic minority gang members.[44] Law enforcement, including the FBI, splits gang membership into three categories: White supremacist prison gangs, outlaw motorcycle clubs, and criminal street gangs. But they don't include majority White gangs as a major group under the criminal street gangs. The gangs that are most tightly policed are criminal street gangs and, more specifically, Black and Latino gangs in urban areas.[45] Law enforcement data suggests that criminal street gangs are primarily an urban phenomenon, and especially in large cities. However, the National Longitudinal Survey of Youth (NLSY) shows that youth in urban and rural areas are equally likely to be a lifetime member of a gang.[46]

As a result of the focus on urban racial-ethnic minority criminal street gangs, White gangs are not policed as much and their members are typically not classified as gang members when they get arrested for a crime. Because they are not classified as gang members, they are more likely to receive an intervention, such as job and life skills training in an effort to change their life.[47]

The "Royals"

One of the largest majority White gangs are the "Royals," who are active in urban and rural areas, including Chicago and Mississippi. The Royals are one of the oldest gangs in the United States and many of their members are White, typically recruited as youth from poor urban and rural areas. One of these White youth recruits from Florence, Mississippi, was Benny Ivey. He grew up in a trailer park in what he describes as a chaotic family. His parents were addicted to crack and opioids and he wanted to be part of something. By joining a gang, he became part of a group and he proudly wore their black and blue colors. From his own experience, Ivey dismisses the law enforcement picture of what gang members look like. In an interview, he stated: "The world should know there are whites struggling in hoods as well as any other race, and more often than not those kids become gang members or drug addicts." Ivey dropped out of school at age 15 and engaged in numerous crimes, including robbery and breaking into houses. He eventually became the president of the Simon City Royals. He spent time in prison—a normal part of life as a gang member. However, when he met his future wife, he started to turn his life around. He stressed that the Royals, although majority White, were not racist or a gang that would fall under the category of White supremacists. Most of the gang members were looking for a group, a sort of family, that would support them and give them a purpose for their life. Recently, Mississippi announced that the Royals are one of the largest criminal street gangs and 53% of gang members are White. Under these circumstances, it is difficult to explain that all 97 prosecutions for gang-related crimes under Mississippi law included racial-ethnic minority offenders. There appears to be a continuing strong belief among law enforcement and prosecutors that the most dangerous criminals are racial-ethnic minority gang members despite the evidence that a large percentage of criminal street gang members are White and that majority of White gangs commit a substantial amount of violent crime.[48]

ASIAN GANGS

The National Gang Center estimates that only 4.6% of all gang members are Asian or Pacific Islanders.[49] There is little research about Asian gangs and they typically also receive very little attention from law enforcement. Asian gangs may have such a very low visibility because of their low rates of criminality and little juvenile delinquency.[50] Chinese and Vietnamese gangs have received the most attention despite the fact that there are many different ethnicities that are involved in gang activity.[51] Most of the research on Asian gangs has been done on gangs in California and New York as those are the states where a large number of Asian immigrants settled. They found that Asian gangs formed during the 1960s because Asian immigrants had little access to services and economic opportunities. The gangs served as support systems and a place to resolve conflicts.[52]

There are numerous Asian gangs, including the Asian Street Walkers, the Tiny Oriental Crips, the Exotic Family City Crips, Asian American Gangs, the Asian Boyz Crips, the Black Dragons Gang, and many others.[53] Even though the FBI does not look at Asian gangs as a major threat, local law enforcement may well experience much violence and crime stemming from Asian gang activity. For instance, Fresno, California, has witnessed major gang activity from Asian gangs. One of the most active Asian gangs in Fresno is the Mongolian Boys Society. On November 17, 2019, two of its members went into a backyard football watch party and started shooting. They killed four people and injured six others. The mass shooting was a retaliation for the death of one of their brother members. He was shot earlier that day and the Mongolian Boys Society believed that he had been killed by its hated rival gang, the Asian Crips. Seven members of the Mongolian Boys Society were arrested and charged with conspiracy to commit murder. The FBI's Safe Streets Taskforce believes that the leaders of the Mongolian Boys Society met and

CASE STUDY 5.2: MS-13 AT WIRT MIDDLE SCHOOL IN RIVERDALE, MARYLAND

According to teachers, parents, and students, gang-related fights are a daily occurrence at William Wirt Middle School in Riverdale, in Prince George's County, Maryland. The school, which is only 10 miles from the White House, has mostly Hispanic students and a small number of them are MS-13 members. Even though there may only be about a dozen MS-13 members, they have become a major force, throwing gang signs, spraying gang signs on the walls with graffiti, selling drugs, and aggressively pursuing students to join the gang. They are also violent, bringing weapons to school and intimidating teachers and students alike. Many of the students at the school immigrated from Central American countries, often countries where gangs, such as MS-13, cause much violence. According to reports, teachers who report gang activity are being ignored by administrators. Parents have also voiced their concerns about the pressure asserted on their kids by MS-13 members attending the school. "Teachers feel threatened but aren't backed up. Students feel threatened but aren't protected," one educator said. "The school is a ticking time bomb." Several teachers reported having been threatened and even sexually harassed. They are afraid that MS-13 members could follow them home and harm them or their family. The school, however, denies the ongoing gang activity and has been ignoring the problems reported by teachers, parents, and students. "The principal is aware of concerns about gangs in the community, but has not experienced any problems in school," John White, a spokesman for the county school system, wrote in an email. During the 2017–18 school year, police were called to the school 74 times and five students were arrested for bringing weapons to school, selling drugs, and assaulting other students. A number of students have apparently banded together to fight the influence of the MS-13 members. This, of course, has resulted in a sort of arms race and more violence and fear. One father stated, "If someone doesn't do something soon, there is going to be a tragedy at that school."[54]

Discussion Questions:

1. If you were the principal at the school, what would you do?

2. If you were a parent of one of the students, what would you do?

3. What strategies could be implemented at schools to decrease gang activity in general?

discussed which homes were hangouts for members and affiliates of the Asian Crips. They then selected a home and posted three members around the neighborhood to look out for police while two other members went into the backyard. During the questioning by police, they stated that they just wanted revenge. These types of revenge killings are common in gang wars and create great risks for the communities.[55]

GANGS AND DRUGS

Urban street gangs have built alliances with drug trafficking organizations to make money. Even though territorial control may be the main objective for gangs, they have to make money to survive. Selling drugs is one of the most lucrative "businesses" and gangs have long taken advantage of such opportunities. Thus, many gangs have drug alliances, especially with Mexican drug cartels. The gangs then fight to control the street-level sales of illegal drugs. This, of course, leads to violence between gangs and with other street drug dealers. It isn't rare that innocent bystanders get killed during these gang fights.[56]

During the 1980s Florida, especially Miami, was the main center of drug trafficking. The drugs came from Colombian drug cartels who smuggled the drugs into the United States and wired the proceeds into bank accounts, real estate, and expensive cars. New laws and stricter drug law enforcement changed the playing field for the Colombian drug cartels and, in the 1990s, they began to focus on drug production, leaving the smuggling and sale of drugs to other trafficking organizations, including the Jamaican, Dominican, and Mexican drug cartels. Mexican drug cartels not only smuggled drugs for the Colombians, they also started to buy drugs from the Colombians and then sold them on the U.S. market. The profits from drug trafficking greatly increased for the Mexican drug cartels and they consistently expanded their territories, resulting in much violence and many deaths. Between 2006 and 2011, 47,000 people died in fights between the drug cartels and government authorities and in fights between the different cartels.[57]

To make more money, street gangs in the United States cooperate with the Mexican drug trafficking organizations (MDTOs). They are mostly active at the southwestern border. At first, these gangs would buy drugs from the MDTOs and sell them in the United States. Later, the street gangs provided protection for drug shipments into the United States, store the drugs in warehouses, and protect shipments across the border. Nowadays, street gangs are trying to cut out the "middle man" by making arrangements with MDTOs, buying larger quantities of drugs, and distributing them directly to their markets. This direct distribution scheme has cut the costs of drug trafficking by about 30% for the gangs, which means their profits have increased by 30% or more.[58]

The partnerships between MDTOs and gangs work really well because they are like-minded in many ways. Both do not cooperate with law enforcement, value loyalty above anything else, enforce discipline of their members, and are strongly committed to making a profit. They are also working to expand their territories and protect the economic area of operations from competitors. Finally, their partnership is mutually beneficial because the MDTOs have a reliable partner in their drug trafficking operations and the gangs are able to increase their profits.[59]

Recruitment of Children

Urban drug gangs are increasingly recruiting children in their neighborhoods as drug runners.

In many cities, including Chicago, gangs have become more aggressive in recruiting teenagers. The gangs need to sell illegal drugs and they are constantly in need of new drug runners. School and police officials have voiced their concerns about the methods used by the gangs. For instance, gangs will make threatening phone calls to teenagers and tell them that there will be consequences if they don't join the gang. They attack students who say "no." And they hang around schools around the time when the kids get out. So, many kids will do extra work for teachers and other things to bide their time until the gang members have left and they can go home. Some teenagers literally fight to stay straight: Lacrista Ewing, an energetic 18-year-old, was warned about her straitlaced ways by the girls of a street gang at Harper High. "They told me: 'You're going to get violated. You're going to get moved on. You're going get bum-rushed,'" she recalled. Lacrista said her first attacker was a gang leader named Mary, who accosted her in a parking lot outside school. When the fighting was over, Mary was hospitalized and needed stitches. For a while, it seemed that Lacrista had to fight every other day. She never lost. Since then, the gangs have left her alone. Lacrista has a B average at school, plays on the volleyball team, and belongs to the African American History Club. She will be the first of the four children in her family to graduate from high school. An older brother just went to jail. Another brother was killed in a gunfight in 1986. "They said it involved drugs, gangs, and a girl," Lacrista said.[60]

Drug Distribution

As noted earlier, gangs are heavily involved in drug trafficking and distribution within the United States. The choice of drugs depends on the geographic area. The prevailing drug distributed by gangs on the West Coast is methamphetamine. In the North Central and South Central regions, gangs mainly distribute cocaine. Heroin is the prevailing drug in the North East and South East regions. The profit from drug distribution is substantial. For instance, the Hoover Crips who distribute drugs in Dallas, Texas, and Tulsa, Oklahoma, obtained $10 million in cocaine from MTCOs.[61] Case Study 5.3 throws light on law enforcement action on a gang.

CASE STUDY 5.3: GANG MEMBERS CHARGED IN SAN DIEGO FOR DRUG AND FIREARMS TRAFFICKING

On February 28, the San Diego County prosecutor charged 37 gang members of the North County street gangs with heroin, methamphetamine, and weapons trafficking. The arrests were made by 100 members of the North County Regional Gang Task Force, the FBI SWAT team, plus other law enforcement agencies. Ten other suspects are still at large. Law enforcement seized numerous items, including heroin, methamphetamine, fentanyl, and firearms, including a semiautomatic pistol, revolvers, and two AR-15 style assault rifles. The gang members operated in North County neighborhoods near a park and several schools.

The investigation began in 2016 after two gang-related homicides in North County caught the attention of law enforcement. A gang task force targeted several gang members known to be major distributors of methamphetamine. They used wiretaps to target the main drug distributors in each gang area. One of the main concerns was the distribution of large amounts of drugs near schools and the trafficking of drugs from Mexico. Following the arrests of the gang members, the FBI stated: "The FBI and our law enforcement partners at the North County Regional Gang Task Force won't accept when gang activity coupled with drugs, firearms and violence infests our communities," said FBI special agent in charge, John A. Brown. "The FBI will continue to pursue violent gang members and work tirelessly with our law enforcement partners to keep our communities safe."[62]

Discussion Questions:

1. What are the dangers gangs such as the North County Street gang pose to the community and residents?

2. Imagine you lived in a neighborhood with gang activity. How do you think that would impact your life and daily routines?

HOW CAN COMMUNITY POLICING REDUCE GANG ACTIVITY AND RECRUITMENT?

A Comprehensive Strategy to Address Gang Violence

Earlier in this chapter, we talked about the prevalence of gang violence in communities across the United States. Scott Decker, who is one of the leading experts on gang recruitment and violence, describes the role of police for preventing gang membership. The role of police is manifold:

1. "Based on their knowledge of youth in their communities—who is in trouble, and who is on the brink of trouble—the police are in a unique position to make an early identification of youth who are at risk of joining a gang.

2. Because they are active in neighborhoods at times when (and in places where) other adults are not, the police can play a vital role in efforts to prevent gang-joining, including referrals to services.

3. SARA—scanning, analysis, response, and assessment, the primary problem-solving model used by law enforcement—and the public health prevention model share complementary data-driven components, which can be used in building initiatives and partnerships that prevent youth from joining gangs.

4. Police legitimacy can be increased through partnerships with community groups and agencies that are trying to reduce the attraction of gangs; when police play a more active, visible role in gang-prevention activities, it builds trust and improves community efficacy.

5. Law enforcement leaders should place more emphasis on recognizing gang-prevention work of patrol officers and making that work more visible to the public."[63]

Based on this description of the role of police, Decker developed an integrated model for gang membership prevention. The integrated model combines parts of the public health model and the SARA model. Table 5.1 shows the basic steps for the two models and the combined model.[64] Case Study 5.4 looks at the activities of a nonprofit organization to bring about a social change.

COMMUNITY POLICING AND JUVENILE GANG MEMBERSHIP

The Office of Juvenile Justice and Delinquency Prevention (OJJDP) developed a comprehensive gang model aiming to reduce gang membership and violence. The model consists of five strategies to address the causes of gang activity. The OJJDP agency operates under the assumption that gang activity is caused by system failures or community dysfunction. The five strategies are:

Table 5.1 Combining the Public Health and SARA Models to Prevent Gang Membership

Public Health Model	SARA Model	Combined Model to Prevent Gang Membership
Using surveillance to better understand the scope, characteristics, and consequences of the issue.	Scanning the environment to identify the problem.	Surveillance or information-gathering via data from law enforcement, emergency room admissions, and surveys of the community. Police should provide an assessment of the gang problem.
Identifying the risk and protective factors.	Analyzing the problem, using multiple sources of information.	Identifying at-risk youth by creating working relationships between community leaders, emergency room staff, and law enforcement. Police should use gang prevention programs for the entire youth population as well as targeted programs for youth at highest risk.
Designing and evaluating prevention strategies.	Developing a response consistent with the information gathered.	For instance, kids who come to the emergency room with a gunshot wound from a gang fight should be blanketed with a therapeutic intervention that includes the entire family, including younger siblings to prevent these younger siblings from becoming gang membership.
Disseminating and implementing the best strategies.	Assessing the effectiveness of the response.	Law enforcement should assess which strategies were successful and share their information with other departments and the public.

1. *"Community Mobilization*: Involvement of local citizens, including former gang-involved youth, community groups, agencies, and coordination of programs and staff functions within and across agencies.

2. *Opportunities Provision*: Development of a variety of specific education, training, and employment programs targeting gang-involved youth.

3. *Social Intervention*: Involving youth-serving agencies, schools, grassroots groups, faith-based organizations, police, and other juvenile/criminal justice organizations in "reaching out" to gang-involved youth and their families, and linking them with the conventional world and needed services.

4. *Suppression*: Formal and informal social control procedures, including close supervision and monitoring of gang-involved youth by agencies of the juvenile/criminal justice system and also by community-based agencies, schools, and grassroots groups.

5. *Organizational Change and Development*: Development and implementation of policies and procedures that result in the most effective use of available and potential resources, within and across agencies, to better address the gang problem."[65]

CASE STUDY 5.4: YOUTH ALIVE!

On March 2–3, 2009, the nonprofit organization Youth ALIVE! organized a symposium for hospital-based violence intervention programs in Oakland, California, to bring together the experts in the field and discuss key program components and best practices in the field. In addition, the symposium was meant to be a first step in establishing a national network of hospital-based programs. At the time, there were a total of nine programs across the United States that had been in operation for at least one year. All nine programs participated in the symposium.

During the symposium, the nine programs developed 10 key components of successful hospital-based violence intervention programs.

1. *Secure Hospital Buy-in*
   The hospital staff and administration must buy into the program before it is implemented. To ensure buy-in, it is important to engage all different types of staff (i.e., doctors, nurses, medical social workers, etc.) and administrators in the planning process.

2. *Select Target Population*
   Most programs focus on adolescents and young adults but often also include siblings and family members. Each program should determine its target population based on the violent injuries that occur in the immediate area where the hospital is located in order to effectively serve the needs of the community.

3. *Establish Goals and Objectives*
   The main goals of the violence prevention programs include reduction of gang violence and especially retaliatory violence.

4. *Streamline Referral Process*
   Programs need an easy-to-follow referral process to refer clients to a case manager for clients to receive the services they need quickly. Clients should be referred to a case manager before they are discharged from the hospital to ensure that clients have a support system before they return to their neighborhood.

5. *Determine Structure of Service Provision*
   Programs should assess the risk factors and needs of each client and provide tiers of services, such as using tiers based on risk factors. For instance, clients with low risk of retaliatory violence and reinjury may be referred to an advocacy group, such as the Victims of Crime, and to other outside agencies. Clients with multiple needs, such as mental health issues, job training needs, etc., and who are not at high risk of retaliatory violence and reinjury, may be referred to a case manager who will assist with mental health services, job training, and a program for three to six months. Clients who are high risk for retaliatory violence and reinjury need intense case management for at least six months.

6. *Engage Resource Networks*
   Clients need a case manager who will coordinate the intervention by selecting appropriate outside agencies. Each client has specific needs and it is imperative that agencies are selected based on these specific needs.

7. *Make Informed Direct Service Staff Hiring Decisions*
   Programs have to select staff who will be responsible for clients. For instance, the case manager should be carefully selected as they will have the closest interaction with the clients. The program must also establish clear guidelines for work expectations and match clients with case managers based on life circumstances and other relevant factors.

8. *Support Direct Service Staff Through Training and Supervision*
   Staff needs to be trained at the beginning and throughout their career. Most trainings include initial program and agency/hospital orientation, professional development, and skills building.

9. *Conduct Effective Evaluations*
   Programs should establish clear guidelines for the assessment of the program. The main

Continued

CASE STUDY 5.4: YOUTH ALIVE! (CONTINUED)

question is whether the program is successful in accomplishing the goals over several years.

10. *Set Funding Goals for Sustainability*
    Programs are in constant need of funding. The program directors should have a clear plan to ensure the sustainability of the program.[66]

Discussion Questions:

1. Discuss the purpose of the hospital-based violence prevention programs.

2. What do you think: Are these programs effective in reducing gang-based violence? Explain your answer.

3. Imagine you would want to build such a program. Pick a hospital in your neighborhood and write down the information for each key component based on what you know about your neighborhood. For instance, what do you think should be the target population for your neighborhood and what should be the goals? Which community programs would be able to assist your hospital? And how would you convince the staff and administration to implement a hospital-based violence prevention program?

There have been several implementations of the model. The OJJDP tested the model at four sites across the United States: Los Angeles, California; Richmond, Virginia; Milwaukee, Wisconsin; and North Miami Beach, Florida. Based on these model programs, the OJJDP identified best practices to be used in such programs. These best practices include the following: The programs are targeting high-crime high-risk communities by providing essential services, including prenatal and infant care, afterschool activities, truancy and dropout prevention, and job programs. The program identified young children aged 7–14 who are at high risk of delinquency and provided special services to these children. These services consisted of community services, services by faith-based groups, and resources provided by schools, such as afterschool programs and early college awareness programs. There are also gang awareness training programs for children, their families, and local businesses. The intervention strategy focuses on active gang members and their associates and provides support services to encourage them to disassociate from the gang activity. For instance, clients receive individual and group therapy after a needs assessment, including drug and alcohol abuse. Clients also have access to job training, tattoo removal, and anger management classes. The suppression strategy targets the most dangerous gang members with the goal to remove them from the community. The reentry strategy identifies serious gang members who are being released after having been confined to provide services and supervision.[67]

EDUCATION AND TRAINING

Combating gang activities is a difficult task for law enforcement. As you have learned in the prior section, schools are a primary recruiting ground for street gangs, such as **MS-13** or **Barrio 18**. The problems faced by school administrators is how to protect teachers and students when some of the students are gang members using their influence and presence to intimidate and recruit new members. One of the strategies used nationwide is educating youth about gangs. The leading program for this purpose is called **G.R.E.A.T.**, which stands for Gang Resistance Education and Training.

Gang Resistance Education and Training (G.R.E.A.T.)

One of the main tools law enforcement officers can use is educating young people and trying to prevent them from joining gangs. The G.R.E.A.T. program, which started in 1991 in Phoenix, Arizona, is an evidence-based gang and violence prevention program built around a school-based, law enforcement officer–instructed classroom curriculum. The main goal of the program is to immunize youth against delinquency and gang membership. The training takes place in the classroom at an age before children are typically recruited by gangs. Schools can contact the program and choose from a variety of components. The program has a 13-lesson middle school program, a 6-lesson elementary school program, a summer program, and a family program. Law enforcement officers who come to the schools are trained and certified by the program. There are currently about 13,000 officers certified to teach the lessons and more than 6 million children have been part of the program.[68]

The G.R.E.A.T. program also works together with other youth programs, including the Boys & Girls Clubs of America, Inc.; Families and Schools Together®; and the National Association of Police Athletic/Activities Leagues, Inc. Since 2011, the OJJDP has provided the funds needed to support the training of G.R.E.A.T. instructors and materials for instructors and students.[69]

SUMMARY

Gangs and gang violence are a great problem in communities across the United States. It has been very difficult to reduce gang violence and gang membership. In fact, according to a National Survey of Law Enforcement Agencies, gang violence has increased in the past years. The FBI and research have focused mainly on White supremacist prison gangs, motorcycle gangs, and racial-ethnic minority-led gangs, especially Hispanic people and Black people. However, there are also many White gang members who are part of a majority White gang not associated with the White supremacists. In addition, there are Asian gangs, especially in California and New York, which cause a significant amount of violence and are of great concern to local law enforcement. There are several community policing strategies that promise success in reducing gang membership and violence. These strategies include hospital-based violence prevention programs, comprehensive strategies aimed at adolescents, and educational programs, such as the G.R.E.A.T. program. Gang violence will continue to be a challenge task for decades to come.

KEY TERMS

Barrio 18 106
Gang 86
G.R.E.A.T., 107
MS-13 106
Outlaw motorcycle gangs (OMG) 88
Prison gangs 87
Street gangs 87

DISCUSSION QUESTIONS

1. Discuss the gang MS-13 and why it has become one of the main targets of law enforcement.

2. Discuss the characteristics of the different types of gangs and the threats they pose to society.

3. Discuss why street gangs have formed partnerships with Mexican drug trafficking organizations. How is this partnership mutually beneficial?

4. Discuss how gangs recruit juveniles and how community policing may prevent the recruitment of juveniles.

5. Discuss the risk factors for juveniles with regard to joining gangs. How can community programs address these risk factors?

6. Discuss the comprehensive strategy to address gang violence. What are the strengths and weaknesses of this strategy?

CHAPTER 6

COMMUNITY POLICING AND HATE CRIME

On an early Sunday morning in Jasper, Texas, in 1998, three men offered to give James Byrd a ride home. Their act was not a gesture of neighborliness, however. Rather, the three men, who were self-proclaimed White supremacists, started to beat Byrd, spray painted his face, and then tied him to the back of their pickup truck. They dragged him for over three miles along a back road. One of his arms was severed and he was decapitated. The police later found his body parts and a trail of blood spread across an isolated back road. The offenders, John William King, Lawrence Russell Brewer, and Shawn Allen Berry, stated that they killed him because he was African American and because he had the right to vote and serve on a jury. King and Brewer became the first White men in modern Texas history to be sentenced to death for the killing of a Black person. The third man, Berry, received life in prison. He will be eligible for parole in 2038. Brewer was executed in 2011 and King was executed in 2019. King had denied having been part of the slaying during his trial and thereafter. However, his codefendants had identified King as the ringleader and confirmed to the police that he had participated in the killing of Byrd.

In 2001, Texas passed the James Byrd Jr. Hate Crime Act and Byrd's siblings started the Byrd Foundation, which "promotes racial healing and cultural diversity through education." The slaying of James Byrd has had profound negative effects on the victims' and offenders' families, people in the community, and the city of Jasper. It has been difficult to attract businesses because of the nationwide publicity the offense and the offenders received. Basically, the pastor had to console the victim of the family and he also to console the family of the offenders because they all suffered the loss of a family member. Despite substantial efforts to heal, the family of the victim and the Black community continue to be confronted by discrimination. Neither King nor Brewer showed any remorse for their crime and Brewer stated before his execution that

Learning Objectives

1. Define the term *hate crime* and discuss the motives of the offenders.
2. Explain how hate crime data is collected.
3. Discuss the extent and trends of hate crime.
4. Explain the main purpose of hate crime legislation and enhanced penalties.
5. Discuss the major arguments for and against hate crime legislation.
6. Describe community policing strategies that are effective in responding to hate crime.

he would do it all over again.[1] In this chapter, we will discuss the extent of hate crimes, hate crime legislation, and how community policing can effectively respond to hate crimes and create the support communities need to overcome hatred toward others.

INTRODUCTION: DEFINING HATE CRIME

Hate crime laws are basically laws that provide prosecutors with enhanced penalties for crimes committed out of prejudice. The Federal Bureau of Investigation (FBI) defines a **Hate crime** as "a criminal offense committed against a person, property, or society that is motivated, in whole or in part, by the offender's bias against a race, religion, disability, sexual orientation, or ethnicity/national origin."[2]

The main concept in this definition is the definition of *hate* as a bias toward something or somebody. Most people think of hate as a form of *anger* or *rage*. However, with regard to crime, the term *hate* means that someone has a bias against another individual based on a specific characteristic defined by the law, such as ethnicity. The *crime* is typically a violent crime, such as assault, murder, vandalism, or similar crimes. It also includes crimes such as conspiracy to commit a hate crime or inciting another person to commit a hate crime.[3] All these crimes aim to destruct something or cause injuries to a person.

Hate crimes are different from hate incidents. **Hate incidents** are acts committed out of prejudice that do not involve a criminal act. Examples of hate incidents are name-calling or distributing pamphlets. Even though hate incidents are not investigated as crimes, police, and especially community police officers, should pay attention to these incidents because they can turn into criminal acts.[4] One of the questions discussed regularly is whether hate speech is also a crime and should be punished. **Hate speech** can be defined as "communication that carries no meaning other than the expression of hatred for some group, especially in circumstances in which the communication is likely to provoke violence. It is an incitement to hatred primarily against a group of persons defined in terms of race, ethnicity, national origin, gender, religion, sexual orientation, and the like."[5]

In 2018, the Lawyers' Committee for Civil Rights Under Law and the Fund for Leadership, Equity, Access and Diversity (LEAD Fund), part of the American Association for Access, Equity and Diversity (AAAED), distributed a survey to U.S. universities and asked about hate crimes, hate speech, and conduct prohibited under an institution's antidiscrimination policy within the last two years. About 84% of the participants indicated that they had observed or heard about conduct prohibited under an institution's antidiscrimination policy and 82% reported that they had encountered hate crimes. In addition, 65% stated that they had observed or encountered hate speech. Many hate incidents consisted of outside groups coming to the campus and distributing papers and other literature. Most of them were geared toward nationalism. Hate messages are also often sent via social media or are posted inside dormitories.[6]

In 2017, a White supremacist, Richard Spencer, toured university campuses and held speeches across the United States. A day after he toured the campus of the University of Virginia with torches and chanting Nazi refrains, a protest in Charlottesville, Virginia, turned violent, resulting in the death of one woman. By then Spencer had already moved on to repeat his speech at the University of Florida (UF). The president of the university, Kent Fuchs, prepared early in anticipation of possible fights between the supporters of Spencer and protesters. The university created a question-and-answer website and informed the public about the visit by Spencer. The president also sent a message via Twitter: "I don't stand behind racist Richard Spencer. I stand with those who reject and condemn Spencer's vile and despicable message." The preparation was successful in that the event did not cause any destruction of the campus and no one got injured or killed.[7]

Other universities had substantially more problems during the visits of controversial speakers. For instance, the University of California, Berkeley, experienced substantial destruction during a visit from Milo Yiannopoulos, a right-wing provocateur and former senior editor of *Breitbart News*. UC-Berkeley was unprepared for the violent protests that Yiannopoulos's speech would spark. It has since spent more than $1.4 million in security costs. The university canceled the "Free Speech Week" following the Yiannopoulos talk. Yiannopoulos accused the university of sabotaging the "Free Speech Week" and his planned talks by making it difficult to make the required logistical arrangements.[8]

HATE CRIME DATA COLLECTION

Data Collection From the General Population

Hate crimes are not well documented across the United States. As discussed earlier, only a limited number of states have reporting laws. Thus, there is only limited verified data. The Hate Crime Statistics Act, passed in 1990, lays out how to identify and collect hate crime statistics. It states that the "Attorney General shall acquire data, for each calendar year, about crimes that manifest evidence of prejudice based on race, gender and gender identity, religion, disability, sexual orientation, or ethnicity, including where appropriate the crimes of murder, nonnegligent manslaughter; forcible rape; aggravated assault, simple assault, intimidation; arson; and destruction, damage or vandalism of property." The data should be collected via the FBI and the Bureau of Justice Statistics (BJS).[9]

At the federal level, the FBI and the BJS collect hate crime data. The FBI collects data voluntarily reported by law enforcement agencies to the Uniform Crime Report (UCR). The first Hate Crime Report was published by the FBI in 1992. Only 81% of police departments report to the UCR however. In addition, 88% of the reporting departments reported zero hate crimes. Media reports demonstrate that this does not reflect the reality of hate crimes across the country.[10] Hate crimes, overall, are greatly underreported due to four main reasons. First, states vary greatly with regard to their hate crime statutes and range of crimes covered, which will be discussed in more detail below. For instance, what is reported by one state as a hate crime may not be counted as a hate crime in other states. This means that comparing data across all states is impossible. Second, states do not have consistent data collection procedures. Some states' procedures are of much higher quality than others and thus their data overall is of higher quality and not comparable to states with a lower quality. Third, states have different training requirements for law enforcement with regard to hate crime reporting. Some law enforcement agencies are much better trained to recognize and report hate crimes than others. Thus, the reliability of the reported data is unknown. Finally, states also vary on the question of whether hate crimes are separate crimes or not. Some states may have a separate category for assault committed due to racial bias. Other states may simply record the assault as a simple or aggravated assault, not as a separate crime. In these states, hate crimes are indistinguishable from other crimes.[11]

The BJS receives its data from the National Crime Victimization Survey (NCVS). The NCVS is a self-report survey where households across the United States are asked to report hate crime victimization. The question about hate crimes was added in 1999. The NCVS asks persons whether they have experienced hate crimes. In order to determine

THINK ABOUT IT: THE FIRST AMENDMENT AND HATE CRIME

The First Amendment guarantees citizens the freedom of speech and prohibits the government from prosecuting people for their beliefs. People are free to express beliefs that are offensive to others, even if these beliefs are untrue or based upon false stereotypes. This also applies to threats that are vague or distant and hateful ideologies and messages from groups such as the Ku Klux Klan or White supremacists. The question is what constitutes free speech and what constitutes a crime. The U.S. Supreme Court (hereafter referred to as the "Court") has heard several cases on this issue over the past century.

One of the earliest cases was Chaplinsky v. New Hampshire (1942), where a member of Jehovah's Witnesses called the city marshal a "damned fascist" and "damned racketeer." The Court held that such speech can cause a breach of peace because these phrases are likely to provoke an average person to retaliate. Thus, New Hampshire had the right under its public law to punish Chaplinsky.[12]

In Virginia v. Black (2003), the Court had to decide whether burning a cross was a crime. The Virginia statute makes it a felony "for any person . . ., with the intent of intimidating any person or group . . ., to burn . . . a cross on the property of another, a highway or other public place." The statute further stated that burning a cross in itself would be sufficient evidence of an intent to intimidate a person or group. The defendant Black objected and argued that the Virginia law violated his right to free speech guaranteed under the First Amendment. The Court decided that Virginia did have the right to ban cross burnings that were intended to intimidate. Sandra Day O'Connor, who wrote the majority opinion, stated that cross burnings are historically intertwined with the reign of terror by the Ku Klux Klan. The Ku Klux Klan used cross burning to intimidate Black people and threaten them with violent acts. Often, the intent of the cross burning is that the recipient fears for their life.

However, the Court also held that the prosecutor must prove that the cross burning was actually meant to intimidate someone. The burning of the cross in itself is not evidence of an intent to intimidate because it could also be an act of political speech, which would be protected under the First Amendment.[13]

In Watts v. U.S. (1969), the Court distinguished between threats and constitutionally protected speech. The Court held that a person must have made a threat with the intent to carry it out. Watts was convicted in the District of Columbia for making a statement during a political discussion. He stated, "They holler at us to get an education. And now I have already received my draft classification as 1-A and I have got to report for my physical this Monday coming. I am not going. If they ever make me carry a rifle the first man I want to get in my sights is L. B. J. They are not going to make me kill my black brothers." The Court held that Watts had basically just stated his political opinion and that in itself is protected speech. This type of speech cannot be punished.[14]

The crux of the problem lies in determining someone's motivation. In Watts, the Court did not believe that the defendant actually intended to harm the president of the United States. That, however, is a judgment call and other courts and justices may have come to a different opinion. Critics also point out that it is questionable whether hate crime legislation can be enforced by the people responsible when many of them are not experts in First Amendment intricacies.[15]

What do yo think?

1. Discuss why it is so difficult to distinguish free speech protected by the First Amendment and speech that would be considered a crime.

2. Discuss whether you think that hate speech should be prohibited. Why or why not?

whether the crime was a hate crime, the person must report at least one of three hate crime indicators: (1) hate language, (2) hate symbols, and (3) police investigators confirmed that the incident was a hate crime. The survey obviously is also not a true reflection of hate crimes and hate incidents because not every household is surveyed and not all people who are surveyed report hate crimes and incidents. The BJS data shows that between 2003 and 2015, only 41% of hate crime victimizations were reported to the police. Between 2003 and 2015, the BJS recorded 104,600 victimizations. But, only 14% of these crimes were classified as hate crimes. At the end of the day, only 6% of hate crimes are represented in the UCR statistics. Thus, the data leaves much to be desired.[16]

Even though the national data is highly problematic due to the reasons discussed above, there are some states that have collected high-quality data for a long period of time. For instance, in 1988, Minnesota passed a law requiring law enforcement officers to collect hate crime data. California has also collected hate crime data since 1995 and the recorded data is more extensive than the data contained in the UCR.[17]

A national coalition of journalists and civil rights groups founded the project Documenting Hate. They gather media reports of hate crimes and hate incidents and verify them. However, the data is limited to media reports and many hate crimes and hate incidents do not come to the attention of the media. Thus, the dark number of hate crimes is high in these statistics, too.[18]

Data Collection From the College Population

Federal law requires all institutions of higher education to report hate crimes either to the U.S. Department of Education (ED), to the Office of Postsecondary Education (OPE), or the UCR. There are about 7,000 institutions of higher education, 6,000 of which reported to the ED and 400 to the UCR. Studies have found that the reported data is highly inaccurate due to reporting errors, different interpretations of the law and which crimes should be reported, different procedures with regard to case processing, differences in college safety departments, and so on. Some colleges that reported to the UCR and the ED reported significantly different numbers to these departments.[19]

Despite the good intent of the Hate Crime Statistics Act, hate crime data is far from accurate, which makes it very difficult to determine the true extent of hate crimes and develop effective policies targeting offenders.

EXTENT AND TRENDS OF HATE CRIME IN THE UNITED STATES

As described above, the FBI collects hate crime data that is reported by law enforcement agencies across the United States. This, of course, leaves out a substantial number of hate crimes because many people to not report hate crimes to the police. Some people don't report hate crimes because they feel embarrassed or because they are concerned that police will not take them seriously. Others may not report the hate crime because they know the offender and are afraid of revenge or because they don't believe that the offender will be arrested and punished.

The Center for the Study of Hate and Extremism at California State University, San Bernardino, examines the extent and trends of hate crimes nationwide and also for a select number of cities. According to the center, hate crimes have increased significantly in the

▶ **Photo 6.1** A memorial outside the Tree of Life synagogue in Pittsburgh, PA, for the 11 victims of a shooting in 2018. It was the deadliest anti-Semitic attack in U.S. history.

past four years. In 2017 alone, hate crimes increased by 12.7%. This trend is contrary to the general crime trends, which show that violent crime decreased by 0.8% and property crime decreased by 2.9% in the first half of 2017. The cities with the highest number of hate crimes were New York (339), Los Angeles (254), Phoenix (230), Washington, D.C. (179), and Boston (140). Some cities had substantial increases from 2016 to 2017. For instance, in Los Angeles, hate crimes increased by 10.8%, in Phoenix by 33%, and in Washington by 67%. New York and Boston registered a small decline in hate crimes by 2%. The increase in hate crimes started in 2016. The FBI reported a spike in hate crimes during the three days after election day. Overall, there was a 25.9% increase in hate crimes in the fourth quarter of 2016. In California, the highest number of hate crimes were recorded during the three days immediately following election day on November 8, 2016.[20]

Nationwide, UCR data and data from the NCVS combined show that the extent of hate crimes changed very little between 2004 and 2015 but has substantially increased since 2016. In 2004, the hate crime rate for reported and unreported crimes was 0.9 per 1,000 persons age 12 or older. In 2015, the hate crime rate was 0.7 per 1,000 persons age 12 or older. This rate means that on average 250,000 people become victims of a hate crime per year.[21]

There are grave differences in hate crime rates reported by the UCR and the NCVS. In 2004, the NCVS data shows 281,670 hate crimes committed, whereas the UCR data suggests that only 7,649 hate crimes had occurred. In 2015, the number of hate crimes reported through the NCVS was 207,880, but through the UCR only 5,850. This reflects the difference between crimes reported to police and a closer estimate of the extent of hate crimes.[22]

According to the NCVS data from 2011 to 2015, the most common reported hate crimes were violent crimes, including simple assault (61.6%), aggravated assault (17.7%), robbery (8.3%), and rape or sexual assault (2.5%). Property crimes were fairly rare with 9.6% and here the most common crime was theft (7.4%). The most common location where hate crimes occurred was at or near the victim's home (38.7%), a parking lot or public space (24%), a commercial place (14.2%), and a school (13.6%). Hate crimes were most often reported to the police by the victim (69.8%). However, about 54% of hate crimes were not reported at all. There are many reasons why victims do not report hate crimes. The most common reasons are that the incident was handled in another way (40.7%), the incident was not important enough (17.5%), the belief that the police would not help (17.5%), and the belief that the police could not do anything (5.1%).

One of the main issues is the classification as a hate crime. The NCVS asks how the victims knew that this was a hate crime. The vast majority (98.7%) of victims cited hate language by the offender as the primary evidence that a hate crime had occurred. Only very few victims (5.4%) stated that the offender had left behind hate symbols or that the police confirmed it was a hate crime (6.9%). More than 63% of hate crimes were committed by single offenders and almost 71% were male offenders. About 37.7% of the offenders were White and 34.3% were Black.[23]

The UCR shows a significant increase since 2015 in hate crimes. According to the FBI, there were 6,121 reported hate crimes in 2016 and 7,175 reported hate crimes in 2017. This is an increase of about 15% within one year. However, the number of law enforcement agencies reporting to the FBI increased by 1,000 during the same time period. Thus, some of the increase might be attributable to the increased reporting. The most common bias categories were race/ethnicity (59.6%), religion (20.6%), and sexual orientation (15.8%). The majority of hate crimes (about 5,000) were committed against a person, including intimidation, assault, and other violent crimes. About 3,000 offenses were property offenses, such as vandalism, robbery, and burglary.[24]

THE EXTENT AND PURPOSE OF FEDERAL AND STATE HATE CRIME LEGISLATION

Federal Hate Crime Legislation

On June 21, 1964, three civil rights workers—Michael Schwerner, James Chaney, and Andrew Goodman—were murdered in Neshoba County, Mississippi, by members of the Ku Klux Klan. The three men were working on voter registration in the area during what became known as the Freedom Summer. The offenders burned down their church. Schwerner, Chaney, and Goodman were arrested by police later that day, but released, and told to leave the town. On their way out of the town, they were stopped by several White men who also included police officers, taken to a remote location, and murdered. Even though it took several more decades until the offenders were punished, this hate crime would change the nation in a way that the offenders could not have foreseen. The incident had significant effects all the way to the White House. In 2005, Mississippi attorney general Jim Hood finally brought murder charges against Edgar Ray Killen. Killen was sentenced to 60 years in prison for the killing of the three men.[25]

In 1968, four years after the murder of the three civil rights workers, the first federal hate crime statute was signed into law by President Lyndon Johnson. The statute made it a crime to "use, or threaten to use, force to willfully interfere with any person because of race, color, religion, or national origin and because the person is participating in a federally protected activity, such as public education, employment jury service, travel, or the enjoyment of public accommodation, or helping another person to do so."[26] For instance, assault is a class D felony. A person convicted of assault may receive up to seven years in prison. If it is a hate crime, however, then the offender can be sentenced up to 15 years because the crime is upgraded to a class C felony.

There are five federal hate crime laws. First, the most comprehensive law was signed in 2009 by President Barack Obama. It is called the Matthew Shepard and James Byrd Jr. Hate Crimes Prevention Act. Matthew Shepard and James Byrd Jr. were victims of two separate hate crimes. We discussed the killing of Byrd at the beginning of the chapter. This crime rattled the nation, not only because of the violence involved but also because the FBI was prohibited by law from investigating and prosecuting the offenders. The offenders were convicted in the state of Texas, but several members of the U.S. Congress believed that these crimes should fall under the jurisdiction of the FBI rather than the state. Thus, Congress enacted the new federal hate crime bill to give the FBI more power.[27] The Act expands the federal definition of hate crimes to include crimes committed due to bias against an "actual or perceived religion, national origin, gender, sexual orientation, gender identity, or disability of any person, only where the crime affected interstate or foreign commerce or occurred within federal special maritime and territorial jurisdiction."[28] It also enhances the legal toolkit available to prosecutors in cases where the hate crime was motivated by the victim's actual or perceived sexual orientation or gender identity. Finally, it provides greater support for local and state partners. In 2017–18, the federal government charged over 50 offenders under this act—for a total of about 300 since 2009.[29]

Second, the Criminal Interference with Right to Fair Housing statute "makes it a crime to use or threaten to use force to interfere with housing rights because of the victim's race, color, religion, sex, disability, familial status, or national origin." This includes the renting, selling, or purchasing of a dwelling, the occupation of a dwelling, the financing of a dwelling, and the opportunity to participate. The statute specifies punishments of imprisonment of not more than one year, or if bodily injury occurs of up to 10 years, or if death occurs of up to life in prison.[30]

Third, the Damage to Religious Property, Church Arson Prevention Act of 1996 "prohibits the intentional defacement, damage, or destruction of religious real property because of the religious nature of the property."[31] The statute also makes it a crime to obstruct a person via force or threat of force in their free exercise of religious beliefs. For instance, on October 17, 2018, Marq Perez was sentenced to 24 years in prison for burning down the Victoria Islamic Center in Texas in January 2017. Perez had told a witness that he wanted to send a signal and that he would burn down the mosque again if the community rebuilt the mosque.[32]

Fourth, the Violent Interference with Federally Protected Rights Statute "makes it a crime to use or threaten to use force to interfere with federally protected activity, such as public education, employment, jury service, travel, or the enjoyment of public accommodations."[33] Almost all criminal codes include enhanced punishments for crimes against older people, very young people, teachers on school grounds, or law enforcement officials.[34]

State Hate Crime Laws

State law legislation varies widely. There are five main differences in hate crime statutes. First, different statutes include different protected groups. Some statutes focus mainly on race, whereas others are more inclusive, adding gender, sexual orientation, ethnicity, religion, etc. Second, statutes vary in whether and how they deal with criminal and civil penalties. Third, statutes vary on the range of crimes covered. Fourth, statutes vary on the reporting criteria of hate crimes by law enforcement. Finally, statutes vary on whether there is a training requirement for law enforcement officers. In the past decade, states have expanded their statutes and included a wider range of protected groups and expanded penalty enhancements.[35]

The first state that enacted hate crime legislation was Oregon with the 1981 Hate Crimes Act. The Act mainly targeted racially motivated crimes and provided for enhanced penalties for offenders. The Act also allowed the victims to collect damages by bringing a civil lawsuit against the offender. The civil action is independent from a criminal case and conviction. This means that even if the offender is not convicted on criminal charges, they may still be awarded damages. Under the 1981 Hate Crimes Act, prosecutors can also invoke injunctions against persons and groups who are believed to engage in hate crime activities.[36] Many other states followed Oregon's approach in the following years. However, there are important differences between states' legislation.

Types of Recording Statutes

There are three categories of states with regard to hate crime statutes and recording of hate crimes: (1) States and territories with hate crime laws that require data collection of hate crimes; (2) States and territories with hate crime laws that do not require data collection of hate crimes; and (3) States and territories without hate crime laws. There are 30 states and territories with hate crime laws that require data collection of hate crimes. Eighteen states and territories have hate crime laws, but they do not require data collection of hate crimes. Thus, a total of 48 states and territories have passed hate crime laws. The remaining eight states and territories do not have hate crime laws: American Samoa, Arkansas, Georgia, Guam, Northern Mariana Islands, South Carolina, U.S. Virgin Islands, and Wyoming. All federal hate crime laws do apply in all states and territories, of course.[37]

Types of Laws

Further, we can distinguish three types of hate crime laws among the 48 states that have passed legislation. First, 45 states and the District of Columbia have passed penalty-enhancement laws. Most of them were modeled after a statute by the Anti-Defamation League of 1981, which provides for more severe penalties when the victim was targeted because of bias. The prosecutor must prove the motivation of the offender beyond a reasonable doubt. Almost all of these states provide for enhanced penalties if the victim was targeted because of their race, religion, or ethnicity. Only 26 states and the District of Columbia include gender-based crimes, and only nine states and the District of Columbia have enhanced penalties for gender identity-based crimes.

Another type of hate crime legislation is institutional vandalism statutes. These statutes apply in 42 states and the District of Columbia. They mainly protect institutions such as religious schools, houses of worship, and cemeteries from destruction or damage.

Finally, hate crime legislation also takes the form of data collection and law enforcement training mandates. A total of 27 states and the District of Columbia require data collection and 14 states require law enforcement officers to complete training that teaches officers to identify, respond, and report hate crimes.[38]

THE PROBLEM OF MOTIVE IN HATE CRIME INVESTIGATIONS: ARE HATE CRIME LAWS EFFECTIVE?

The purpose of hate crime laws is threefold: (1) Deter people from committing hate crimes; (2) Support the victims of hate crimes by punishing offenders more harshly; and (3) Provide law enforcement with the necessary tool to combat hate crimes and especially the violence they cause. Proponents and critics argue about the ability of the current hate crime legislation to accomplish these goals. Arguments in Favor of Hate Crime Legislation

The main argument for hate crime legislation is that this type of violence must be condemned by providing for harsher penalties. Here, proponents believe that when a person is victimized because of their skin color, race, religion, natural origin, gender, sexual orientation, gender identity, or disability, all of the members of that group "feel like potential targets and experience a shared sense of persecution."[39] Thus, the proponents of hate crime laws note that these laws protect all population groups and deter people from harming people out of bias. In addition, these laws help law enforcement fight the type of violence caused by bias and hatred. These types of crimes are worse for the victims because they attack the victim at the core of their identity and the victims often suffer greater psychological trauma than other crime victims.[40]

Finally, enhanced penalties for hate crimes are appropriate and important to deal with low-level criminal assaults, intimidation, and property crimes. These low-level crimes are often not vigorously investigated, especially when police have to deal with more serious crimes. This can have substantial negative consequences for the victims because hate crimes tend to be repetitive and interfere with other constitutional rights. For instance, offenders may target a person because of their race and deny them the right to rent or buy a house in a certain neighborhood. This would be a violation of the Fair Housing Act. If these crimes are not investigated and the offender is not punished, then the victim suffers not only the original injury but also continuously because they are unable to rent or buy a house in their desired neighborhood simply because of their race. Thus, proponents of hate crime legislation argue that these laws are imperative to enforce the rights of all citizens. In addition, there is much concern that these low-level offenses can spiral into violent offenses if police do not vigorously investigate and punish the offenders. Law enforcement has to demonstrate that this type of behavior will not be tolerated in order to deter more violent crimes.[41]

Arguments Against Hate Crime Legislation

Critics argue that the current hate crime statutes fail to accomplish either of these goals. The main problem is the issue of proving the motive of the offender. As discussed earlier, in order to prove that a person committed a hate crime, the prosecutor must be able to prove that the motive of the offender was biased toward against a race, religion, disability, sexual orientation,

▶ **Photo 6.2** While this may look like a historical photo, this image was taken in 2019 as active members of the Ku Klux Klan burned a cross in the suburbs of Madison, Indiana.

or ethnicity/national origin. This problem of proving the motive has resulted in a heated discussion between proponents and opponents of hate crime legislation. The motive may be confused with the intent, but they are different. **Intent** is the "purpose to use a particular means to achieve some definite result."[42] For instance, the offender may use a knife to injure a person. In contrast, **motive** is defined as the "cause or moving power that impels action to achieve that result."[43] For instance, the offender may feel hatred toward immigrants and takes the knife to injure the person who they believe is an immigrant. But, it's also possible that the offender injured the person out of revenge or for some reason other than bias.

How can police officers, prosecutors, judges, and juries determine the true motivation behind a person's statement or what the person's true motives are? Critics of hate crime laws argue that people may be convicted of a hate crime when their motivation was not actually prejudice toward a certain population group. For instance, a White man might kill a Jew during a bar fight and he could be charged with a hate crime even though he was simply drunk and got into a fight with another patron. He may not have even been aware that he was getting into a fight with a Jew. The requirements for prosecutors to establish a hate crime are therefore high. First, the prosecutor has the burden of proof and must prove beyond a reasonable doubt that the defendant committed the crime. Second, the prosecutor must prove that the defendant had the mens rea or a guilty mind to commit the crime. Finally, the prosecutor must prove beyond a reasonable doubt that the defendant attacked the victim "because of" or "by reason" of that person's race, ethnicity, gender, sexual orientation, or religion. This requirement is very difficult to meet because we can't exactly know what a person was thinking. For instance, a person may have attacked a Black person with the intent to kill the person, but the motive for the crime was to take the purse with the money. In this case, the purpose of the act and the motive can be distinguished. The purpose was to kill the Black man, but the motive was to take the purse with the money. The issue of making inferences about motive is especially problematic in cases where the defendant invokes their Fifth Amendment right to be silent and has not revealed their motive for the crime and in cases where there are no witnesses who can speak to the motive. In addition, an offender may have multiple motives. The prosecutor has to prove that the bias, such as racism, was the sole or main motive. In the absence of actual knowledge of the motive, the prosecutor's case often relies on circumstantial evidence. People have been convicted of crimes they did not commit and it seems likely that people will receive harsher punishments because the jury concluded that they had committed a hate crime even if that is not actually true.[44]

A related problem is the conclusion by the jury. A jury may find that the offender had a racist motive, but what if the jurors themselves were biased? We all have unconscious biases that affect our behavior. **Unconscious biases** are typically defined as "prejudice or unsupported judgments in favor of or against one thing, person, or group as compared to another, in a way that is usually considered unfair."[45] Many jury instructions now include statements that educate the jurors about unconscious biases. However, research suggests that unconscious bias occurs automatically. People don't actually think about their biases when they hear evidence, listen to witnesses, and deliberate about the verdict. Rather, they use their past experiences and background in their decision-making. Most of these unconscious biases include factors, such as class, gender, race, ethnicity, religious beliefs, age, disabilities, sexual orientation, and similar factors.[46] These factors may benefit some offenders and disadvantage others. Research on jury decision-making across jurisdictions found that offenders who are committing crimes that fit the stereotype of a hate crime are the most likely to be convicted. These crimes are violent incidents committed by White offenders on Black victims. This, of course, leaves out a substantial number of victims, and especially victims who were targeted because of their sexual orientation or gender identity, such as gay people or transgender people. Jurors may have biases against these victims and may not punish the offender.[47]

Critics have also pointed to the fact that it is very difficult for prosecutors to obtain a conviction under a hate crime statute due to the problems outlined above. In addition, prosecutors may not charge offenders under a hate crime statute because they fear that they will lose the case if they can't prove the motive. The offender would go free and that is not in the interest of justice. Thus, prosecutors prefer to charge an offender under the traditional criminal statutes. Finally, inferring the motive by jurors leads to inconsistent verdicts because jurors have to use their subjective judgments. The criminal justice system becomes more arbitrary and deterrence cannot be achieved when criminal convictions are random and unpredictable.[48]

Low conviction rates and low rates of success in civil lawsuits are one problem. Another problem is that these victims are often ignored by police who fail to properly investigate crimes.[49] A study by Wolff and Cokley (2007) found that in 31% of violent crimes against these victims, police refused to file an incident report.[50] In addition, victims also experience abusive treatment by law enforcement officers. For instance, between 2007 and 2008, the reports of physical abuse increased by 150% from 10 to 25, the number of

THINK ABOUT IT: REFORMING HATE CRIME STATUTES

Some of the critics have proposed changes to the current hate crime statutes. The main focus is on easing the burden of proof of prosecutors that the motive of the crime was prejudice. The proponents suggest that instead of putting the burden of proof of the motive on the prosecutor, the burden of proof should be shifted to the defendant. In this case, the defendant would have to prove that the motive was not prejudice. For instance, in cases of interracial violence, it would be presumed that the offender was driven by a racist motivation. This approach would likely result in a greater number of convictions under hate crime statutes and enable victims to be successful in civil lawsuits[51] and work toward sustainability.

What do you think?

1. If you were a defense lawyer, what problems would you point out in this approach of shifting the burden of proof to the defendant?

2. Discuss the issues raised by this approach with regard to the defendants' constitutional rights that guarantee equal protection under the law and the standard of "innocent until found guilty."

incidents of verbal abuse rose by 50% from 34 to 51. And in 2008, law enforcement officers were the primary offenders in 196 cases—an increase of 11% compared to 2007.[52] When victims not only receive no support from law enforcement in a significant number of violent incidents but are also victimized by law enforcement officers, the goals of hate crime statutes are not accomplished.

Minority rights groups have also voiced their concerns, especially people who are part of the LGBTQ community who have a low socioeconomic status and/or are of color. These people are very vulnerable to hate crimes, but they are too often not protected by current hate crime statutes, law enforcement, and the criminal justice system in general. A study by the National Coalition Against Hate Violence shows that most violent hate crimes against this group occur at home, in a public space, and in the workplace. The public spaces where most of the violence occur are gendered spaces, such as restrooms, locker rooms, jails, and shelters. The largest group of offenders are strangers, followed by landlords and neighbors.[53]

COMMUNITY POLICING STRATEGIES TO RESPOND TO HATE CRIME INCIDENTS

Community policing is especially important during responses to hate crime incidents. Community police officers know their neighborhood and the people who live there. They often have knowledge of people who belong to a gang or who commit offenses out of prejudice. Hate crimes have very serious impacts on the victim and the community. The willingness of police and prosecutors to pursue hate crimes and punish offenders is very important for police–community relations. Many groups, including human rights groups, civic leaders, and law enforcement, have highlighted the importance of being tough on hate crime offenders. Case Study 6.1 features an interview with the director of the U.S. Department of Justice of COPS.

CASE STUDY 6.1: INTERVIEW WITH BERNARD MALEKIAN, DIRECTOR OF THE U.S. DEPARTMENT OF JUSTICE OFFICE OF COMMUNITY ORIENTED POLICING SERVICES (COPS)

How can community policing help address hate crime?

It is imperative to build relationships with the community by meeting with community members and communicating about important issues. We must provide a venue where people can come together and talk because it is much easier to hate someone you don't know. By providing opportunities for people from different communities and different ethnic and religious backgrounds to meet and converse about their concerns, people get to know each other and get to understand each other's beliefs and concerns better. It also puts a human face to police, which helps build trust, and as a result a greater reporting of incidents of hate crimes.

Why is hate crime data so difficult to collect?

Hate crime data is difficult to collect because in order to classify an act as a hate crime, the police have to identify the motive of the offender. That can be very challenging. For instance, the police have to determine whether a crime, such as a robbery, occurred because the offender wanted the money, or whether the offender was motivated by hatred toward the victim due to ethnicity, religious beliefs, sexual orientation, etc.

Why does hate crime classification matter?

It matters because the number of hate crimes is an indicator of community health. It is important to use the data to develop strategies to improve the community health.

How can police departments make it easier for people to report hate crimes?

Police departments need to build relationships with community groups and encourage people to report hate crimes. People have to be confident that when they come to the police to report a hate crime, they will not be trivialized or dealt with in a condescending fashion. Crime in general, and hate crime in particular, is underreported and without a fair estimation of the actual crimes committed, it is very difficult for police to address hate crimes.

How does community policing serve victims?

Victims of hate crimes sometimes feel like things, rather than human beings. In order to make the victims whole again, we have to help them reintegrate into the community and make sure they don't feel that they have been reduced to a particular characteristic, such as gay or Black.

What message do you have for police department leaders?

The police have to be part of a larger community effort. The police have to find a common place where they can interact with community members on a regular basis. These common places can be community gatherings, neighborhood block parties, and other events. Police should not only be present during a crisis but rather interact with the community at social events that are part of everyday life. The community members must know the police officers and vice versa. Knowing the other creates trust, which results in more hate crimes being reported. In addition, police must take hate crimes very seriously or otherwise the victim may feel dismissed. Victims who feel dismissed are less likely to finish the process of filing a police report and providing useful information about the crime. Thus, police must strengthen the victims' resolve.[54]

Discussion Questions:

1. Which type of violent acts would you classify as hate crimes?
2. How can police better identify the motive of an offender to determine whether a crime was a hate crime?

3. Should punishments be harsher if a crime was motivated by hate? For instance, if an offender assaults a person motivated by hate, should the penalty be enhanced because of the offender's motive? Make arguments for and against enhanced punishments.

COMMUNITY POLICING STRATEGIES IN RESPONSE TO HATE CRIME INCIDENTS

Law enforcement officers have a special task in responding to hate crime incidents. One of the main issues is the possible escalation of hate crimes. Stated differently, low-level crimes could turn into higher-level crimes. When offenders are not punished, they may believe that they can continue and escalate their behaviors. First, police must prioritize hate crimes to send a message to the community that hate crimes will not be tolerated and will be investigated vigorously. Second, training on hate crimes is crucial for the investigation and reporting of hate crimes. Third, creating a special task force among police departments should include members of the community. The partnership between the community and the police depends on coordinating hate crime law enforcement, victim services, and community partners.[55] See Case Study 6.2 to learn how the Oak Creek, Wisconsin, Police Department developed a response to a hate crime incident in which four people were killed at a Sikh temple.

CASE STUDY 6.2: OAK CREEK—COMMUNITY POLICING IN THE AFTERMATH OF A TRAGEDY

Oak Creek is a suburb of Milwaukee, in Wisconsin, with a population of about 35,000. On August 5, 2012, a White supremacist shot and killed four people and injured four others at a Sikh temple. People had gathered in the temple for Sunday services. Two children ran outside the temple to play when they saw a man, Wade Michael Page, pull up with his car and start shooting. While the children ran into the temple to warn everyone, the shooter killed two people in the parking lot. He then went into the temple and started shooting. One of the injured victims was police lieutenant Brian Murphy. When Lieutenant Murphy arrived at the scene, the shooter left the temple and ran toward his car where he was confronted by Murphy. During the violent encounter, Murphy was shot 17 times. At that moment, another officer arrived at the parking lot and opened fire at Page. Page was hit, and he eventually killed himself with a gunshot in the head.

After the incident, the Oak Creek Police Department met with members of the community to ensure them that they were safe and that the police were focusing on proactive training in the event of active shooters. The police department also spent a significant amount of time on building strong community relations. For instance, the police department

Continued

CASE STUDY 6.2: OAK CREEK—COMMUNITY POLICING IN THE AFTERMATH OF A TRAGEDY (CONTINUED)

organized a vigil for the victims and 5,000 residents attended.

An important part of building strong relations is communication of important information to build trust. During the months after the attack on the temple, the police provided much information about the attack and the results of the investigation. They also held numerous meetings where citizens could ask questions and voice their concerns. Together with the community, the police implemented new public safety measures. The community and police, in collaboration with the U.S. Department of Justice of Community Oriented Policing Services (COPS), produced a film that is part of the Not in Our Town series. One of the main goals is to engage community partners to work with the police department on preventing hate crimes, increasing hate crime reporting, and addressing underlying tensions within the community that could result in hate crime.

In addition, the police department identified local White supremacist groups and other groups in which the shooter was a member. They started sharing intelligence data with the community to get more information about these groups, their members, and hateful activities. They worked together with the Sikh community to create stronger security measures and ensure that everyone felt safe. This includes police officers stopping by the Sikh community center every morning and having police present at all major events at the Sikh temple. The attack resulted in the community taking was a stronger stance against hate crimes, including the accurate reporting of every incident that might be a hate crime to the Federal Bureau of Investigation (FBI).[56]

Discussion Questions:

1. Why is it important to build strong relationships with the community to prevent hate crime?

2. What can police departments do to build trust with members of the community?

3. How does intelligence assist police in preventing hate crime, and what role does the community play in the process of gathering intelligence?

SUMMARY

In this chapter, we defined the term *hate crime* as a crime motivated by bias. These types of crimes often target ethnic minorities, race, or religion, or are based on a disability, gender, gender identity, or sexual orientation. The number of hate crimes in the United States has greatly increased in the past four years. The main explanations of this rise in hate crimes are complex.

We also discussed the five main federal hate crime laws, the most comprehensive of which is the Matthew Shepard and James Byrd Jr. Hate Crime Prevention Act of 2009. Hate crime laws and recording of hate crimes vary widely across the states and a few states do not have specific state statutes for hate crimes. However, all federal laws apply to the states and thus hate crime perpetrators can receive enhanced punishments regardless of where they commit the crime. Community policing has a great impact on hate crime offenses because law enforcement officers must demonstrate to the community that they are taking a strong stance against crimes committed based on prejudice. In the aftermath of a hate crime incident, strong police–community relations can be very helpful in supporting the victims, arresting the offenders, and bringing the offender to justice. In the absence of a combined police–community effort, hate crimes can flourish, creating more tensions because the community believes that the police, who are supposed to protect them, are either ineffective or have no interest in deterring the offenders. As you learned, deterrence is one of the main goals of enhanced punishments. But this goal depends greatly on the ability to arrest the offender and prove that their motive was based on prejudice.

KEY TERMS

Hate crime 110
Hate crime laws 110
Hate incidents 110
Hate speech 110
Intent 119
Motive 119
Unconscious biases 120

DISCUSSION QUESTIONS

1. Discuss the definition of hate crime. What types of crimes fall under this category?
2. Debate whether the First Amendment of the U.S. Constitution protects hate speech.
3. Debate the pros and cons of hate crime laws.
4. Discuss the extent and trends of hate crime over the past decade. What explanations have been advanced for these trends?
5. Discuss the problems associated with the collection of data on hate crimes and hate incidents.
6. Explain how community policing can effectively reduce hate crimes.

CHAPTER 7

COMMUNITY POLICING AND MILITARISM

Learning Objectives

1. Explain the development of military tactics and weapons in police departments across the United States.

2. Discuss why growing militarization of police may be a threat to community policing.

3. Discuss the impact of battle dress uniforms (BDUs) on community policing.

4. Explain how the growing militarization is influencing police training and vice versa.

5. Explain the importance and benefits of military equipment for police agencies.

Imagine that you are at home with your family, sleeping soundly in the early morning hours. You awaken suddenly to a loud explosion and the sound of glass shattering. A bright light blinds you and there is a terrible ringing in your ears. You cannot see anything, but through the ringing you hear the harrowing sound of your front door being broken down as your children begin to scream in the next room. As you come to your senses, you look outside your window and see what appears to be a tank in your driveway. Suddenly, people—you have no idea how many—break through your bedroom door. In the darkness, all you can see is that they are wearing black and carrying assault rifles, and their faces are masked. You hear people yelling at you and your partner to get on the floor and put your hands behind your back. Your children are still screaming in the next room and your dog is barking loudly. The people lead you, wearing whatever you wore to sleep that night, into the living room, pointing assault rifles at you the entire time. You are ordered to sit, and someone quickly handcuffs you to the chair. More people then bring your partner and your children into the living room at gunpoint. Your dog is still barking, and one of the people shoots it, killing it instantly, in front of you and your children. They then proceed to ransack your home, breaking down doors and shattering windows. You can see that the explosion you heard earlier came from a grenade that now lies near your feet, scorch marks covering the floor from the blast. They hold you and your family at gunpoint for the next several hours, refusing to answer any questions about why they are there or what they are looking for. Once they have finally left, you find your home in shambles. Broken glass litters the floor, and doors are broken from where the police kicked holes in them. Your dog lies breathless in a pool of its own blood. Tables are overturned, papers are strewn about, and electronic equipment has been ripped from the walls and left on the floor. Your partner is desperately trying to calm your hysterical children.

Unfortunately, this is not a scene from an action movie, and it did not happen during the course of a protracted battle in an overseas war. This is the militarization of our state and local police, and events like this are happening every day in homes throughout America[1,2] (p. 12).

INTRODUCTION

In this chapter students will learn how policing has changed and incorporated more weapons and tactics that were traditionally used by the military. We will discuss the use of Specialized Weapons and Tactics (SWAT) teams and their pros and cons. We will also detail the impact of police militarization on community policing. Students will also learn about how police militarization has changed police training. Finally, we will detail recent developments and increasing public scrutiny of military police tactics and equipment.

DEATH BY SWATTING

On December 28, 2017, police in Wichita, Kansas, received a 911 call from a man claiming to have shot his father and to have taken his mother and siblings hostage. The caller also told police that he had poured gas over the house and wanted to set the house on fire. The police surrounded the house of the suspect. Before police could make contact, Andrew Finch, the resident, came to the door. The police told him to keep his hands up. When Finch failed to comply with the orders of the police and moved his hands toward his waist, a police officer fired a single shot. Finch was pronounced dead at the hospital. Following the incident, the police searched the house. They found four of his family members inside the house. All were alive and unharmed. Finch's body was also searched for weapons. He was unarmed. The investigation showed that Finch had not called the police nor had he shot his father. He had become the victim of "swatting." Tyler Barriss, an online gamer from California, playing the game "Call of Duty," had made a 911 call falsely claiming to have shot his father and to have taken his mother and siblings hostage, holding them at gunpoint. Barriss was arrested and charged with involuntary manslaughter. This was not the first time that Barriss had engaged in **swatting**, the false reporting of emergencies to police in order to get a SWAT team sent to the address. The gamers get credit in the game for every successful call, raising their online reputation. Swatting has become a major problem for police across the United States and many celebrities, including actor Tom Cruise and singers Rihanna and Chris Brown, have become victims of swatting.[3]

In fact, swatting has become such a problem that several states and the federal government are working on bills that would increase the penalties for swatting. In April 2018, Kansas governor Jeff Colyer signed an "anti-swatting bill" into law imposing stiffer penalties for "swatting" or false 911 calls that result in an injury to a third party.[4]

THE HISTORY OF POLICE MILITARIZATION

The Civil Rights Riots: How It All Began

In racial-ethnic minority communities, the militarization of police is more pronounced than anywhere else and has become part of the daily life of the residents. The militarization of police is closely related to the troubled racial history of the United States. In the 1960s, many cities, including Detroit, experienced riots as part of the civil rights movement. Between 1964 and 1971, more than 700 civil disturbances resulted in many deaths, injuries, and property damage in predominantly Black communities. As a response, President Lyndon Johnson signed the Omnibus Crime Control and Safe Streets Act in 1968. This Act provided local law enforcement with funding for military resources in preparation for the riots. The first Special Weapons and Tactics (SWAT) team was developed in the city of Los Angeles in the 1960s. The Act also provided funding to small cities and today 90% of all cities with a population of more than 25,000 have SWAT teams.[5]

The rise of **SWAT teams** during the 1960s and 1970s was not matched, however, by the appropriate training of officers. Larger cities, such as Los Angeles and New York, had developed training procedures for SWAT officers. The training emphasized the appropriate use of the military equipment and tactics. SWAT teams were typically only deployed in emergency situations and SWAT officers were trained to deescalate violent situations through negotiations first and use paramilitary tactics and weapons only as a last resort. This was not the case in small cities and police departments. There was little to no training for officers and many of the SWAT officers were part-timers who had a full-time job with more conventional police work. Instead of emergencies, SWAT teams in small towns were often used against protest groups and civil rights activists.[6] One police chief voiced the following concern to the *New York Times*:

> Some of these men have lost perspective of their role in society and are playing mental games with firearms. . . . And if you set yourself up to use heavy firepower, then the danger exists that you will use it at the first opportunity, and over-reaction—the opposite of what the [SWAT] concept is about—becomes a real danger.[7]

The War on Drugs and Police Militarization

In June 1971, President Richard Nixon declared the "war on drugs." Over the past decades, the war on drugs has substantially contributed to the militarization of police. Myles Ambrose, head of the Office of Drug Abuse and Enforcement under President Nixon, stated: "Drug people are the very vermin of humanity. They are dangerous. Occasionally we must adopt their dress and tactics."[8] This statement very clearly demonstrates the "tough on crime" approach and especially the tough on drug crimes approach that would dominate the philosophy of policing. Nixon substantially increased the size and power of federal drug control agencies. The growing militarization of police is directly related to this "tough on crime" and "tough on drugs" policy. During this time, numerous bills and Supreme Court decisions gave police greater power and eroded the civil liberties of citizens. For instance, a new crime bill was passed by Congress that allowed no-knock drug raids, that is, police did not have to knock and announce themselves before breaking into someone's home who was suspected of drug possession and/or drug dealing. As you learned in the beginning of the chapter, these no-knock raids had serious consequences for many people as SWAT teams raided house in which children and other innocent residents were present and suffered traumatic experiences, were injured, and sometimes killed.[9]

Nixon also created the Office of Drug Abuse Law Enforcement (ODALE) teams led by Myles Ambrose. The office was later called the "Office of Fear" by Edward Jay Epstein in his book with the same title. One of the police raids that made national news at the time and still stands for the warrior cop mentality used by SWAT teams today occurred in Collinsville, Illinois (see Case Study 7.1).[10] There are many examples of raids that "went wrong," where innocent people were harmed and killed, wrong houses were raided, and where officers planted evidence after the raid because they didn't find drugs. But Nixon's war on crime was a successful political strategy at a time when crime rates were surging

CASE STUDY 7.1: THE COLLINSVILLE, ILLINOIS, DRUG RAID

Herbert Giglotto made good money as a boilermaker. That made the job's early mornings more tolerable. He and his wife Evelyn went to bed each night at 8:00 p.m. to be sure he was up by 5:00 a.m. to get to his job. The couple lived in Collinsville, Illinois, a small suburban town of about 20,000 people, 15 minutes outside of St. Louis, Missouri. At a little after 9:30 p.m. on April 23, 1972, the Giglottos woke to a crash. And then another. The couple's inner and outer doors were being ripped from their hinges. Someone was breaking into their home. "I got out of bed; I took about three steps, looked down the hall and [saw] armed men running up the hall dressed like hippies with pistols, yelling and screeching." Giglotto turned to his wife, who was still in bed, and said, "God, honey, we're dead." "That's right, you motherfucker!" one of the men screamed. The men—15 of them—then stormed the bedroom. One of them threw Giglotto to the bed, bound his hands behind his back, and put a gun to his head. "Move and you're dead," the man said. He then motioned in the direction of Evelyn Giglotto. "Who is that bitch lying there?" "That's my wife." Evelyn Giglotto cried out, "Please don't kill him!" "Shut up!" the man snapped. The man with the gun at Herbert's head quickly flashed a badge, though he didn't give Herbert time to read it. These were cops. The man then read a list of names and asked Herbert if he knew any of them. He knew none of them. "You're going to die if you don't tell us where the drugs are."

Giglotto pled with the man, "Please, please, before you shoot us, check my wallet for my identification. Because I know you're at the wrong place." Seconds later, someone shouted from the stairs. "We've made a mistake!" The men unbound the Giglottos and began to filter out.

Herbert struggled to put on his pants to chase after them for more information. He shouted, "Why did you do that?" The man who'd just held a gun to his head answered, "Boy, you shut your mouth." Evelyn Giglotto was most upset that the police had also thrown the couple's animals—three dogs and a cat—outside. (Given the frequency of dog-shooting during raids in the coming years, the Giglottos' pets got off easy.) "When you don't have children, your pets sort of become your children," she later explained in a newspaper interview. When she asked the police if her pets had been harmed, one of them replied, "Fuck your animals." And with that they left.[11]

and there was a widespread fear of crime among citizens and politicians across all party lines. Indeed, there was overwhelming bipartisan support from Congress and Senate for the crime bills and the war on drugs.[12]

During the second part of the 1970s, under President Jimmy Carter, policing turned from the hard-liner power cop to a softer approach in line with community policing. Paramilitary police tactics and methods declined and so did the use of SWAT teams. This did not last very long, however. When Ronald Reagan became president in 1981, he revived the hard-liner approach to crime and the war on drugs expanded substantially. In the process, police militarization also expanded and SWAT teams were often used for drug raids. During his presidency from 1981 until 1989, Reagan signed several laws that expanded the role of the military in the war on drugs. One of the most important acts was the Military Cooperation with Law Enforcement Act, a proposed amendment to the Posse Comitatus Act, which was signed into law in 1981. This Act provided for a much larger role for the military in the war on drugs. It basically allowed military officers to work with drug enforcement officers on interdiction of drugs, searches, and arrests. This included especially the use of military radar systems to search for drug smugglers. During the Nixon presidency, the Posse Comitatus Act had been interpreted by the courts to allow the military to "indirectly" support state and federal law enforcement agencies, but the amendment signed by Reagan further expanded the role of the military in the war on drugs. The "indirect" support consisted of intelligence, military equipment, research, and access to military bases. On April 8, 1986, Reagan issued the National Security Decision Directive 221, which stated that drugs are a threat to national security because drug monies subsidize terrorist organizations. As a result, the military was encouraged to more actively engage in drug enforcement.[13] The same year, on October 27, Congress passed the Anti-Drug Abuse Act of 1986[14] (HR 5484) and two years later on November 18, the Anti-Drug Abuse Act of 1988[15] (HR5210). These bills were partially a response to the growing drug problem caused by crack cocaine. Crack cocaine had become a major epidemic and was especially widespread among African American communities. The crack epidemic was associated with violence and death, and many people, including community leaders of African American neighborhoods, supported the harsh punishments implemented by the drug bills. There was, however, also much criticism and concern that the crime bills targeted mainly African American neighborhoods because the punishment for possession and sale of crack cocaine was much harsher than for powder cocaine used mainly by White people. Even more disconcerting was the expansive use of SWAT teams to dismantle "crack houses." In Los Angeles, SWAT teams were notorious for breaking into people's houses using an armored military vehicle with a 14-foot battering ram attached. It wasn't rare that the SWAT team would find children inside the houses and only small amounts of crack or other drugs. African American communities certainly faced an unprecedented amount of paramilitary style drug raids, arrests, and incarceration.[16]

Presidents George H. W. Bush and Bill Clinton continued the militarization of police in the 1990s. During the Bush administration, the government created "joint task forces" between the military and the drug enforcement officers. Drug raids by SWAT teams increased greatly and by the end of Bush's term, there were more than 40,000 paramilitary style raids of American homes per year, the vast majority of which were for nonviolent drug offenses. The American Civil Liberties Union (ACLU) estimates that every day (or more likely every night) there are about 124 drug raids.[17]

For several decades now, SWAT teams have been raiding mostly racial-ethnic minority communities in the war on drugs. They use flash bang grenades before they storm into a house and point assault weapons at the residents and their children. They use military weapons and tactics to serve warrants and search for drugs. These SWAT raids typically include about 20 police officers, heavily armed with military-style weapons and often with no-knock warrants. They come in the middle of the night, break into houses, and leave after creating much fear and chaos.[18] Undoubtedly, SWAT teams have destroyed drug labs and arrested drug dealers, but many people question whether these tactics are necessary and whether the same result could be accomplished in a different way, such as community policing. But the war on drugs was not the only "war" that has greatly contributed to the militarization of police. The "war on terrorism," which began under George W. Bush, has also been a major contributor.

The War on Terrorism and the Militarization of Police

On September 14, 2001, three days after the 9/11 terrorist attacks, President George W. Bush declared the "War on Terrorism." The newly created Department of Homeland Security (DHS) poured money into law enforcement agencies allowing them to purchase military equipment to combat terrorism. Special DHS grants provided additional funds for the war on terror by police and, as a result, military equipment. Police departments that applied for the two grant programs, the State Homeland Security Program (SHSP) and the Urban Areas Security Initiative (UASI), were required to dedicate at least 25% of the funds to "terrorism prevention–related law enforcement activities." However, police departments were free to use the funds for other police-related activities.[19] Undoubtedly, counterterrorism duties today have a much higher priority than prior to the 9/11 terror attacks. The training of police officers became increasingly similar to military forces, including tactical training and building the mentality of a soldier. With a higher priority on counterterrorism, police also needed new equipment to adequately perform their duties to catch terrorists and prevent terrorist attacks. The new equipment consists mainly of military-style equipment, including body armors, armored vehicles, assault rifles, night vision goggles, and even attack helicopters. The idea was that if police are more involved in duties that used to be performed by the military, then the police also need the military equipment. That is a logical conclusion. However, police quickly began using military-style equipment that was purchased for counterterrorism tasks for other police work, including serving search warrants to people who were not suspected terrorists. Many legal experts have raised serious concerns about the use of military-style weapons by police because it blurs the line between police and military forces. Police officers may also start to see themselves as soldiers rather than peace officers. That shift has very important consequences for the mindset of the officer. Peace officers are trained to solve problems and lethal violence is only used as a last resort. Military officers are trained to kill the enemy. If police officers act in ways that are consistent with the mindset of a soldier, they are more likely to see a suspect as an enemy and instead of thinking about solving the problem with the help of the legal system, they are likely to use violence, including killing the suspect. People who are raided basically see police who look like soldiers, are trained like soldiers, use weapons used by soldiers, and who may act like soldiers rather than police or what people would expect from police.[20]

Police Perceptions of Militarization and SWAT Developments in the 21st Century

The increase in active shooter and terrorist attacks has led many police departments to change the training and equipment of officers. While the public expects police to properly handle such incidents, there is also much criticism of how police have responded to civil protests and the use of SWAT teams in general.

Law enforcement agencies in general have increased their military-style training, equipment, and tactics. Officer safety is one of the major reasons for the increasing militarism. The quality of the equipment has greatly increased and led to better outcomes for officers, hostages, and suspects during violent encounters. Especially important has been the development of body armors, which have become lighter, more flexible, and provide more coverage. In addition, weaponry has changed and improved the selection of weapons, especially less-lethal weapons. SWAT teams also get support from robots, some of which can be thrown into a window and provide eyes and ears for the officers outside. The officers are better able to assess the situation and make tactical decisions. SWAT teams are also changing tactical decisions based on prior experience. For instance, SWAT teams that were serving search warrants for drugs would often hit a structure hard and fast to preserve the evidence (i.e., drugs). This was very problematic because people got injured frequently. Now, SWAT teams typically breach and hold out while they are calling the suspect. If there is no answer, they send a robot. If the suspect refuses to come out, then they treat the situation as a barricade and they use special resolution tactics.[21]

One of the areas in which the SWAT response has changed entirely is with regard to calls for service for suicides. SWAT teams used to intervene regularly when individuals attempt to commit suicide. Many police departments decided to stop using SWAT teams in these situations due to frequent injuries of the subject and court rulings not favoring SWAT teams and patrol officers who injured the subject. Thus, in cases where the subjects are by themselves and do not pose a danger to anybody else, SWAT teams do not interfere. Lt. Joe Dietrich of the Maricopa County (AZ) Sheriff's Office (MCSO) explains: "It comes down to how much pain are we willing to put on somebody that doesn't want help and is not committing a crime. If they don't want to seek treatment and don't want any help, why are we going there?" SWAT teams do respond if other people are in danger, of course.[22]

SWAT teams also respond in situations of public protest and civil unrest, mainly because most police agencies do not have the officers or equipment necessary to respond effectively. SWAT teams are filling the void, but they are not meant for this purpose. The public has perceived SWAT teams as threatening during peaceful demonstrations and it has created a perception that police have become too militarized.[23]

There has been little research about the question of how police feel about the increasing militarization. Scott Phillips conducted a survey of police leaders to learn about the use of SWAT teams, tactical equipment, and changes to policing related to the increase in militarization. The survey was conducted during the 2015 FBI National Academy in Quantico, Virginia. A total of 370 police chiefs, upper level administrators, and managers completed the survey. One of the survey questions asked about the use of SWAT teams during events involving narcotics and arrest warrants. The respondents stated that the use of SWAT teams during serious or violent incidents, such as high-risk narcotics incidents, felony arrests, hostage incidents, barricaded suspects, and building searches, would not change. These types of situations are dangerous for the officers and victims and SWAT teams are better equipped to arrest suspects without endangering the lives of others. However, 70% of police leaders also answered that they will continue to use SWAT teams during public protest and unrest despite the fact that the public has great concerns about the use of SWAT teams in such situations. About 5% of police leaders stated that they would use SWAT teams more often during public protest or civil unrest because these situations are dangerous for the officers and officer safety is a major concern for any law enforcement agency. The survey also asked about uniforms and potentially changing the uniforms to make police officers look more like civil police rather than military personnel. The vast majority of police leaders (about 78%) did not believe that changes to the uniform were necessary. Similarly, the vast majority of police leaders also indicted that they will continue the use of armored vehicles that very much resemble military-style vehicles.[24]

The survey also asked about respondents, view of militarization within police agencies. More than 90% of the police leaders agreed that armored vehicles, use of SWAT teams in routine patrol, military-style training, use of SWAT teams for warrants, the SWAT uniforms, and patrol rifles are all factors that have contributed to the militarization of police. Police believe that these tools are imperative to keep the public and police officers safe during certain tasks that carry a high risk of violence. Police did also acknowledge though that the military-style appearance of SWAT teams is unacceptable for many citizens.[25]

The Department of Defense Excess Program

In 1989, a series of laws established what is referred to as the Military Program 1033 or the Department of Defense (DOD) Excess Property Program. The purpose of the program was to "enhance the effectiveness of domestic law enforcement agencies through direct or material support."[26] The program is administered by the **Defense Logistics Agency (DLA)**. DLA is responsible for the transfer of controlled and noncontrolled (or uncontrolled) military equipment. **Uncontrolled items** are items that are not tightly controlled, are of small monetary value, and have a short life span (i.e., machine paper). Police departments can also purchase **controlled items**, which require a high degree of accountability. Examples of controlled items are weapons and vehicles. The DLA policy states that

> to participate in the program, a law enforcement agency must be a federal, state, or local government agency whose primary function is the enforcement of applicable federal, state, and local laws and whose sworn compensated law enforcement officers have powers of arrest and apprehension.[27] (p. 6)

In 2016, a total of 8,600 federal, state, and local law enforcement agencies have transferred some type of military surplus equipment from the DOD. The vast majority of equipment (96%) was transferred to state and local departments. The logistics of the program are straightforward. The DOD transfers excess items to the DLA Disposition Service, which makes the item available on their request website. DLA has a list of authorized individuals who may view and request property. Authorized individuals from police departments can go online and look up which items are available. The department then submits a request with detailed justifications to the Law Enforcement Support Office (**LESO**). Officials from LESO review the request and either approve or deny it. Some of the controlled items, including aircrafts, vehicles, and weapons, require a greater degree of supervision, and departments who request such items must answer additional questions and provide documentation. According to LESO, the main reason for a denial is that the requested item has already been allocated to some other department. It is rare that LESO denies a request because it seems unnecessary for the department or the justification for the request was not sufficient.[28]

THINK ABOUT IT: GENERAL ACCOUNTING OFFICE (GAO) STUDY OF THE DOD EXCESS PROPERTY PROGRAM—A FICTITIOUS POLICE DEPARTMENT GETS ACCESS TO LESO

The DLA is responsible for the correct implementation of the LESO program. The procedures established by the DLA to ensure that only authorized and legitimate agencies receive access to the military property did not prevent a fictitious police department from accessing the military excess property. In 2016–2017, the GAO conducted a study of the DOD Excess Property Program. The GAO created a fictitious police department with a fictitious agency name, number of officers, contact information, and address. Several GAO investigators posed as authorized federal law enforcement officers. In 2016, the fictitious police department submitted an application to the LESO program office and was granted access to excess property in early 2017. No one had noticed that the department did not actually exist, had no employees, and had fictitious contact information, email, and a fictitious address. All communications between the fictitious department and the LESO officials was conducted via email. There was no attempt by the LESO office to contact the fictitious police department via telephone or other verbal means of communication. There was also no attempt to visit the police department to confirm its legitimacy. The LESO office simply requested confirmation of the police department's authorizing statute in compliance with U.S. Code. The fictitious police department sent fictitious authorizing provisions as laid out in the U.S. Code.

The GAO investigators and their fictitious police department had gained access to the online inventory system and began ordering excess military property. In total, the fictitious police department ordered over 100 controlled items worth about $1.2 million. The ordered items included "night vision goggles, reflex (also known as reflector) sights, infrared illuminators, simulated pipe bombs, and simulated rifles"[29] (p. 22).

In March 2017, the GAO investigators met with DLA officials and gave them the report with their findings. The DLA immediately began visiting police departments that were participating in the LESO program. In April 2017, the DLA office visited a total of 13 federal law enforcement agencies. However, the DLA did not revise its procedures on the verification process of law enforcement agency applications. The GAO reported concluded that

> DLA has not ensured compliance that on-site officials routinely request and verify valid identification of the individual(s) authorized to pick up allocated property from the LESO program, as required by the guidance. If DLA does not ensure that Disposition Services on-site officials routinely request and verify valid identification, then DLA will lack reasonable assurance that controlled property is transferred to authorized individuals.[30] (p. 25)

Discussion Questions:

1. Which procedures would ensure that fictitious departments and people cannot access the excess military equipment?

2. How could the LESO program be abused by people who are planning terrorist or other violent attacks?

▶ **Photo 7.1** SWAT team.

MILITARIZATION AS A THREAT TO COMMUNITY POLICING

As we have discussed previously, the original idea of the British prime minister and creator of the Metropolitan Police, Sir Robert Peel, was that police should look different from the military and police would use different weapons and methods. Community policing is based on building relationships and trust with the residents of a community so that crimes can be prevented and people who are planning terrorist acts or other serious crimes can be arrested before they can act. Community policing is based on the willingness of the public to cooperate with police and use of force by police should be minimal. It is meant to reduce fear of crime and fear of the police and create working relationships between police and citizens. It has been argued that the militarization of police threatens these goals.

In contrast to police, military officers rely on the use of force to accomplish their mission and they use different types of weapons, including flash grenades, bombs, assault rifles, and night vision goggles. They also use the surprise factor by starting a mission in the middle of the night whereas community police act during the day when they can interact with people.[31]

SWAT teams look like military units rather than police officers and the weapons and tactics they use are military in nature. For instance, SWAT teams often conduct drug raids when most people are asleep. They break into the house without giving notice via no-knock warrants, and they behave very much like soldiers while inside the house. The Community Policing Services Office (COPS) and other organizations have argued that the increased deployment of paramilitary SWAT teams to incidents of domestic disputes does not further the goal of community policing; instead the military dress patterns and military-style training of the SWAT team officers negatively impact the relationship between key community members and police.[32]

Thus, whereas community policing is supposed to reduce conflicts and fear of police, SWAT team raids of community members exacerbate conflicts and increase the fear of police. The success of police departments depends on the cooperation of citizens to prevent and solve crimes. Citizens who have a positive attitude and trust toward police are much more likely to provide information to police that helps them. But when citizens are fearful of police, distrust police, and perceive them as authoritarian, they are less likely to interact with police and provide important information. For instance, traditional police uniforms tend to instill a sense of the "serve and protect" theme, whereas battle dress uniforms, which are military-style uniforms, imply a state of heightened aggression toward the citizens, suggesting that extreme measures are necessary.[33]

For police departments, it is imperative to consider how the officers will be perceived by the community and how these perceptions will impact their success of preventing and solving crime. Case Study 7.2 lists the weapons arsenal of the Maricopa County Police Department and Case Study 7.3 showcases a SWAT team raid made in error.

The Impact of Battle Dress Uniforms on Community Policing

In addition to the increased use of SWAT teams, there has also been a trend toward an increased use of **battle dress uniforms (BDUs)** for police officers, including patrol officers. The color of the uniform may impact how police are perceived by community members. Some research has found that the military-style uniforms, and especially black

CASE STUDY 7.2: MARICOPA COUNTY, ARIZONA, WEAPONS ARSENAL

The police department of Maricopa County has compiled a massive weapons arsenal. The following is a list of the military-style weapons.[34]

- 32 bombs
- 704 night vision goggles
- 1,034 guns, including 712 rifles
- 42 forced entry tools
- 830 units of surveillance and reconnaissance equipment
- 120 utility trucks
- 64 armored vehicles
- GPS devices
- 17 helicopters
- 21,211 other types of military equipment

CASE STUDY 7.3: THE CASE OF BABY BOU BOU

On May 30, 2014, a SWAT team from Cornelia, Georgia, with a no-knock warrant, went to the house of Wanis Thoneteva for the purpose of a drug raid. The door was locked and appeared to be blocked. To gain entry, one of the officers threw a flash grenade into the house. The flash grenade landed in the playpen of 19-month-old Bounkham Phonesvanh, called Baby Bou Bou by his family. Baby Bou Bou suffered life-threatening injuries to his face and chest, third-degree burn wounds, and was put into a medically induced coma. He had to undergo several surgeries to reconstruct his face and chest. The SWAT team who had raided the house found no drugs, no suspect, and made no arrests. The no-knock search warrant was based on testimony by Deputy Nikki Autry that Wanis Thoneteva had sold methamphetamine worth $50 to an undercover agent. This information turned out to be false. Autry resigned and was later indicted on charges of making false statements to a judge in order to obtain a no-knock warrant. The no-knock search warrant was based on a $50 drug deal that never happened. The officers who were part of the SWAT team said they did not know that there were four children inside the house. Autry was not convicted and the officers involved in the raid were later cleared of any wrongdoing. The county shifted the blame for the incident to Bou Bou's family, claiming that they should have known that they were staying with a relative who was dealing in drugs. Bou Bou's family was staying with their relative because their house in Wisconsin had been destroyed in a fire. The medical costs for the treatment of Bou Bou's injuries had added up to $1 million by 2015. The county refused to help with the medical bills. However, in 2015, a judge ordered that the family of Bou Bou receive a settlement of $3.6 million to pay the medical bills and provide for Bou Bou in the future.[35]

uniforms, create more aggression, exacerbating the level of conflict between police and residents.[36]

When SWAT teams break down the doors to someone's house in the middle of the night, it is not rare that the residents believe they are the victims of a burglary attempt or an armed robbery. This is especially true in no-knock warrants where police do not announce themselves as police. These situations can result in a shootout in which residents, children, and police officers get killed. Examples of shootouts and resulting deaths and injuries are widely available and often covered in the news. In December 2013, a SWAT team broke into the trailer house of Henry Magee, who was asleep with his pregnant girlfriend. Magee quickly got his gun, believing that criminals were trying to burglarize his house. He shot and killed one of the officers and was charged with capital murder. The grand jury refused to issue an indictment, however, stating that Magee had acted in self-defense. The SWAT team was looking for a marijuana-growing operation. What they found were a few marijuana plants. Based on a tip from an informant who was looking to make a deal with the prosecutor, a judge had signed a no-knock warrant. The events occurred as follows:

> By about 5:40 a.m., the deputies were in place. They wore uniforms as dark as the night sky, with lettering that revealed their identities—"Sheriff's Office"—only on the back. Then came a crashing thud at the door. "Hank, what was that?" Ms. White asked. "Who is it?" Mr. Magee shouted, according to Ms. White. "Who's there?" No answer, then another thud at the door. Mr. Magee scrambled into his bedroom and

retrieved an AR-10 semiautomatic rifle from a closet. As he re-entered the living room, the front door burst open, followed by a deafening explosion. Ms. White screamed as a dark figure crossed the threshold. "I thought we were being robbed," she said. "It was my worst fear, that it was like on TV with people kicking in the door and coming in." Mr. Magee raised the rifle and fired several times toward the door, just above Ms. White on the couch. She jumped up and dashed toward him, brushing her neck against the steaming barrel, then dropped to the floor, and crawled into the bedroom. "I still didn't know who it was or what was going to happen," she said. "I just knew my best bet was getting as low as possible and getting out of that room." Only then, she said, did she hear the announcement: "Burleson County Sheriff's Office! Come out with your hands in the air." Mr. Magee dropped the rifle and complied. Ms. White followed him out the door, stepping over a broad-shouldered body as blood pooled on the wooden flooring. Although wearing body armor, Adam Sowders had been struck in the head.[37]

The jury found that the black BDUs made it impossible for Magee or his girlfriend to identify the intruders as police. These uniforms in combination with the no-knock warrant create highly volatile situations, resulting in unnecessary violence and deaths. Dale Stroud, who was working at the sheriff's office at the time, stated that they stopped the no-knock tactics after that incident.[38]

Researchers have found that the color of the uniform can create trust or cooperation but also fear and aggression.[39] For instance, Johnson (2005) found that lighter color uniforms are generally associated with a calming and pleasant reaction, whereas black uniforms tend to increase aggression and hostility toward police. Johnson's study asked the following question: "Is it possible that some uniforms promote a negative impression formation, thus making it more difficult than necessary for officers to overcome citizen apprehensions and anxieties?" (p. 59). Johnson found that officers with black uniforms were perceived as "cold, mean, forceful, unfriendly, aggressive, and corrupt," whereas the officers with the lighter blue uniforms were perceived as "good, nice, warm, gentle, friendly, passive, and honest" (p. 61). These perceptions of officers are very important to the goals of community policing. Citizens who are intimated by police are much less likely to approach officers when help is needed, such as in situations of domestic violence. They are also less likely to approach police when they have information about a crime or a possible terrorist attack.[40]

Others contend, however, that these no-knock tactics are for officer safety. Officers are concerned that people who own weapons would put their lives in danger if they would knock and announce themselves. There are also other arguments in support of the BDUs. The BDUs allow police officers to carry additional equipment that would not fit into the traditional police uniform. In addition, the uniform is a form of representation and a way to demonstrate their training and responsibilities. In this sense, the BDUs are consistent with the increasing military-style training at police academies. The uniforms and gear promote the impression that the officers are well trained, well armed, and ready to respond to any situation.[41] Wearing these uniforms is a strategy to deal with certain issues. A statement by Springfield, Massachusetts, police commissioner William Fitchet reflects this way of thinking: "Members of the department's Street Crime Unit will again don black, military-style uniforms as part of his strategy to deal with youth violence"[42] (p. 1). The black uniforms are a symbol of authority and that is exactly the purpose of these uniforms.

Another advantage of the BDUs is that they provide additional safety due to their wearability and comfort. They have no buttons or pins or other accessories that could become tangled or get ripped during stressful situations. They are also tougher and more durable because they were made for military purposes, which saves police departments money over the long run.[43]

The question remains, however, whether these BDUs are suitable for police, especially within the context of building relationships with communities. The idea that SWAT teams raid the neighborhoods at night with which they are trying to build a working relationship during the day seems contradictory. How can citizens trust police when they appear in the middle of the night raiding houses in paramilitary uniforms and using military weapons for the purpose of finding drugs? How can residents trust police who put the lives of innocent people, including children, at risk? At the same time, of course, SWAT teams have saved the lives of many people during terrorist attacks, school shootings, hostage situations, and other violent encounters. The main issue revolves around the type of situations in which SWAT teams are appropriate to use. This will continue to be a controversial question.

The Impact of Military-Style Training of Police Officers on Community Policing

The organization of police has many similarities to the military, including the hierarchical structure with the decision-making power concentrated on the top level of administration, such as the sheriff or police chief. The behavior of police officers is constrained by the organization's expectations of how things ought to be accomplished. Mundane behavior by officers is closely regulated and monitored. The police chief and other top administrators make the decisions for the police officers, which impair the ability of officers to interact with citizens, participate in important decisions, and communicate with citizens because officers are expected to simply follow orders rather than finding solutions to problems. The community policing model, however, requires police officers to communicate consistently and effectively with citizens, and often in racial-ethnic minority communities that are not likely to assist police. The police recruits learn early on that they are supposed to follow orders and that goals are best accomplished by the use of their authority, rather than the building of relationships, a very necessary component of community policing.[44]

Another concern for many supporters of community policing is the increase of military-style training in police academies, and especially stress training and the "warrior-like orientation." This type of training tends to create an "us-versus-them" mindset. Stress training in particular creates significant hurdles for an effective police–citizen relationship, which in turn is contra-productive for community policing.[45] A police trainer stated, "We trainers have spent the past decade trying to ingrain in our students the concept that the American police officer works a battlefield every day he patrols his sector."[46]

Karl Bickel, who is the senior analyst of the COPS office, has asked an interesting question: Are police academies training new recruits to be problem solvers or are they training new recruits in a way that is counterproductive to community policing? Are police academies creating obstacles to the development of partnerships between police and citizens and preventing crimes and terrorist acts by militarizing the training of new police officers?[47]

▶ **Photo 7.2** Police officers training in martial arts techniques.

There are two general models of police officer training: (1) stress training and (2) nonstress training. The main benefits of the militaristic stress environment are believed to be self-discipline, following orders, self-confidence, and command presence. The benefits of nonstress training have been found to be the ability to identify and solve problems and work collaboratively with coworkers and community partners. Officers who completed the nonstress training also showed higher levels of job satisfaction and field performance.[48] In addition, stress training can have negative psychological and physiological consequences. In 1988, Massachusetts governor Michael Dukakis ordered an investigation of the training of recruits at Agawam Police Academy after doctors had found widespread kidney problems among the recruits. The panel of investigators concluded: "The so-called drill instructor approach to training that includes indiscriminate verbal abuse, debasement, humiliation, confrontation, harassment, hazing, shouting, and physical exercise as punishment has no place in police training."[49] The panel also found that stress training was "not conducive to training men and women in a manner that will best enable them to serve society."[50] Several other studies have also found the negative effects of stress training on police officers and their ability to fulfill the needs of community policing. For instance, Conti (2009) found that police recruits who go through stress training were more likely to act defensive and depersonalized.[51] Others have noted that the stress training isolates officers from citizens because the recruits identify more and more with their peers as they go through the tough training and they become more separated from the population who are perceived as outsiders.[52] It also adds to the "us-versus-them" mentality, which can become a great obstacle to building partnerships with community members. Finally, stress training may lead to maladaptive coping strategies, such as excessive use of force during stressful situations, alcohol use and abuse, an aggressive attitude, distrust of outsiders, and conforming to militaristic behaviors.[53] These maladaptive behaviors inhibit officers' ability to build relationships with community members based on trust.

THE IMPORTANCE AND USE OF MILITARY EQUIPMENT

The militarization of police has led to much concern and criticism, especially with regard to its use during drug raids in racial-ethnic minority communities. However, there are also many other situations, especially emergency situations, during which police departments need military equipment to save people's lives. All police buy military surplus, which includes armored vehicles and assault rifles, but also small items such as sleeping bags and mosquito nets. The majority of items purchased by police departments fall into the category of "uncontrolled items." But police departments also get "controlled items," which include weapons and vehicles. There are different types of military vehicles police agencies may acquire, including assault vehicles, mine-resistant vehicles, and armored vehicles. These vehicles are imperative to insert rescue teams in flooded areas or areas where there was much destruction due to a natural disaster. For instance, after Hurricane Katrina, the city of New Orleans and surrounding areas were completely flooded and destroyed. Patrol cars were unable to drive and get to people who were injured and stuck on the rooftops of their houses. Police officers used military trucks to pull boats through fallen power lines and trees until they reached the Lower 9th Ward that was completely flooded. For 14 hours straight, they saved hundreds of lives. Without the military trucks, they would not have been able to get the boats into these districts.[54] More recently, in 2017, after severe flooding in parts of Texas, the sheriff's county office in Harris, Texas, received two armored vehicles for high-risk operations and high-water rescues.[55]

These military vehicles are also imperative to apprehend terrorists after an attack. For instance, during the San Bernardino terrorist attack on December 3, 2015, the San Bernardino Police Department used armored vehicles at the scene of the shooting and to apprehend the two shooters, Syed Rizwan Farook and Tashfeen Malik, who were heavily armed with semiautomatic weapons. Both were shot by police during a shootout. Without armored vehicles, police officers would have been very vulnerable to the shooters.[56]

Uses of the Military Excess Equipment

The LESO program is being used by state and federal police departments for a variety of purposes. Table 7.1 shows the different types of uses of the military excess equipment. The GAO report identified nine main uses: (1) counterdrug operations; (2) counterterrorism operations; (3) border security; (4) rescue and search operations; (5) active shooter or hostage situation; (6) building search or examination of suspicious packages; (7) tracking fugitives; (8) protection of officers and public; and (9) natural disasters.[57]

▶ **Photo 7.3** Armored vehicles have many uses.

Benefits of the LESO Program

According to police departments, the LESO program has a variety of benefits. Federal agencies reported the following benefits: (1) obtaining specialized gear and equipment in a cost-effective manner; (2) obtaining critical items without committing additional government funds; and (3) obtaining equipment that increases efficiency, saves officers time, increases operational capabilities, and increases the safety of the officers. Similarly, state agencies reported: (1) ability to order equipment that would not be able to be purchased with limited budget; (2) enhanced law enforcement activities; and (3) improved ability to react to active shooter situations. Table 7.2 shows the results of the GAO study from 2017, in which 50 police departments responded to questions about the LESO program. The study included self-report survey data and case study data from state and federal agencies.[58]

Table 7.1 Examples of Reported Uses of the Department of Defense Excess Controlled Property in the Law Enforcement Support Office Program (by Type of Law Enforcement Activity)

Law enforcement activity	Reported example of use
Counterdrug	• Helicopters were used daily daily to patrol drug use. • Vehicles were used as undercover vehicles to monitor drug house activities and to make controlled buys. • High Mobility Multipurpose Wheeled Vehicles were used to patrol and enforce in rugged areas that are not accessible by two-wheel drive vehicles. • Mine Resistant Ambush Protected vehicles were used to conduct numerous drug raids in the rural countryside, sometimes several in a single day. • Weapons were used to protect citizen and officers while conducting court-approved drug search warrants. • Night-vision equipment was used to maintain surveillance of drug activities in low light conditions. • All-terrain vehicles and other small vehicles allowed a police department to patrol off-road, wooded areas where known drug trafficking activities have taken place for many years. • Tactical gear was used during drug raids and search and seizure. • Optics, night vision, and thermal Imaging were used during investigations and surveillance of marijuana cultivation on public lands.
Counterterrorism	• Helicopter was flown on several Homeland Security and counterterrorism missions each year, such as for 9/11 anniversaries or times of increased security concerns, and on coastline patrols and Metro-North Railroads patrols. • Helicopter was used for 9/11 coastline patrols. • Mine Resistant Ambush Protected vehicles were ready for deployment and on standby status for such high-profile events as the Super Bowl and other dignitary protection details, such as visits by the president and vice president of the United States and by presidential candidates. • Boat was used to provide better water patrols of a dam from a homeland-security standpoint.
Border security	• Night-vision equipment was used on border crossing locations to detect violators (northern and southern borders). • High Mobility Multipurpose Wheeled Vehicles were used to patrol the border for illegal activity. • Mine Resistant Ambush Protected vehicles were used to patrol the border for illegal activity. • Fixed, aerial, and mobile capabilities were used to enhance border surveillance and detection capabilities on the southwest border in key areas.
Search and rescue	• Aircraft provided law enforcement agencies the ability to conduct searches of vast areas for search and rescue. • Helicopters were used for searching for missing persons. • Mine Resistant Ambush Protected vehicles, High Mobility Multipurpose Wheeled Vehicles, and cargo carriers were used to respond to flooding and blizzard situations. • Watercraft were used for search/rescue on several occasions.

(Continued)

Table 7.1 (Continued)

Law enforcement activity	Reported example of use
Active shooter or hostage situation	• Mine Resistant Ambush Protected vehicles were used in situations involving an active shooter and another armed person threatening violence with a weapon. • Armored vehicles were used to safely rescue citizens and officers from active shooter situations. • Armored vehicles were used to respond to active shooter situations and in high-risk warrant situations. The armored vehicles received fire and protected the officers inside. • A Mine Resistant Ambush Protected vehicle was used when a gunman barricaded himself. Once the vehicle arrived, the gunman surrendered without incident.
Building search or examination of suspicious packages	• Robots were used to enter areas that might not be safe for officers. For example, robots were sent in to investigate bombs or in environments that are risky due to the presence of drugs, hazardous materials, or violent individuals. • Robots were used to search buildings to eliminate the risk of sending in officers to conduct dangerous searches.
Tracking fugitives	• Helicopters were used for tracking fugitives. • Periscopes were used as a safer way to clear an attic. A periscope aided in the protection of officers instead of sending a person into an attic where an armed suspect could have been hiding.
Serving warrants	• Tactical vehicles and gear were used for the safety of officers serving high-risk warrants, especially when served at nighttime.
Protection of officers and public	• Radio communications equipment helped meet the primary mission of the agency in serving the public. • Enhanced sights on weapons during various types of tactical activities increased the capabilities and safety of the officers. • Aircraft provided a force multiplier during critical mission where officer safely was at risk.
Natural disasters	• High Mobility Multipurpose Wheeled Vehicle and tactical vehicle were used during hurricanes and snowstorms to patrol when the streets were not passable and flooded. • High Mobility Multipurpose Wheeled Vehicle was used to rescue homeowners trapped at their flooded residences. • High Mobility Multipurpose Wheeled Vehicle was used after tornado to get around when the patrol cars kept getting flat tires. • Helicopters were used to fight fires.

Sources: GAO survey and case study data. | GAO-17-532

Credit Line: US Government Accountability Office. Report to Congressional Committees, "DOD Excess Property: Enhanced Controls Needed for Access to Excess Controlled Property" (July 2017).

Table 7.2 Examples of Benefits That Law Enforcement Agencies Reported from the Receipt of Department of Defense Excess Controlled Property

Reported example of benefit by:

Federal law enforcement agency survey respondents

- The Bureau of Alcohol, Tobacco, Firearms and Explosives, U.S. Department of Justice, reported that the Law Enforcement Support Officer (LESO) program allowed it to obtain specialized gear and equipment in a cost-effective manner that tremendously enhanced and aided its readiness to carry out its mission. Further, in a time of budget constraints, the officials stated that the LESO program allowed them to obtain critical items.
- The Drug Enforcement Administration, U.S. Department of Justice, reported that the LESO program is a key source of supply without committing additional government funds.
- The Bureau of Indian Affairs, U.S. Department of the Interior, reported that the LESO program saved the agency hundreds of thousands of dollars over the last three years because it had the ability to acquire equipment at no cost.
- The U.S. Forest Service, U.S. Department of Agriculture, reported that much of this equipment would not be affordable to most law enforcement agencies and that this equipment has made them more efficient, saved money, greatly increased its operational capabilities, and made its officers safer.

Federal law enforcement case study respondents

- The U.S. Marshals Service, U.S. Department of Justice, Atlanta office, reported it was creating a Special Response Team in 2014, and considered cost-savings measures and how to acquire items to aid its mission instead of buying items out of its budget. Officials at U.S. Marshals Service Atlanta stated that the LESO program allowed it to obtain equipment, such as High Mobility Multipurpose Wheeled Vehicles, that were beneficial to its district and mission, and helped save money and acquire better technology.
- Officials from the Federal Bureau of Investigation, U.S. Department of Justice, Detroit office, reported that overall, the LESO program operates as a budgetary enhancement, as it cannot afford the majority of the equipment it is authorized to use. For instance, it said it would cost $35,000 to outfit the entire team with infrared lasers; however, its budget for this type of item is only $12,000.

State coordinator survey respondents

- More than 85 percent of state coordinators reported that controlled property received through the LESO program enhanced law enforcement activities in their state.
- More than one-third reported that law enforcement agencies within their state could not afford purchasing their own equipment or would purchase their own equipment piecemeal over a long period.

State and local law enforcement case study respondents

- A local law enforcement official from Michigan reported that the program allowed smaller departments to get items that they would not normally be able to acquire and as a result, his department saved $5,000 to $10,000.
- A local law enforcement official in Maryland reported that his sheriff's department saved over $200,000 through the LESO program.
- A local law enforcement official in Texas reported that 96 percent of the department budget goes to salaries and that the LESO program helped the department acquire items that it would otherwise not be able to afford; the official estimated that the program saved the department $2 million to $3 million.
- University law enforcement officials reported the threat of active shooters on campuses, and reported that the equipment they received from the LESO program aids their ability to respond to such threats.

Source: GAO survey and case study data. | GAO-17-532

Credit Line: US Government Accountability Office. Report to Congressional Committees, "DOD Excess Property: Enchanced Controls Needed for Access to Excess Controlled Property" (July 2017).

THINK ABOUT IT: THE POLITICS OF MILITARIZATION

President Barack Obama

On August 9, 2014, Michael Brown, a Black 18-year old was shot dead by a White police officer, Darren Wilson, in Ferguson, Missouri. Following the death of Michael Brown, African Americans showed their disgust and protest with the police actions. The city of Ferguson experienced peaceful protests and serious riots lasting several weeks. On November 24, a grand jury decided not to indict Darren Wilson on criminal charges for the killing of Michael Brown. This decision by the grand jury set off another round of protests and riots. Police used armored vehicles, snipers, and military gear to gain control of the riots. The pictures of police in their military gear quickly went around the country on TV and social media. Many people felt appalled by the way the police looked—like soldiers in a war zone. Many people also felt outrage over the pattern of police brutality that seemed to target African Americans. The riots grew more severe. Upset citizens threw rocks and bottles at police officers, burned police cars, smashed the windows of stores, set buildings on fire, and engaged in the looting of businesses. There were also reports of automatic gunfire in parts of the city. Police used smoke and gas to gain control of the riots. Lambert, Airport in St. Louis, was closed Monday evening for incoming flights. Schools were closed for a week and the governor declared a state of emergency. The governor also called for help from the National Guard as police was overwhelmed by the chaos of the riots. After the riots had stopped, there was much talk about police use of force, especially against African Americans. In March 2015, the U.S. Justice Department ordered the city of Ferguson to overhaul its criminal justice system. The Justice Department in its review of the city's procedures had found several violations of constitutional rights.[59]

As a reaction to the riots in Ferguson and the use of military equipment by police in Ferguson, President Obama ordered a review of the Pentagon military program 1033 that allowed police departments to purchase military surplus equipment. Following the review, President Obama issued an executive order restricting the ability of police departments to buy military equipment, including military surplus gear, explosives, battering rams, riot shields, and helmets. In addition, the presidential order also prohibited the transfer of such equipment between departments. The purpose of the restrictions was to create a mentality of guardianship by police rather than a warrior mentality.[60]

President Donald Trump

On August 28, 2017, President Trump rolled back the limits placed on police departments to purchase and transfer military surplus equipment. The presidential order restored the military program 1033 that began in 1990. Police unions had lobbied the administration to restore the program and allow them to purchase equipment from the military. The Trump administration has defended the restoration of the program as a means to ensure that police can get the equipment they need to do their job. Attorney General Jeff Sessions said:

> We will not put superficial concerns over public safety. . . . The executive order that President Trump will sign today will ensure that you can get the lifesaving gear that you need to do your job and send a strong message that we will not allow criminal activity, violence, and lawlessness to become the normal. And we will save taxpayers money in the meanwhile.[61]

Vanita Gupta, the former head of the Justice Department's civil right division under President Obama, voiced her concerns about the reversal of the restrictions. Gupta said that she found it "especially troubling that some of this equipment can now again be used in schools where our children go to learn."[62]

Article

United States Government Accountability Office. Report to Congressional Committees.
"DOD Excess Property: Enhanced Control Needed for Access to Excess Controlled Property."

Discussion Question:

What do you think? Should the police department be able to purchase as much military surplus equipment as it thinks it needs, or should there be restrictions on the purchase and transfer of such equipment? Debate the pros and cons.

SUMMARY

The militarization of police dates back to the civil rights riots and the Omnibus Crime Control Act under Lyndon Johnson and has continued since, fueled by the war on drugs. After the terrorist attacks in 2001, militarization of police further increased as police departments were tasked with homeland security efforts. Today, the vast majority of police departments have military-style equipment and a SWAT team. Overall, the war on terrorism and the war on drugs have led to an unprecedented level of police militarization. The vast majority of SWAT team raids of private homes are for drug raids, however. These raids have received much media attention, especially cases in which innocent people or police officers were severely injured or killed. The Office of Community Oriented Policing Services (COPS) has repeatedly warned that militarized police and raids by SWAT teams of officers with battle dress uniforms who seem to be at war with citizens create serious problems for community policing. Aggressive militarized tactics can destroy the relationships between police and community leaders and negatively impact the ability of police departments to prevent and solve crimes.[63]

When people lose trust in an institution that is supposed to protect them, the goals of community policing are difficult, if not impossible, to accomplish. Many police departments are struggling to implement community policing when they are tasked to focus on drugs and homeland security issues. Some researchers have suggested that the goals of homeland security and community policing can be integrated, however. They believe that community policing is very valuable for homeland security. As we have discussed in Chapter 4, building relationships with community leaders is imperative to preventing terrorist attacks. This is only possible if citizens have trust in police as the institution that protects them. If citizens feel that police are at war with their community, this goal becomes impossible to reach.[64]

KEY TERMS

Battle dress uniforms (BDUs) 136
Controlled items 133
Defense logistics agency (DLA) 133
LESO 133
SWAT teams 128
Swatting 127
Uncontrolled items 133

DISCUSSION QUESTIONS

1. Explain what swatting is and the danger it creates for citizens. How can swatting be reduced?

2. Discuss the impact of SWAT teams and the growing number of paramilitary tactics on community policing.

3. Discuss how battle dress uniforms impact community policing and conflicts between police and citizens.

4. Discuss the argument by Karl Bickel from the COPS office that police academy training may be counterproductive to community policing. What type of training would be beneficial to community policing?

5. Discuss how police departments use armored vehicles and other types of military equipment.

6. Discuss under what circumstances paramilitary weapons are appropriate to be used by law enforcement.

CHAPTER 8

COMMUNITY POLICING AND DRUGS

Learning Objectives

1. Describe the extent of the current drug problem in the United States.
2. Discuss the changing role of police in the face of the current drug crisis.
3. Explain the challenges open-air drug markets pose for communities and law enforcement.
4. Discuss which programs have been successful in disrupting drug markets.

A DEADLY COMBINATION: HEROIN LACED WITH FENTANYL

Washington, DC, has become ground zero for a new drug overdose epidemic among African Americans. In 2017 alone, 279 people died of an opioid overdose, three times as many as in 2014. More than 80% percent of the victims were Black and in more than 70% of the cases, the opioid fentanyl was involved. Just to put these numbers in context, the number of opioid deaths is now higher than the number of homicides. While the media has focused mainly on the impact of the opioid crisis via prescription drugs on White neighborhoods, African Americans, and especially African American long-time heroin users, are facing a substantial risk of death every time they use heroin.[1] The danger stems from the lacing of heroin with the opioid fentanyl. Fentanyl is a prescription opioid that is prescribed to patients for end-of-life pain management, such as in cancer patients. However, it is also used for patients in severe pain after surgeries, burn victims, and people with severe chronic pain. As a prescription drug, fentanyl is known under the names "Actiq," "Duragesic," and "Sublimaze." It has been used for this purpose since 1959 and is typically given as a shot, a patch on the skin, or as lozenges, that can be sucked like cough drops. Fentanyl is much more potent than heroin and is also called "China-White," "China Girl," "Apache," "Dance Fever," and "Tango & Cash." It is sold as a powder that can be mixed with other drugs, including heroin, cocaine, and methamphetamine. In fact, fentanyl is 50 to 100 times stronger than morphine, which makes it inherently dangerous in itself as a very small amount can slow or stop a person's breathing, which then results in **hypoxia**—a condition where the brain does not receive enough oxygen.[2]

Heroin is also an opioid, derived from morphine, which is derived from poppy plants. The street names for heroin include "Big

H," "Horse," "Dust," and "Smack." Heroin is illegal and has no medical purpose, but it produces similar effects to those of prescription opioid painkillers, such as fentanyl or oxycontin. Heroin is generally cheaper and easier to get than opioid prescription drugs and thus people who are addicted to opioids may switch to heroin. This is also true for people who cannot get another prescription from their doctor.[3]

Both heroin and fentanyl are fast acting, which means that people get addicted very quickly and they build a tolerance to the drugs quickly. **Drug tolerance** means that users need a higher amount of the drug or need to take the drug more often to get the same effect from the drug. Thus, over time, drug users increase the amount of drugs they are taking and an overdose becomes ever more likely. In addition, when users are unaware that the heroin is laced with fentanyl, they may take too much of the heroin and overdose.[4] Fentanyl has made heroin much more potent—and deadly. Drug users are well aware of the danger and buy from dealers whom they trust. However, even experienced drug users cannot see the difference between clean heroin and heroin laced with fentanyl. Thus, many injection drug users have suffered an overdose, sometimes with deadly outcomes.[5]

A fairly new drug, called **Naloxone** or Narcane, can reverse an overdose of heroin or other painkillers. It was approved by the Federal Drug Administration in 2015. Naloxone blocks the effects of opioids and basically reverses the effects of an overdose. People who receive Naloxone may experience withdrawal symptoms. Naloxone can be injected, auto-injected via a prefilled device similar to an EpiPen, or it can be used as a nasal spray. It has been used for many years by emergency room doctors and paramedics. Police officers are also increasingly carrying Naloxone. More recently, family members and friends of people who are addicted to heroin or other opioids are allowed to carry Naloxone. Some states require patients to bring a prescription for Naloxone, but in other states, it is available without a prescription.[6] In some states, including Texas, Naloxone is now also available online via the website www.NaloxoneExchange.com. This was spurred by the high number of overdose deaths. In 2017, 2,998 people died of an overdose in Texas alone. In light of the current opioid epidemic, it seems very likely that many other states will follow and make naloxone more widely accessible. Currently, the cost of two doses of naloxone are around $83 online, but there is also a generic version at a reduced cost.[7] However, this does not solve the broader problem. Opioid drugs such as heroin and fentanyl are cheap and freely available and there is a high demand for opioid drugs. As long as this trend continues, there will be a high number of deaths due to drug overdoses. Police are on the forefront in combating the current opioid epidemic and drug markets in general. This chapter will discuss the extent of the problem, the role of police, and effective community policing strategies.

INTRODUCTION

In this chapter, students will learn about community policing efforts to combat drug use and abuse as well as the consequences of such behaviors. Students will learn about dangerous trends in drug use and abuse, with a focus on the current opioid crisis. Drug use and abuse are widespread across the country and police agencies have to adjust constantly to the changing drug prevalence and patterns of use and distribution. For instance,

▶ **Photo 8.1** Some organizations distribute Naloxone kits to individuals who are homeless. What role should the police play in this crisis?

combating open crack markets requires a different response than hidden drug markets such as prescription opioids. Finally, students will also learn which community policing strategies have been successful.

EXTENT OF THE CURRENT DRUG PROBLEM IN THE UNITED STATES

The National Drug Threat Assessment (NDTA) (2018) details the extent of drug abuse reported by local, state, federal, and tribal law enforcement agencies. The NDTA also includes data from public health agencies, open sources, and other government agencies. One of the main questions the report intends to answer is which drugs pose the greatest threat to the United States and its citizens. Since 2011, the number of deaths caused by drug poisoning has been rising every year. In fact, deaths due to drug poisoning are outnumbering the number of deaths due to firearms and motor vehicle crashes.

Opioids

Opioids are currently the biggest threat and have also received the greatest attention in the media and among politicians, doctors, and researchers. The highest number of overdose deaths are caused by opioid drugs. A substantial number of people are taking prescription opioid medications and some of them become addicted. Prescription drug abusers typically cite physical pain as the reason for the addiction and the opioids are typically fairly easily obtained from a doctor, friend, or relative. The most often prescribed opioid prescription drugs are hydrocodone and oxycodone. In 2017, doctors prescribed 5.5 billion dosage units of hydrocodone and 4.5 billion dosage units of oxycodone. As described in the introduction, drug abusers of prescription opioid drugs may turn to heroin if they can't get prescription drugs or because heroin is much cheaper.[8]

In the past few years, there has been much talk about the "opioid epidemic." The Center for Disease Control and prevention has identified three waves of the opioid epidemic. First, since 1999, overdose deaths have steadily increased due to the rise in prescription opioids. This first wave was closely related to the emergence of oxycontin as a potent opioid pain medication. Oxycontin can work for 12 hours and was labeled as nonaddictive, which led many doctors to use oxycontin instead of morphine and other opioid pain medications that were labeled addictive. The so-called pill mills popped up across the United States, and especially in Florida, where doctors prescribed high quantities of oxycontin. Oxycontin, of course, was similarly addictive as other opioid pain medications and this led to a rising number of people addicted to opioids and a resurgence of heroin as a replacement for oxycontin. The second wave of the epidemic was marked by a surge in heroin overdose deaths. Federal legislation had closed down the pill mills and made it more difficult to buy oxycontin with a prescription. Thus, people with drug addictions replaced opioid pain medications with heroin. The third wave involved the emergence of illicit fentanyl and synthetic opioids. By 2017, more than 70% of overdose deaths were caused by fentanyl or fentanyl analog drugs. As a result, the number of overdose deaths steadily increased from 23,166 in 2012 to 47,600 in 2017, an increase of 49% within five years.[9]

Most of the heroin comes to the United States from Mexico. As the demand has continued to increase, so has the supply. Fentanyl is the second most common opioid causing many overdose deaths. This is especially true when heroin is laced with fentanyl as we have discussed in the introduction. There are several ways in which fentanyl is sold on the drug market. First, most of the fentanyl sold in the United States comes from Mexico and China. Chinese sellers typically send fentanyl via the U.S. Postal Service and international express mail carriers. Some sources, including Thomas Overacker, the executive director of the Office of Field Operations at the U.S. Customs and Border Patrol, suggest that this type of trafficking has greatly decreased. Between January and July 2019, drug seizures of high-purity fentanyl have dropped substantially, which indicates that less fentanyl is being trafficked via mail. Border patrol only seized a few pounds in 2019. In the past two years, the United States has been working with China on a drug control strategy, and in May 2019 China criminalized possession of all fentanyl-related substances. As a result, many Chinese dealers are not mailing fentanyl any longer. In addition, the U.S. Postal Service now requires data on all senders, recipients, and contents of all international parcel with special emphasis on Chinese mail. The U.S. Postal Service also installed new scanners to detect drugs. These changes were highly successful as the agency reported a 1,000% increase in the number of parcels seized containing synthetic opioids between 2016 and 2018.[10]

However, trafficking of fentanyl from Mexico has greatly increased as border patrol is unable to scan the vast amount of private and commercial vehicles that cross the border every day. The reality is that about $1 billion worth of legitimate freight comes across the

▶ **Photo 8.2** What do you think could be done to slow the spread of illegal drugs across the U.S.–Mexico border?

border daily via tens of thousands of vehicles and only 2% of private and 16% of commercial vehicles can be scanned. Thus, there is much opportunity to traffic fentanyl into the United States.[11]

Second, fentanyl may be counterfeited and sold as pills, which has also greatly contributed to the number of overdose deaths as people underestimate the potency of these pills. The counterfeiters can produce tens of thousands of pills using pill machines and flood the drug market. For instance, in February 2018, DEA agents arrested a person in San Diego who was counterfeiting Xanax pills. The DEA seized a pill machine capable of producing 10,000 pills. The same pill machine could also have produced fentanyl or other prescription drugs. Finally, fentanyl can also be diverted to the illegal drug markets from legitimate sources. For instance, one such method is referred to as "lost in transit." **"Lost in transit"** is defined as prescription drugs being misplaced while they are being transported from one place to another. This includes theft by employees and customers, burglaries, armed robberies, and losses in transit. Even though the overall number of prescription drugs "lost in transit" was reduced in 2017, there has also been a substantial increase in 22 states in controlled prescription drugs. This type of drug diversion has great impacts on crime rates. For instance, in Washington, the number of armed robberies increased by 200% and in Texas and Wisconsin by 100%. This, of course, creates problems in the communities and for police who are working to reduce crime.[12]

The problems related to the opioid epidemic are not limited to the rising number of overdose deaths and violent crime, but also the increase of infectious diseases, especially HIV and hepatitis. Drug users who share needles to inject themselves with methamphetamine, heroin, or other drugs are at a great risk of getting infected. Another type of infection that threatens the health of drug users is an infection of the heart, referred to as *Staphylococcus aureus*, which causes the heart valves to stop functioning properly. People who have this heart infection may need a heart transplant. A study from North Carolina found a 13-fold increase in heart infections in people who misuse drugs as compared to people who don't. There is also some evidence that drug users who use high doses of opioid pain medications have a significantly higher risk of getting an infection of the lungs, also called pneumonia. This phenomenon is still being explored, but it appears that opioid pain medications suppress the natural immune system of our body. Without a properly functioning immune system, people are susceptible to viral, bacterial, and fungal infections, increasing their risk of death.[13]

In addition to the adverse consequences of drug abuse for people with drug addictions, drug abuse also has significant adverse economic impacts, including increasing costs for treatment, health care, and the criminal justice system. The admissions to treatment for heroin addiction has increased from 270,564 in 2010 to 401,743 in 2017. Similarly, emergency room admissions have also gone up significantly. People who are addicted to drugs not only put a burden on the health care system, they are also not as productive at their job. They are more likely to skip work and underperform. They may also put their coworkers at risk for injury.[14]

The federal government has contributed $11 billion in 2017 to combat drug abuse via prevention, treatment, and recovery programs as well as research programs, criminal justice programs, health surveillance programs, and supply reduction programs. Overall, there were 57 programs aimed to address the current drug epidemic. Of the $11 billion, 2.12 billion are set aside to specifically target the opioid epidemic.[15]

Community–Police Partnerships Help Combat the Opioid Epidemic

Dayton, Ohio, has been one of the epicenters of the national opioid epidemic. The police began working with the East End neighborhood in 2012 via a grant sponsored by the Department of Justice, called the Community Based Crime Reduction initiative. The grant was meant to tackle high-crime neighborhoods. The partnership involved community organizations, law enforcement, public health, and addiction services. The police first looked at the crime data and the causes of crime. They determined that more than 90% of the crime, mainly robberies and burglaries, were driven by opioid abuse. Most crimes were committed to support the offenders of drug abuse. The partnership developed a comprehensive strategy to combat drug abuse and crime. One part of the strategy involved adding treatment beds. At the outset of the partnership, there were 13 treatment beds for the entire city. By 2018, East Dayton had its own treatment facility. In addition, it provided training to officers, community residents, and family members on how to use naloxone to save lives in case of an opioid overdose. The partnership also created a program, called *Conversations for Change*, to reduce violence. The program was intended to provide an opportunity for drug abusers and their families to meet, have a meal, and talk. Trained mediators gave advice about treatment options and service providers and

CASE STUDY 8.1: WASHINGTON STATE SYRINGE EXCHANGE PROGRAM—PREVENTING HARM OR ENABLING DRUG USE?

Syringe exchange programs (SEP) aim to reduce harm to injection drug users by providing clean syringes. They provide five main services. First, the exchange used syringes for new ones free of charge. Second, they provide injection equipment, including alcoholic wipes and cottons, to enable drug users to use sterile practices and prevent viral and bacterial infection. Third, they distribute information about preventing overdoses. They also distribute naloxone, a drug that can reverse an opioid overdose, such as an overdose with heroin and fentanyl. Fourth, they provide free tests for Hepatitis C and HIV and they assist with wound care. They also help drug users enroll in health insurance. Finally, they refer drug users to treatment services.

Washington has an extensive syringe exchange program with 25 programs in 18 counties.

The programs are carried out by local health departments, community organizations, and tribal entities. It is estimated that Washington State has a total of 33,318 injection drug users. In 2018, the Alcohol and Drug Abuse Institute published a survey where 1,079 drug users from 18 SEPs of the 18 counties answered questions about their drug use, routes of administration, and other health problems, including infectious diseases. The two most common drugs injected were heroin and methamphetamine. Eighty percent of the respondents answered that they had injected heroin by itself in the past three months, 82% stated that they had injected methamphetamine by itself in the past three months, and 46% reported that they had injected both, heroin and methamphetamine mixed together. This combination is also called a "goofball." Prescription opioids play an important role as 53% of the respondents indicated that they got hooked on prescription-type opioids, such as oxycontin, vicodin, or fentanyl, prior to turning to heroin. The average age of first injection was 24 years and overall 70% reported that they had injected drugs every day in the last seven days. The syringe exchange program targets this population and the respondents stated that they were using the program on average two times in the past 30 days. They also exchanged syringes and other materials on behalf of someone else. About 19% of the respondents had suffered an overdose from opioids in the past 12 months and 64% had witnessed an overdose. It is not surprising that the ownership of the drug naloxone has greatly increased. Whereas in 2015 only 47% of the respondents owned a naloxone kit, in 2017 this had increased to 66%. The SEP was critical in this increase as 83% of the respondents received their kit from the SEP. Other respondents had received the kit from family, friends, or a drug-using partner.[16]

A study by Nguyen (2014) suggests that from an economic perspective, SEPs could save millions of dollars every year as it is much cheaper to provide clean syringes for no cost compared to the costs of medical treatment for infectious diseases due to needle sharing. Needle sharing is the second most common cause of HIV infection and each year there are about 2,575 new infections. The lifetime treatment costs for HIV are about $391,223. Thus, each year the total treatment costs are $1.01 billion. Nguyen (2014) estimated that with a $50 million increase in funding for syringe exchange programs, the United States would reduce the number of new HIV infections due to drug use to 1,759 people and save a total of $269.1 million in treatment costs.[17]

Within the United States, 39 states have SEPs and the Centers for Disease Control and Prevention has endorsed the practice. The number of SEPs has substantially increased since 2016 as the opioid crisis has hit the country and has been making the news constantly. The rising death toll has spurred much of this movement. However, critics are concerned that SEPs may actually increase drug use, making the opioid drug crisis even worse. Some people believe that there has been a rise in injection drug abuse because they are finding more needles in public parks and on the street. Patty Pastore and several other volunteers from Port Angeles search the parks and street for needles. They have great problems getting rid of the needles. They are not drug users and thus are not eligible to use the syringe exchange to get rid of the needles. They have also found that the number of needles they find keeps rising. In an interview, Pastore stated what many others may be thinking: "Why don't they just hand the addicts a gun and let them shoot themselves? Because that's what they're doing. No citizens in this town were ever given a right to vote on something, yet our taxpayer dollars were used for the syringe exchange." Some cities have responded to the criticism by closing the syringe exchange. Others, such as Kitsap County, are attempting to move the program from a mobile van to local health care centers. There is, however, no scientific evidence that the SEPs are in fact bringing more needles to the community and increasing drug use. Quite the contrary, in Miami, Florida, which does not have an SEP, people found eight times more needles than in San Francisco, which has had an SEP for many years.[18]

Discussion Questions:

1. Discuss the pros and cons of syringe exchange programs. If you were the mayor of a city with a high number of injection drug users, would you fund a syringe exchange programs? Explain your answer.

2. Discuss the ethical pros and cons of syringe exchange programs.

educated attendees on opioid abuse and drug addiction. The police are also targeting drug users who overdosed previously and referred them to the *Conversations for Change* program and other harm-reduction programs. The partnership proved to be fruitful. In Montgomery County, the number of overdose deaths declined from 566 in 2017 to 294 in 2018. In addition, the crime rates for property crimes, such as theft and burglary, also decreased by 18% between 2016 and 2017.[19]

Research has shown that the family is not only important for the recovery from drug abuse, but that the family also has a great impact on future drug abuse. A study by de Vaan and Stuart (2019) shows that opioid prescription drug use spreads within families. More specifically, if one member of the family uses prescription opioids, it is very likely that other family members will also use prescription opioids. The authors examined hundreds of millions of medical claims and 14 million opioid prescriptions in one state between 2010 and 2015. They concluded that household exposure to opioids increased the demand for these drugs within the family. The use of prescription opioids is a social behavior similar to smoking and drinking alcohol in which family members influence each other. Family members also serve as examples for "what works." A family member whose pain was successfully treated with an opioid will promote the product to other family members and friends who complain about pain and doctors will be asked for a prescription even if an opioid may not be necessary or appropriate for their medical condition. Thus, it is imperative to involve the family in the drug treatment program to limit exposure to the drugs and to create a deeper understanding of the consequences of opioid prescription drug use.[20]

THE CHANGING ROLE OF POLICE IN THE FACE OF THE CURRENT DRUG CRISIS

Police are on the forefront of the current overdose death crisis caused by the overprescription of opioid painkillers, such as oxycontin and fentanyl, the rising use of heroin, and the spiking of heroin with fentanyl. Police departments are ill equipped to handle the growing influx of drug users. Serious problems related to withdrawal symptoms can occur when drug users are arrested and put in a holding cell until they can make bail or until their court hearing. This process can take several days and withdrawal symptoms can take many different forms, including hallucinations, aggression, and depression. Withdrawal syndromes, also referred to as discontinuation symptoms, occur when a person has become physically dependent on a drug and vary by drug. For instance, withdrawal syndrome for people who are physically dependent on alcohol includes tremor, sweating, anxiety, agitation, depression, and nausea. Heroin withdrawal syndrome typically consists of aching muscles, chills, and abdominal cramps. Methamphetamine withdrawal syndrome includes fatigue, depression, and psychosis, such as hallucinations of animals crawling under their skin. Some of these withdrawal syndromes can cause serious risk of injury to themselves or others. There are also symptoms that can directly lead to death, such as seizures and heart attack, which can occur when people withdraw from alcohol or tranquilizers.[21]

Police departments do not typically have mental health and drug addiction specialists to screen all of the arrestees for possible withdrawal symptoms. In addition, the arrestees may continue to use drugs while they are in custody and an overdose can happen inside a holding cell just as quickly as on the street. The Boston Police Department has experienced several overdose deaths of people who were in custody. Even though the officers carry the drug naloxone that can reverse an overdose, overdose death can still occur because the department does not have the resources to supervise all inmates 24/7. Police departments are at high risk of civil lawsuits when people die while in custody. The main issue is typically whether the death was preventable and whether the inmate was treated humanely. On September 28, 2018, police in Vermont arrested Madelyn Linsenmeir. She had been addicted to opioid drugs for many years and weighed only 90 pounds. She had called her mother the day before the arrest and told her she needed medical help. While she was in custody, she asked for medical care. According to a lawsuit filed on behalf of the family by the American Civil Liberties Union, officers denied her medical attention and made sarcastic comments while she was telephoning with her mother. On October 7, Linsenmeir was transferred to the Hampton County's Sheriff Department, which then took her to the hospital. She died the same day from a staph infection. With proper medical treatment, she probably would not have died. This case is only one of many around the country. Police departments and officers are trained to patrol the streets and protect the public; they are not trained to protect people with substance abuse disorders. In response to the growing problem, some police chiefs have banded together. In Middlesex County, New Jersey, 54 police chiefs are supporting legislation to create a special facility for detainees. The facility would include medical professionals, mental health counselors, and other staff trained to identify people with substance abuse disorders, detainees at risk for overdose, suicide, and withdrawal syndromes. The police chiefs all agree that this type of service cannot be provided by their departments.[22]

The Police Executive Research Forum has published a report that outlines the issues police departments and officers are facing due to the heroin epidemic and the legalization of marijuana in several states in the United States. The main question that the report addresses is whether police must change their attitude toward drugs and drug users in order to effectively combat drug abuse. Police have typically focused on arresting the drug sellers and drug traffickers, and on confiscating the drugs. This strategy has not been very successful. There are more drug users than ever, there are more drugs available on the street, and the purity of the drugs has also greatly improved. For instance, the purity of heroin went from 2% to 3% in the 1970s to 70% in 2014. Some cities, including in Minnesota, have heroin that is 97% to 98% pure. Heroin is also very cheap, with prices as low as $3–$5 per bag. This makes heroin much more accessible and more addictive, and it also increases the likelihood of an overdose. Heroin is a problem all across the United States, that is, it is not limited to urban areas any longer. The reason for the increased spread of the supply of heroin is a change in drug trafficking. During the last 30 years, Mexican and Colombian drug cartels have established themselves as the major heroin traffickers. They are trafficking multiple drugs and have established trafficking routes into the United States. During the 1970s and 1980s, heroin came mainly from Southeast and Southwest Asia. They only distributed heroin to certain large cities. That has changed toady where about 50% of the heroin is trafficked to the United States, with Mexico supplying heroin to the entire country including rural areas. Finally, prescription opioid pain medications play a substantial role in the heroin epidemic. The Drug Enforcement Agency (DEA) estimates that 80% of people who began using heroin in the past few years started with prescription drugs.[23]

Police in different cities are taking matters into their hands, developing strategies, and publishing their data. For instance, the police chief of Morwood, Massachusetts, implemented a comprehensive strategy targeting drug dealers as well as drug users. Most drug dealers live in rental properties. Every time police arrest a dealer and raid the property, they notify the landlord and provide information on how to evict a drug dealer. The purpose is to remove the dealer from the property by the time the case gets to trial. The police also notify the drug buyers they have identified and provide information about treatment options. Overall, police are increasingly recognizing that drug addiction is a medical illness and cannot be addressed by simply incarcerating the offender. Drug users need treatment to address their drug abuse issues that have physical and psychological consequences. Thus, the role of police is changing to include being a service provider for drug users and often also their families. Police departments have been asking public health agencies to take the lead on the issue, but they understand that they are the first responders when people suffer an overdose or are involved in criminal behavior. Across the country, police departments are developing their own approaches to the heroin epidemic, such as carrying naloxone to reverse an overdose. They are also working with community health agencies to refer low-level drug offenders into treatment programs. In fact, in some communities, police have become the number one source for referrals to drug treatment.[24]

Marijuana

In addition to the heroin epidemic, police are also facing the changing attitude among the population toward marijuana and the extent of marijuana consumption. Similar to heroin, most of the marijuana consumed in the United States comes from Mexico. In 2017, the U.S. Customs and Border Protection confiscated more than 500,000 kilograms of marijuana in more than 20,000 incidents. Given that only a very small fraction of the freight from Mexico is being inspected, one can only imagine the extent of marijuana smuggling. Not surprisingly, marijuana is the most popular illicit drug among adults and juveniles. In 2017, 8.9% of the population were regular marijuana users. Marijuana use has steadily increased in the past decade. Among people 12 years and older, marijuana use increased from 5.8% in 2007 to 8.9% in 2017.[25]

There is also a substantial amount of marijuana produced within the United States. This development is closely related to the legalization of marijuana in some states. By 2019, 33 of the 50 states have legalized medical marijuana. In addition, 11 states plus the District of Columbia have legalized small amounts of marijuana and 26 states plus the District of Columbia have decriminalized small amounts of marijuana.[26] This development creates unique challenges for law enforcement, especially given the fact that marijuana use is still a federal crime and federal law supersedes state law. This means that even though the use and sale of certain amounts of marijuana may be legal in certain states, federal agents can still arrest people for possession and/or sale of an illegal substance. Many people are buying marijuana legally but then drive across the border to their home state where possession of marijuana is illegal. Police officers regularly make drug busts on highways coming from Colorado and other states in which marijuana is legal. People do not understand that even if they buy the marijuana legally in one state, it is still illegal to take it into a state where marijuana is illegal. The critics of the marijuana legalization argue that marijuana is still a gateway drug and one in six marijuana users will also use other drugs. A substantial number of people are admitted to treatment centers for marijuana addiction. Research also suggests that marijuana use has adverse effects on brain functioning, especially in young people. The differences in state laws also create considerable confusion among people, especially young people, as there is a mixed message about drug use. Most people understand the negative consequences of using heroin, cocaine, methamphetamine, and other drugs, but they are not sure whether marijuana is harmless or dangerous.[27]

Retail stores have popped up in many locations selling marijuana plants or edibles, such as brownies that include the tetrahydrocannabinol (THC) extract.[28] These products, especially the edibles, have caused significant problems. For instance, marijuana edibles include brownies, chocolate, gummy bears, and other candy. These edibles are often more potent than the marijuana plant, which is smoked or vaped. A study from Denver shows that edibles caused a substantially higher number of medical crises, severe intoxication, cardiovascular problems, and acute psychiatric symptoms in people with no history of mental illness as compared to inhaled pot. The study, led by Dr. Andrew Monte, tracked 2,600 marijuana-related patient visits to the emergency room over a five-year period.[29]

There have been several deaths associated with edibles, all of which were surprisingly violent. In 2014, Richard Kirk, a father of three boys, shot his wife in the head, killing her. He blamed his ingestion of marijuana edibles for his behavior. He was convicted of second-degree murder and sentenced to 30 years in prison in 2017. In an interview with

PBS, Kirk stated that he was nibbling on a THC-infused orange gummy candy when he started to feel very weird: "It was like I was in a different place. It was like I didn't know," he said. "I remember being out on our deck. I remember jumping through my youngest son's window, trying to go to the screen when there is a door right outside of his room. I remember scraping my shins and being sore." He then started to fumble with his gun: "I didn't know it was my wife," he sobbed when asked why he pulled the trigger. "I thought it was somebody else, I guess. That's the only way I could have done it. I never ever once thought about even hurting my wife or pushing her or anything—let alone taking her life, taking her away from her three boys." Dr. Monte has confirmed that edibles can lead to severe psychosis in people who are mentally healthy. However, in 2014, Kirk was addicted to pain medications due to back pain and family members reported that he had a short temper and that they had marital and financial problems. We may never know for sure whether the edibles caused a psychosis that led Richard Kirk to kill his wife, but the sheer extent of edibles and the rise in consumption are certainly cause for concern. In 2017, Colorado alone had about 6.5 million marijuana user visits and sold more than 7.2 million in edibles laced with THC. Colorado and other states have made substantial changes to marijuana regulations after this highly publicized incident.[30] In 2019, the Colorado legislature implemented new labeling, packaging, and product safety regulations to inform consumers of the possible dangers.[31]

There are other risks and problems caused by the marijuana business. First, in order to extract the THC from the marijuana plant, butane must be filtered through the marijuana and then burned off. Butane is highly explosive and if done inside, the house or apartment can explode. Police officers experience these explosions and the consequences thereof on a regular basis. Another issue for police is the seizure of marijuana plants in the case of an alleged violation. People may grow marijuana with certain limits, but it is very difficult for police to enforce these laws. For instance, in Colorado the police may seize the plants, but they must return them alive or pay a fine of $10,000 per plant if a court rules that the plants must be returned to the owner. Police departments do not have places to grow marijuana and keep seized plants alive. Thus, they do not seize marijuana plants anymore. Instead they take pictures or small clippings of the plants.[32]

The Department of Justice expects an increase in marijuana production and use as more states are changing their marijuana laws. In addition, traffickers will increasingly penetrate domestic markets where marijuana is illegal. The legal uncertainties will continue to be problematic for law enforcement and there will be increasing discussion about the different state laws and their impact on law enforcement, banks, and medical professionals (Department of Justice, 2018). For instance, in Colorado, doctors have found profound impacts of marijuana use on patient treatment. Patients who are being sedated may need two or three times as much sedation medication if they are regular marijuana users. If patients don't tell their doctor about their marijuana use, this can have great negative effects on their sedation and treatment. However, very little is known about the impact of marijuana use on medical treatment and federal grants are not available to study the impacts because marijuana is illegal under federal law. This is very problematic for medical professionals who need to study the impact of marijuana to treat patients effectively.[33]

Cocaine

About 93% of cocaine sold in the United States stems from Colombia. In 2016, about 1.8 million people 12 years or older said they used powder and/or crack cocaine within the last month. Powder cocaine is by far the more popular drug. Similar to marijuana and heroin, cocaine availability has continued to increase together with the quality of the cocaine. The average purity of cocaine sold in the United States is 84.4%. At the same time, prices have decreased, which appears to be driven by greater competition within the drug market. Specifically, between 2012 and 2017, prices decreased by 8% and cocaine purity increased by 22%. Thus, it is not surprising that more people are using cocaine and more people die from cocaine poisoning. Between 2010 and 2017, the death rate went up significantly from 4,183 in 2,010 to 10,375 in 2016. In New York City, cocaine poisoning was the cause of death in 46% of all overdose deaths. And in Ohio, overdose deaths from cocaine poisoning increased by 61.9% (from 685 to 1,109) between 2015 and 2016. Other states have experienced even greater increases.[34]

Cocaine is also increasingly laced with fentanyl, which is called a "speedball." The sale of "speedballs" has increased by about 297%. Some dealers are also selling "super speedballs,"—a combination of cocaine, heroin, and fentanyl. The combination of very potent opioids with cocaine is very dangerous and greatly increases the risk of an overdose. It is also very difficult to reverse the overdose since it is not clear which drug is the principal drug responsible for the overdose. This combination of cocaine with fentanyl is a main driver of the rising number of overdose deaths. Between 2007 and 2016, overdose deaths due to multidrug use involving fentanyl increased from 3% to 40% of all cocaine poisoning deaths. The Department of Justice expects an increase in cocaine supply from Colombia as production has been surging. Most likely, cocaine-related deaths will also increase due to the rise of fentanyl and the lacing of cocaine with fentanyl and heroin.[35]

Methamphetamine

Most of the methamphetamine in the United States comes from Mexico and its availability increased by 7.48% between 2015 and 2016. Domestic production of methamphetamine has greatly decreased. Since 2012, methamphetamine lab seizures have declined by 78%. As a comparison, at the height of the methamphetamine epidemic in 2004, police seized 23,703 meth labs. In 2017, police seized only 3,036 labs. However, the methamphetamine production outside the United States has increased sharply and border patrol agents have been seizing more methamphetamine on the southern border every year. Whereas in 2012 border patrol seized about 9,000 kg, by 2017 that number had increased to nearly 30,000 kg. The drugs are typically hidden in vehicle tires, cargo, and hidden car compartments. Methamphetamine can also be transported as a liquid in bottles, concealed as water, alcoholic beverages, oil, etc. For instance, the Atlanta FD seized several bottles labeled as aloe vera, which contained methamphetamine. Methamphetamine is also smuggled and sold as pills. Traffickers have also used drones to transport methamphetamine across the border. For instance, in August 2018, a drone steered by a 25-year-old U.S. citizen dropped 13 pounds of methamphetamine in the United States. Even though drones can only transport a limited amount of the drug, they have some great advantages. The trafficker can be far away, avoiding getting caught by border patrol. Also, some drones have autopilot options that make them programmable. Methamphetamine abuse continues to be very high. Similar to other drugs, prices have been declining, but the potency has risen. The purity level of methamphetamine is very high with an average purity of over 96% in 2017. There were about 667,000 people aged 12 or older who were users in 2017. Most of the users (594,000) were older than 26 years. There was an increase in current treatment admissions and a substantial increase in methamphetamine-related deaths from 1,608 in 2005 to 7,542 in 2016. The death rate is much higher on the East Coast of the United States as compared to any other geographic area. The Department of Justice expects that traffickers from Mexico will continue the supply of the United States with high-purity methamphetamine. Because precursor drugs, such as pseudoephedrine, are becoming more difficult to obtain, they will use other chemicals and adjust their production. Consumption will likely continue to be high because prices are decreasing and purity and access to the drug is increasing.[36]

OPEN-AIR DRUG MARKETS

In 2019, 27 defendants were charged by the U.S. Attorney's Office with drug trafficking of heroin and crack cocaine in Newark and surrounding areas. The U.S. Attorney's Office stated that they pumped massive amounts of drugs into the area and turned a profit of about $10,000 per day. The defendants had occupied several abandoned buildings for the illegal enterprise. One of the abandoned buildings was turned it into an impenetrable fortress that they could only access via a ladder to the second-floor window. The ladder was pulled into the house once they had entered. They used firearms to protect their building. The drugs were sold through a small hole in the wall against cash. They also used other abandoned buildings for their drug dealing. All the buildings were located in residential areas. Some of them were near an elementary school. In 2017, the U.S. Attorney's Office formed the Violent Crime Initiative (VCI), which specifically focused on this type of group of drug traffickers. The leaders of the trafficking organization were Shaheed Blake, aka "Sha," aka "Sha Gotti," aka "Bruh," and Anderson Hutchinson, aka "Murda Rah." Several members of the group were also members of the gang "CKarter Boys," which is an affiliate of the "Bloods." The VCI used wiretaps and other surveillance methods to gather evidence about the organization and eventually arrested 27 members.[37]

This type of drug dealing is part of open-air drug markets. **Open-air drug markets** "tend to be visible public settings where few barriers to access exist, as individuals unknown to dealers are able to purchase drugs."[38] These drug market areas are characterized by a high number of drug sales and purchases within a geographic area. Even though the characteristics of drug markets vary depending on the drugs, they are typically located within cities and they have four common features: (1) they are located in areas with low socioeconomic status households; (2) the sellers stay in the same place; (3) they are located near a transportation hub for ease of access, and (4) they are highly visible to the size and concentration of drug activity. Open-air drug markets are very convenient for buyers and sellers. The buyers can access several drug dealers and weigh the quality against the price. That means that buyers can compare quality and prices of drugs without much effort. In addition, buyers can choose from different types of drugs

CASE STUDY 8.2: HOW THE OPIOID EPIDEMIC HAS CHANGED THE ROLE OF POLICE

In the wake of the current opioid epidemic and the staggering number of overdose deaths, police officers are often working as caretakers rather than just making arrests. They are regularly called to situations where drug users have overdosed. Sometimes the overdose death occurs in the presence of children, friends, or family members. In these situations, police officers have to respond to both, the person who has overdosed and the family who may be angry, desperate, or in shock.

A police captain from Chillicothe, Ohio, describes his experiences from a perspective of a town of about 20,000 where overdoses are a daily occurrence. He states that police increasingly have to take the role of caretakers rather than enforcing laws and solving crimes. They act as nurses and social workers. A major problem is the decreased budget and availability of mental health services. The captain has lived in Chillicothe for 27 years and has worked in law enforcement for 21 years. It is not rare that police arrive on the scene of an overdose and find children and other family members left without anyone to care for them.[39]

But it's not just the overdoses of adult drug users that are a major concern. These drugs are also a deadly hazard for children. Two days after Christmas in 2018, one-year-old Darwin Santana-Gonzalez accidentally ingested fentanyl in his parents' house. His parents had been preparing packages with heroin laced with fentanyl to sell to other drug users. Darwin died shortly after ingesting the drugs. His father fled the United States and the mother has been charged with murder. Special prosecutor Bridget Brennan highlighted the deadly potency of fentanyl: "The amount of fentanyl it would take to kill you or me would fit on the tip of your baby finger, and a small child would be much more susceptible." Two days earlier, an 18-month-old girl from Michigan, Ava Floyd, also died after ingesting fentanyl. Her parents were manufacturing and distributing drugs inside the family home. Both parents were charged with murder.[40]

A study by researchers from Yale University states that between 1999 and 2016 more than 9,000 children and adolescents died as a consequence of an opioid overdose. About 7% of these victims were younger than five years (Gaither, et al., 2018). The fact that children are often the victims of their parents' drug addiction or drug manufacturing habits is nothing new or specific to opioids, however. During the methamphetamine epidemic, which started on the U.S. West Coast in the 1990s and worked its way through the country, the number of abandoned, injured, and dead children was staggering.

For instance, by 2006, Oregon had experienced an increase of 45% in foster children due to the methamphetamine epidemic. The district attorney for Marion County, Oregon, Walt Beglau, stated at the time: "Meth has emerged as nothing short of a weapon of mass destruction in our community, leaving in its wake its greatest casualty, and that is our children, one child at a time, by the hundreds."[41]

In an attempt to curb the production of methamphetamine, Oregon made pseudoephedrine a prescription drug. Pseudoephedrine, a cold medicine, is one of the main drugs necessary to produce methamphetamine. Thus, the idea was that without access to pseudoephedrine, people could not make methamphetamine. Many other states also passed laws that made it more difficult to buy pseudoephedrine. These laws could not curb the production of methamphetamine in Mexico's superlabs however, and thus methamphetamine continues to be a great problem within the United States.

As a result, police officers are left to take care of people who are dying and who have children and other relatives who need care. In these situations, police officers become nurses, caretakers, and social workers—roles they weren't trained for. Officers may feel overwhelmed, especially when children are present. They have to attend to the person who has overdosed, asking them questions about what drugs they were taking, or if the person is unconscious, they have to decide whether to administer naloxone. Officers are not nurses and they are concerned whether they are administering the naloxone correctly. When children are present, officers have to decide what do with the children: Should they drop them off with a family member or take them to social services? Without any further information about the family, officers are not sure whether the children will be safe, whether the family members are also using drugs, and so on. This high-stress environment takes an emotional toll and officers wonder what they can do to fix the problem. Should they try and arrest more drug dealers? Should they focus more on educating the public about the dangers of drugs? Should they team up with other local public health and community agencies to tackle the problem? There is no quick fix to this problem and police agencies across the country struggle to find the most effective solutions.[42]

Discussion Questions:

1. Do you think that charging the parents with murder is just considering that the deaths of the children were accidental? Debate the issue.

2. How could police officer training be modified to better prepare them for the reality of their job in the midst of the persistent drug problems?

3. Which strategies would you use if you were the police chief in response to the drug problem?

in one location. Drug markets of highly addictive drugs, such as heroin and cocaine, are also often open 24/7 because people who are physically dependent on drugs needs access to the dealers at all times. Drug dealers benefit from open-air markets as they have a high visibility and can therefore maximize buyer access. But open-air markets also create risks for dealers and buyers. Both parties are highly visible to everyone, including police. Thus, there is always a risk of being arrested. In addition, buyers may be purchasing drugs from strangers, which encourages rip-offs and robberies. Also, buyers who are not satisfied with the drugs have little recourse, which can also result in violent encounters, including assaults and shootings. The lack of conflict resolution tools is especially problematic in high-value markets and often results in systemic violence; that is, violence is the normal means of settling conflicts.[43] Due to its characteristics, open-air drug markets create unique problems for communities that impact the quality of life of residents. The ever-present violence and open drug consumption create an environment where life is difficult for residents and law enforcement. Neighborhoods with open drug markets experience several adverse effects:

1. Crime hot spots due to open-air drug activities

2. Unusable public spaces, such as parks due to drug dealing and use

3. Prostitution

4. Transients

5. Decrease in property values

6. Decrease in business owners via displacement or failure

THINK ABOUT IT: THE LONG-LASTING EFFECTS OF THE OPEN-AIR CRACK COCAINE MARKET

During the 1980s and 1990s, open-air crack markets were associated with a high rate of violence, especially gun violence. Young Black men were not only often the victims of the gun violence, but this violence also greatly increased gun possession among this population group. Crack cocaine became popular in the 1980s because it was cheap (a single dose costs $2 to $3) and caused an immediate high. This high was short-lived, however, and drug users would buy crack several times a day. Crack became popular first in 1982 in Miami, New York, and Los Angeles where the open-air drug markets were run mostly by young Black men. Because open-air drug sellers depend on a specific location, there was much competition for locations that were profitable. Confrontations were common and often culminated in gun violence and deaths. As the open-air crack markets spread across the United States, so did the gun violence and murder rates of young Black men, which peaked at 129 in the 1990s. As crack cocaine sales slowed down, so did murder rates. However, the murder rates of young Black men have stayed elevated until today.

Evans et al. (2018) examined the question of why the murder rates for young Black men did not drop as much as for other population groups. Their hypothesis was that the higher murder rates were related to higher rates of gun possession. Thus, they studied changes in gun possession during the crack epidemic and after the crack epidemic. They found that young Black men were the primary perpetrators of crack-related gun violence during the crack epidemic in the 1980s and 1990s. The authors specifically found that 17 years after the first crack markets emerged, young Black men were still committing a significantly higher number of murders. The authors believe that gun possession among young Black men remained high after the crack epidemic, leading to a higher number of lethal violent encounters. The research done by Evans et al. (2018) showed that the increase in gun possession was due in part to many Black males seeking guns for their own protection, and lack of trust in the police to be able to protect them from those who were participating in the open-air drug markets. Without guns, many violent encounters may end in injuries, but when guns are part of the encounter, the probability that someone dies is much higher.[44]

Discussion Questions:

1. Discuss why young Black men may have been so heavily involved in open-air crack markets and related violence.

2. What could police departments do to reduce the number of murders related to open-air drug markets?

7. Enhanced crime opportunities for young adults
8. Increase in drug use and related crimes (Kennedy and Wong, 2006)

These problems, of course, result in a high number of calls for service to police. For police, the open-air drug markets are a considerable problem and numerous strategies have been employed in an attempt to eliminate them.

Police Strategies to Eliminate Open-Air Drug Markets

Police departments have generally approached the problem of open-air drug markets by increasing law enforcement efforts to limit the supply and use of drugs. One tactic was called **"inconvenience policing,"** which aimed at creating obstacles for buyers and sellers and discouraging novice and casual buyers from buying drugs at the open-air drug markets. The goal is to disrupt the drug market and reduce public disorder. The basic enforcement strategies include greater police presence at the site, test purchases, and reverse sting operations, where undercover officers pose as drug dealers. In addition, police also use more traditional methods, including arresting dealers and warrant services.[45] These strategies have not proven to be effective in eliminating open-air drug markets because these drug markets may simply be displaced to a different location, causing the same problems in a different neighborhood (Lawton et al. 2005). Other research has found a diffusion effect of these police tactics. Diffusion means that the target site of the enforcement strategies and the surrounding areas are experiencing a reduction in drug crimes.[46] However, these police tactics are also often perceived as aggressive, racially motivated, and unfair. Open-air drug markets are located in racial-ethnic minority neighborhoods with a low socioeconomic status and traditional policing tactics have great negative effects on the community. These communities often have a high number of residents with an arrest and felony conviction record and limited access to education and job opportunities. The criminal record exacerbates the problem as many employers do not want to hire people with a criminal record, especially at a time when there are many qualified workers looking for a job. Thus, the police are perceived as contributing to the problem of poverty in racial-ethnic minority neighborhoods and discrimination based on race.

Research also suggests that these policing strategies can lead to negative public health and social consequences. Theses consequences vary by drug and user behavior. Injection drug users may change their behavior in an effort to avoid police and purchase and consume the drugs before they can be confiscated. They may rush the injection process, which can be harmful. First, users may not clean the injection site properly, which can lead to infections. Second, they may inject the false dose of the drugs, which could result in an overdose. Third, they are more likely to skip important steps in the preparation of the drugs. For instance, they may not heat the drugs to kill bacteria and filter them. This can result in illnesses and infections. Fourth, they may suffer vascular damage if they rush the insertion of the needle. Finally, they are more likely to share injection equipment either because they are accidentally using somebody else's needle or to speed up the injection. All these consequences can have very serious negative effects, including deaths. In addition, users may store their drugs orally or in their nose to hide them from police. If they are swallowed accidentally or purposefully, death from overdose can occur.[47]

Harm can also result from physical displacement of the drugs. Drug users may store the drugs in remote and private indoor locations where they feel safe. One of these types of locations is "**shooting galleries**." Shooting galleries are hidden indoor locations used by drug dealers and users. They are often associated with high-risk behaviors, including professional injectors who inject drug users—often with the same syringe. This, of course, increases the likelihood of infection with HIV, hepatitis, and other diseases. This problem is exacerbated by an interruption of accessing health services. Drug users will stay away from service providers if there is an increased police presence. They may also move to a different area and not have any contacts with health care providers in that area. It has proven to be very difficult for health care services to reach injection drug users in general, and increased law enforcement presence adds to the problem. This makes it difficult to distribute clean syringes, provide treatment opportunities, and distribute educational materials. Finally, injection drug users create risks for others in the community. They may dispose of their used needles and drop them on the street and in parks where children and other people may find them. An increased police presence always carries the risk of being stopped with injection equipment, which could result in an arrest, and as a result drug users are more likely to dispose of their injection equipment on the street.[48]

The benefits and costs of these traditional law enforcement strategies can be difficult to weigh when determining whether these strategies are appropriate and effective in accomplishing the goal of eliminating open-air drug markets. On the benefits side, there is some evidence that these strategies reduce drug activities in the targeted area and increase drug prices, which may deter casual and novice drug users. In addition, these strategies have increased a sense of public safety and public order. However, research suggests that these successes may be time limited, that is, the drug activities only decrease as long as law enforcement is present. Also, the costs associated with these successes can be high. These costs include the harm to injection drug users discussed above, as well as other negative consequences with regard to access to drug users by health services. Also, it is not clear whether drug activities are simply displaced to other locations rather than being reduced or eliminated.[49] Thus, many police departments have started to look at problem-oriented policing strategies to eliminate open-air drug markets.

Problem-Oriented Policing Strategies: The Drug Market Intervention Model

The problems associated with traditional policing strategies and the lack of success in eliminating open-air drug markets have led many police departments to explore alternative strategies based on problem-oriented policing. These strategies include the members of the community and community organizations as stakeholders in the process, which has proven to be an effective way to disrupt drug markets, reduce crime, and improve the relationship between the community and the police. The drug market intervention model consists of three stages: (1) immediate activities; (2) intermediate activities, and (3) long-term activities. Immediate activities focus on arresting drug dealers and incapacitating them. Police show an increased presence in the area by prioritizing calls that relate to drug market activities, increased patrol, and arrests. At the same time, police begin a consistent dialogue with the community to build trust and support. This is important because the increased police presence and arrests in itself do not have lasting effects as arrested dealers are replaced quickly and the drug market activities continue. Thus, following the immediate response, police must focus on intermediate activities. These intermediate activities prioritize community engagement and empowerment by establishing neighborhood groups and continuing an open communication between the community and the police. At this point, the increase in supplemental police can be reduced because the empowerment of the community leads to a higher degree of increased formal social control. It also creates more positive relationships between the community and police. Finally, long-term activities emphasize a regular police presence and building a strong and resilient community. This leads to an increased legitimacy of police and cooperation of residents with police. It also supports informal social control and collective efficacy.[50] Collective efficacy means that if a community shares the belief that they can overcome obstacles then they are more effective in accomplishing their goals.[51]

These strategies combined have shown to successfully eliminate drug markets, reduce drug-related crime and disorder, and improve the neighborhood quality of life.[52]

CASE STUDY 8.3: DRUG MARKET INTERVENTION IN HIGH POINT—A MODEL PROGRAM

High Point is a city of about 95,000 residents located in North Carolina. In 2002, the police department was led by James Fealy, a former narcotics officer from Austin, Texas. Fealy recognized quickly that High Point had a long history of open-air drug markets dating back 40 years.

The High Point Police Department was facing all the typical problems caused by open-air drug markets, such as a high rate of violent crimes, prostitution, and drive-through drug buyers—all of which led to great negative impacts on the quality of life in the city. The police department had employed the traditional strategies, including sweeps, buy-bust operations, arresting dealers, and warrant services, to eliminate the drug markets—but to no avail. There was another negative consequence that was often overlooked. The policing strategies had created racial tensions between the communities and the police. Police were perceived as racist. The police believed that they were just doing their job and that they were not targeting a specific ethnic or racial group. The residents had a very different perception, however. Though police may not have been intentionally targeting a specific ethnic or racial group, by targeting this community that was the end result.

The new police chief, the mayor of the city, and other stakeholders decided to implement a new strategy based on problem-oriented policing. The strategy consisted of having swift sanctions for drug dealers, working with members and organizations in the community to address racial tensions between residents and the police, creating antidrug standards within the community by directly addressing drug dealers and their families, and offering education, job training, and placement to offenders. The new strategy was implemented in May 2004 in the city's West End Open-Air Drug Market. In the following months, the police expanded the strategy to three additional neighborhoods. For three years, the police department together with its community partners followed through on their strategy—with much success. Within these three years, the open-air drug markets in High Point were almost entirely eliminated and violent crime decreased significantly by 20% overall in the city. In the West End neighborhood, crime decreased by 57% within five years. The neighborhoods surrounding the targeted communities also benefited. Community conditions overall improved, including the relationship between residents and police officers. The racial tensions greatly decreased as the typical harm associated with street sweeps was eliminated and a dialogue between the residents, community organizations, and police emerged. Reverend James Summey summarized the accomplishments: "The strategy gave the community a way to confront these people who had been a terror in the community. But at the same time we embraced them, by saying at the same time, you're worth something. It's redemptive. So many in the police and the community don't see eye to eye, but on this we would. We're working together like we never have in our lives. This is the most fantastic thing I have ever seen."[53]

Since 2008, the strategy used by the High Point Police Department has been expanded to 18 sites in the United States under the Drug Market Intervention Program. It is now sponsored by the U.S. Department of Justice through the Bureau of Justice Assistance.[54]

Discussion Questions:

1. Discuss why the new strategy of the High Point Police Department was more successful in eliminating open-air drug markets and crime as compared to traditional policing strategies.

2. Discuss which other types of community problems discussed in this book so far may be reduced using this type of strategy? Explain why you believe this strategy would be successful.

SUMMARY

In this chapter, we discussed the extent and consequences of drug use. More specifically, we focused on the abuse of prescription drugs, such as fentanyl and oxycontin. There has been a substantial increase in the abuse of prescription opioids, and, as a result, in the abuse of heroin. Many people who are abusing prescription opioids eventually use heroin because it is cheaper and readily available. The rising number of overdose deaths, which has made so many headlines, is closely related to these problems. Not only are more people using heroin that is more potent, but dealers are mixing fentanyl into heroin and unsuspecting users may overdose because fentanyl is 50 to 100 times stronger than morphine. Police have reacted to the opioid epidemic and the high number of overdose deaths by carrying naloxone, a drug that can reverse an opioid overdose. Many police departments have experienced a shift in their responsibilities from law enforcers to providing more services to people in need. Police departments have formed relationships with community health providers and other drug addiction specialists to combat the drug abuse and help addicts find treatment options. Police are also actively combating open-air drug markets by using strategies that build trust with the community, reduce racial tension, and target drug offenders to provide treatment, education, job training, and placement. Open-air drug markets create unique problems for communities that impact the quality of life of residents, such as high rates of shootings, prostitution, robberies, assaults, and other crimes. Thus, many police departments expand much effort on eliminating these drug markets. One of these programs, implemented by the High Point Police Department, has become a model for other departments across the United States.

KEY TERMS

Drug tolerance 149
Hypoxia 148
Inconvenience policing 164
Lost in transit 152
Naloxone 149
Open-air drug markets 161
Shooting galleries 165

DISCUSSION QUESTIONS

1. Discuss the problems caused by the opioid epidemic.
2. Discuss the extent of the current drug problem and problems caused to the community.
3. Discuss the changing role of police amid the current opioid crisis.
4. Discuss the dangers of opioids, cocaine, methamphetamine, and marijuana and how community policing strategies can successfully reduce drug abuse.
5. Discuss the challenges that drug markets pose for communities and police. Which strategies are successful in combating these drug markets?

PART III

STRATEGIES AND TACTICS

CHAPTER 9

NET WIDENING AND SOCIAL CONTROL

Learning Objectives

1. Explain the concept of net widening.
2. Explain the connection between diversion programs and net widening.
3. Describe the purpose of quality-of-life policing and its impact on net widening.
4. Asset forfeiture statutes and net widening.
5. Solutions to net widening practices—restorative conferencing programs

SCARED STRAIGHT

Early intervention programs for youth at risk have been popular for decades. The main idea is that early intervention will reduce the risk of serious crime later in life. One of the most well-known early intervention programs is called *Scared Straight*. The Scared Straight program, established in the 1970s, targets juveniles who have committed delinquent acts, including serious crimes, such as assault and robbery. In addition, the program also targets youth who are at risk of becoming delinquent in the future. The juveniles are taken to a prison or jail and confronted with the realities of being incarcerated. Inmates give aggressive presentations, intimidating the juveniles and literally trying to scare them straight. The juveniles are also taken on a prison or jail tour to observe prison life firsthand. The goal of the program is to deter the juveniles from committing further delinquent acts. The program, in various forms, remains popular despite evidence that this strategy has very little positive effects on youth at risk. Two meta-analyses have found that youth who participated in this program had a higher risk of committing future crimes.[1]

The Scared Straight program is an example of early intervention programs that are based on the idea that intervening during the early stages of delinquent behavior is cheaper and more effective than waiting until a juvenile has committed a serious crime. In 1996, San Francisco invested $20 million in the reform of the city's juvenile justice system. The city implemented several prevention and intervention programs. It created a centralized intake system that assessed and referred juvenile offenders to community-based programs with the goal to reduce detention rates. This goal was not accomplished, however. Detention rates actually remained the same despite a shrinking youth population and decreasing youth crime rates. But not only did the program fail to reduce juvenile detention rates, it also drew more juveniles into the criminal justice system to ensure sufficient enrolment in the newly created community-based programs. In effect, more low-risk juveniles were exposed to the criminal justice system.[2]

Early intervention programs, such as the Scared Straight program and boot camps, have increasingly come under scrutiny as they may draw juveniles into the criminal justice program who would otherwise not have been included. This is referred to as net widening. This chapter discusses the concept and effects of net widening.

INTRODUCTION

In this chapter, students will learn what net widening is and how it impacts offenders and the criminal justice system. Students will also learn how different policies and programs are related to net widening. Specifically, the chapter discusses the impact of quality-of-life policing and diversion programs on net widening. In the following, the authors will discuss civil asset forfeiture statutes in the context of net widening. Finally, students will learn which practices have proven successful in addressing the problems associated with net widening.

THE NET WIDENING EFFECT

What Is Net Widening?

The term **net widening** refers to the "increase in the number of people having contact with the criminal justice system as the unintentional result of a new practice"[3] (p. 113). The term *net widening* is most often used in the context of diversion programs, such as drug courts; decarceration programs; deterrence-based programs, such as short-term confinements; and community-based corrections.[4] The definition of net widening has also been expanded to include wider, stronger, and new nets.[5] **Wide nets** are created by "reforms that increase the proportion of subgroups in society differentiated by such factors as age, sex, class, and ethnicity"[6] (p. 169). **Stronger nets** are the result of "reforms that increase the state's capacity to control individuals through intensifying state intervention"[7] (p. 169). Finally, **new nets** are created by "reforms that transfer intervention authority or jurisdiction from one agency or control system to another"[8] (p. 169).

Several programs have been criticized for widening the net, strengthening the net, or creating new nets that draw more people into the criminal justice system and draw people deeper into the criminal justice system.

DIVERSION PROGRAMS AND NET WIDENING

Diversion programs have become a popular part of restorative justice. Diversion programs have been widely implemented, especially for juvenile offenders, drug offenders, and offenders with mental health disorders. Examples of diversion programs include drug courts, mental health courts, and police diversion projects for youth. The intention of these programs was to divert offenders away from conviction and incarceration. They had five main goals: (1) avoidance of negative labeling; (2) reduction of unnecessary social control; (3) reduction of recidivism; (4) provision of service; and (5) reduction of justice system costs.[9]

The assumption was that if offenders were diverted at the front end of the criminal justice process, they would be less likely to re-offend and be incarcerated later. This assumption only holds true if juveniles who would typically be moved into detention are instead moved into prevention and intervention programs. That is not what has occurred, however. Research suggests that the offenders eligible for diversion programs were often offenders who would not have been prosecuted or convicted of a crime in the first place. For instance, a study by the U.S. Department of Justice showed that the juvenile justice system increased the number of juveniles processed. Theoretically, if the system processed 1,000 juveniles, then about 300 juveniles should be placed in a prevention or intervention program. This would reduce the number of juveniles held in detention to 700. Instead, the number of juveniles in detention continued to be 1,000 and an additional 300 juveniles were placed in prevention and intervention programs. Thus, there was a net increase from 1,000 to 1,300 juveniles drawn into the system because cases that would have been dismissed were now moved into the diversion programs and the offenders who would not have been punished or barely punished were now under the control of the criminal justice system. Thus, diversion programs have widened the net of people drawn into the criminal justice system.[10]

Diversion programs have also strengthened the net and created new nets by creating criminal justice programs that did not exist previously and by formalizing criminal justice practices that used to be informal. For instance, diversion programs typically require offenders to successfully complete the entire program to graduate. Offenders sign a "contract," which specifies the consequences of not completing the program. These consequences can include jail time. In these cases, offenders who drop out of the program are drawn deeper into the criminal justice system. Stated differently, offenders who would not have received any punishment or a very small punishment are now facing time in jail.

Proponents of diversion programs argue that these programs reduce recidivism and reduce the occurrence of more serious crimes by intervening early. This saves the state money and, keeps families together, and offenders keep their jobs, and have an opportunity to be productive citizens. Critics counter these arguments saying they may not be correct. For instance, a study by Austin (1980) shows that adult diversion programs had little effect on recidivism rates. This is likely due to the fact that the offenders had committed minor offenses and were not part of the population of offenders who are at high risk of reoffending. Thus, by moving these offenders into diversion programs, the state actually spent more money for administration and supervision—with very little benefit to the state.[11]

In sum, the main concern of critics of these programs is the increasing number of offenders under the control of the criminal justice system. This has great negative consequences for the offenders, including a criminal record and the stigma of having been in a diversion program.

The Negative Effects of Net Widening

Loss of Human Capital. People are generally expected to live a productive life and contribute to the economy. The concept of **human capital** relates to this expectation. Human capital was defined by Adam Smith (1776) as "The acquisition of . . . talents during

THINK ABOUT IT: DRUG COURTS—THE GOOD AND THE BAD

Drugs courts, which are part of therapeutic justice, were implemented in Miami in 1989 to dampen the collateral consequences of the war on drugs. At the time, the increased punishments for drug offenders were overcrowding jails and prisons, prison budgets were skyrocketing, and the private prison industry and its profits grew exponentially. A high number of these incarcerated offenders were low-level drug offenders, including people who had been caught with small amounts of drugs. The drug court movement was also driven by the growing concerns about racial disparities in incarceration rates of drug offenders. For instance, in Milwaukee, a study by the Commission on Reducing Racial Disparities found that Black drug offenders were 42 times more likely to receive a prison sentence as compared to White drug offenders. Drug courts became a popular response to these issues across the United States. By 2004, there were about 1,600 drug courts operating across the country.[12] This response to the collateral consequences of the war on drugs was supported by many academics and institutions that studied the effectiveness of drug courts. Most studies suggested that drugs courts are effective in treating drug abuse disorders, reducing recidivism, keeping families together, and reintegrating drug offenders into society. Drug courts diverted a large number of drug offenders from jails and prison into rehabilitation programs, thus reducing the strain on the criminal justice system and especially the correctional facilities.[13]

However, there is also a persistent criticism of drug courts that relates to the question of whether the war on drugs was a response to a drug crisis or whether the war on drugs was a political tool of racial control. Colorado District Court Judge Morris B. Hoffman, for example, cites a major flaw in the many studies celebrating the low recidivism rates of drug courts in noting that reductions in recidivism are very small and there has been no reduction in incarceration rates. The net widening effects far outweigh the benefits of drug courts.[14]

Even though only 12% of the population are Black, they make up about 40% of the state and prison population. Thus, Black people are greatly overrepresented among the incarcerated population. The war on drugs is widely perceived as a main contributor to this racial disparity in incarceration rates. There are several hypotheses as to why Black people are overrepresented in jails and prisons. First, some people argue that Black people simply commit more drug offenses than White people. Official crime data shows that this hypothesis is not correct. In fact, the proportion of drug offenses committed by Black people is equal to their share of the population, that is, about 12%. Second, Black people commit more serious drug offenses as compared to White people and therefore they are more likely to receive a long prison sentence. Several studies suggest that this hypothesis can also not explain the racial disparities. For instance, higher arrest rates have been contributed to the police practice of racial profiling. Black people are specifically targeted by police and thus caught more often. In addition, studies also show that even though White people are responsible for the majority of drug distribution, especially heroin and prescription drugs, Black people were arrested at much higher rates because law enforcement had focused on crack cocaine, a drug much more often used by Black people. Third, racial disparities have also been explained with the notion that drug offenses by Black people are more likely to include aggravating circumstances justifying a harsher punishment. There are two main aggravating factors: (1) they cause greater social harm and (2) they present a greater threat to public safety. Indeed, some research supports this notion because Black people are more likely to be involved in open-air drug markets, which tend to increase violence within the community, mainly due to turf wars. However, other studies suggest that police are more likely to target open-air drug markets in racial-ethnic minority neighborhoods as opposed to White neighborhoods.[15]

Drug courts have been proposed as a tool to divert drug offenders from prison and reintegrate them into society. However, drug courts generally only accept drug offenders who have committed minor crimes and

Continued

(Continued)

who are capable of completing a rigorous therapy. Drug court participants who fail can be sentenced and sent to jail. These drug court eligible offenders are the least dangerous and would likely not have received a jail or prison sentence to begin with. By being sent to a drug court, they are drawn into the criminal justice system and face a significant punishment if they don't complete the treatment that typically lasts for about 12 months.[16]

Discussion Questions:

1. Do some research on drug courts and discuss the pros and cons of drug courts.

2. What do you think: Did drug courts accomplish the goal of decreasing racial disparities with regard to the prison population?

. . . education, study, or apprenticeship, costs a real expense, which is capital in [a] person. Those talents [are] part of his fortune [and] likewise that of society"[17] (p. 61). In other words, human capital consists of skills, such as reading, writing, critical thinking, engineering a software, and knowing how to build and repair a car that allow people to find a job that enables them to rent an apartment or buy a house, take care of their children and other dependents, and contribute to the economy. Education and training are crucial to acquiring human capital.[18] However, the ability to get an education and training, and find a job that pays enough to live comfortably, is greatly diminished for people who get arrested and convicted. For instance, most students rely on financial aid to go to college. But people with a felony conviction are typically not eligible for financial aid, which prevents them from getting a college education and build human capital. In addition, job applications ask for arrests and felony convictions, which are often reasons not to hire a person, especially in times of high unemployment. Arrests and convictions therefore have a great negative impact on the individual but also their family by disrupting their family and work life. Arrestees cannot go to work for a certain amount of time and this puts great strain on the family. The employer may fire a person who has been arrested, especially if the arrestee cannot post bail and has to stay in jail in pretrial detention. It is often unknown for how long the arrestee will be detained. Arrest and conviction also impact the family because they may not be able to find housing and especially subsidized housing. For instance, public housing regulations may make a drug arrest or conviction an evictible offense. Without a home and stable environment, the family and the children also lose their ability to go to school and work. In addition, the family loses income and support, such as childcare, when a family member is incarcerated.[19]

Of course, serious criminals should be incarcerated to protect the public and punish the offender, but there are also many offenders who do not present a danger to the public and who might be better served with a punishment that allows them to stay in the community, keep their, job, support their family, and contribute to the economy overall. Incarcerating low-level offenders costs much money and wastes human capital.[20]

Group Stigma. Whereas the loss of human capital has direct negative consequences mainly for the offender and their family, group stigma has profound negative consequences for all Black communities. O'Hear discusses three types of harm as a consequence of racial disparities arising throughout the criminal justice system and net widening, including police arrests and placement in pretrial diversion programs and other criminal justice programs. First, the increasing number of Black people drawn into the criminal justice system reinforces and exacerbates the existing perception that Black people are criminals. This perception may result in an even greater number of Black people being stopped and searched by police, drawing more minor Black offenders into the criminal justice system.[21]

Second, net widening results in the loss of confidence among Black community members. People who lose confidence in authority figures are more likely to disobey the law and refuse to cooperate with police and criminal justice agencies.[22]

Finally, neighborhoods with high rates of arrests and people under the control of the criminal justice system create the impression that they are dangerous neighborhoods with high crime rates and disorder. They are not desirable to live in and residents who have the financial ability will move to safer neighborhoods. Negative perceptions about the neighborhood also lead to less contact and association with other residents. This lack of association negatively impacts the ability of the neighborhood to solve crime and social problems.[23]

The harms caused by group stigma have negative effects on community policing. As we have discussed throughout this book, community police officers rely on strong relationships with community leaders and residents. Alienating Black communities leads to strained relationships with the police, which also hurts victims of crime. A lack of cooperation within the community means that victims of crime are less likely to receive justice because the offenders are less likely to get caught.

QUALITY-OF-LIFE POLICING

Quality-of-life policing describes a type of policing that targets highly visible crimes, even if they are minor. These crimes can greatly impact the quality of life of the residents in the community. One of the main underlying theories of quality-of-life policing is the "broken windows theory," which purports that physical decay, such as broken windows, beat-up cars, loitering, and public disorderly behavior can lead to disorder and the decline of the livability of a city. Residents avoid places with obvious signs of disorder because it instills a sense of fear of crime. In addition, obvious signs of disorder encourage more disorder because it appears that it is being tolerated by law enforcement and the community.[24]

The evidence for this thesis is mixed. One of the most cited studies supporting the broken windows theory is the one by Skogan (1990), who conducted interviews with 13,000 individuals from six cities. The results suggest that robberies were significantly higher in neighborhoods with visible signs of disorder.[25] However, in 2001, Harcourt reanalyzed the data and found that the results were driven by one specific city: Newark. Newark had the highest disorder scores and robbery rates. When Newark was removed from the analysis, the relationship between robberies and visible signs of disorder disappeared. In addition, the reanalysis also showed that neighborhood disorder did not significantly impact any other crime.[26]

Several studies have tested the direct impact of quality-of-life policing on disorder and crime. A study by Braga et al. (1999) used an experimental approach. The police implemented quality-of-life policing in 12 neighborhoods throughout Jersey City and tracked disorder and crime. The researchers then compared the disorder and crime rates of these 12 neighborhoods with 12 neighborhoods having similar characteristics that did not have quality-of-life policing. The results demonstrated that disorder and crime substantially

▶ **Photo 9.1** Would you expect to find more crime, less crime, or about the same rate of crime in this neighborhood compared to the neighborhood you live in?

declined in the neighborhoods where police had implemented the quality-of-life policing practice.[27] However, other studies have found no effect in other cities and especially with regard to minor crimes.[28]

Another theory that guides quality-of-life policing is deterrence theory, which goes back to Cesare Beccaria and Jeremy Bentham. Deterrence theory posits that people make decisions that will bring them pleasure and avoid pain. In so doing, they seek to satisfy their own desires regardless of the costs and pain they may cause for others. For instance, a person might steal an expensive car to go on a joyride. According to the theory, the most effective way to prevent people from committing crimes is to deter their behavior with punishments that are swift, certain, and proportionate to the crime. Swift and certain punishment is important because if people believe that they can avoid punishment, they will not be deterred. Despite the massive number of quantitative and qualitative studies testing deterrence theory in a variety of settings at the individual and group levels, there is little consistency in the findings. Overall, the evidence that people avoid criminal behaviors due to a fear of punishment is rather limited.[29]

Elliott et al. tested the question of whether quality-of-life policing had deterrent effects on marijuana possession and use in New York City. They recruited participants from low-income neighborhoods and collected data for five years from 2003 until 2008. The researchers examined several questions related to the underlying assumptions of deterrence theory. First, they asked participants about their perception of the risk and consequences of using marijuana. The participants' responses very clearly indicate that they were well aware of the new arrest policies and that they avoided smoking marijuana in public because they did not want to get arrested and be held overnight at the central booking station. More than 90% of all participants commented on the new arrest policies. One of the participants, Roy Earl (26), stated, "Smoking outside is such a hassle, because the consequences are just too high. Preposterous. Ridiculous"[30] (p. 7). This was a common sentiment among the participants. The researchers also asked about the impact of other social pressures, such as familial and social reputation, on their decision about whether and where to smoke marijuana. Overall, social pressure from friends and family was an important factor in the decision not to smoke marijuana at all or to smoke marijuana outside the house. The teenage participants, especially, were wary of how their behavior reflected on the family. One of the respondents (16 years old) explained:

> Because, like, like, my mother, she has friends who are like always outside. So, like say if I'm standing outside smoking, they're gonna think a bad thing about my mother: "Oh, she doesn't care about her child. She lets her child do whatever she wants," or whatever.[31] (p. 8)

At the same time, for many participants smoking marijuana is part of their social life and hanging out with their friends. In these contexts, they would smoke in public places despite the risk of being arrested. In addition, if the option of smoking inside simply did not exist, then the participants stated that they did smoke in public places and watched out for police. For some participants who had been arrested, the deterrent effect was small because of the experience during arraignment in court. Sunshine noted:

> Yes [I was arrested]. And that's why I no longer smoke in the street... [I was] walking along on a public street. And I was smoking a blunt. And I had a beer in my hand. And I did not see the cop walk past me until he started to call me. A plainclothes [officer] arrested me [at] like three o'clock in the afternoon... It was a blunt. It was

just…down to the…last part of it.… I went to a desk appearance and the judge told the officer to stop bringing bull in his court. [Laughter] They just let me go.

These types of experiences were very common, effectively undermining the deterrent effect of the new policing strategies.[32]

Participants also reported using different spaces, referred to as "**The Cut**," which are "in-between" spaces, "indoor–outdoor" spaces, and controlled public spaces. Some examples are staircases, rooftops, lobbies, alleys, garages, cars, and so on. Most of the participants had several cuts where they would smoke marijuana. Often, the good cuts would be kept secret so that others would not use them. It appears that the deterrent effect was often a displacement effect to spaces where it was safer to smoke. The new quality-of-life policing strategy did result in the removal of disorderly behavior in the city's cultural, commercial, and tourism centers. This was not due to a reduction in marijuana smoking or other disorderly behaviors but rather the displacement of these behaviors to other places less visible to the public. Of course, residents and tourists may very well be mostly concerned about the cleanliness and safety of public spaces, such as parks and streets. This in itself may justify a continuation of the new arrest policies.[33] Case Study 9.1 explains how quality-of-life policing in New York City brought down crime rates and increased tourism.

ASSET FORFEITURE

Civil asset forfeiture allows police to seize property of a person if they believe that the property was used in the commission of a crime and was purchased with money obtained through criminal behavior. The person does not have to be convicted of a crime or even be charged by a prosecutor. Stated differently, people who have not been charged with a crime or been found guilty of a crime can lose their property and have to appeal to have it returned. They have to carry the legal costs of the appeal and wait for months or years to have their case heard by a judge. Until then, their property is held by police. Property that can be seized includes money, cars, houses, trailers, and anything else that is of value.

The federal government as well as each state separately have civil asset forfeiture laws. Under federal law, the Asset Forfeiture Program is part of the Money Laundering and Asset Recovery Section of the Criminal Division of the U.S. Department of Justice. Asset recovery or asset forfeiture is an important tool for criminal law enforcement and mainly targets criminal enterprises, such as drug traffickers, money launderers, and other criminals who make significant amounts of money through a criminal enterprise.[34]

The Asset Forfeiture Program has several goals. First, punish the offender by depriving them of the fruit of their crime. By taking away property from criminals, the government takes away the proceeds gained from the crime. Often, criminals are more concerned about losing their property than about a jail or prison sentence because the property ensures a good life for them after they have served their sentence. It is also important for their family while they serve their sentence. Some criminals only commit crimes to provide resources to their family. Thus, taking away their assets ensures that they actually feel the punishment.

Second, deter other people from committing crime with the goal to make money. If criminals would be able to keep the money made from criminal activity and live a lavish

CASE STUDY 9.1: QUALITY-OF-LIFE POLICING IN NEW YORK CITY—NET WIDENING AND RACIAL BIAS?

In the 1990s, the New York City Police Department (NYPD), under Mayor Rudolph Giuliani, implemented a new policy where police officers arrested individuals who had committed offenses that impinged on the quality of life of its citizens. William Bratton, the police commissioner, believed that small but highly visible offenses, such as smoking marijuana in public and riding the train without a ticket, should be discouraged because they reduce the quality of life of New York's residents. Until then, police had typically ignored such small offenses. Under the new policies, police were mandated to arrest anybody who committed disorderly offenses in public settings, including panhandling, prostitution, sleeping on public benches, graffiti writing, riding the train without paying, smoking marijuana in public, and other misdemeanors and violations. The offenders would be held in jail for about 24 hours until their arraignment before a judge and the judge would impose a swift penalty, such as a fine or community service. The goal of the new policy was deterrence of such behaviors by informing the public of the mandatory arrest policy and by imposing a swift and certain punishment.

Proponents argue that quality-of-life policing has cleaned up the streets, decreased serious crime, increased tourism, and decreased the visibility of homelessness. Some studies found that arrestees were well aware of the mandatory arrest policies and reported that they had cut back on those behaviors for which they could get arrested. The greatest reductions were observed in riding a train without a ticket, disorderly conduct, and traffic violations. In addition, the Tactical Narcotics Team aggressively targeted drug markets and swept entire neighborhoods to remove drug dealers and users. Finally, New York also authorized the Business Improvement Districts to serve as a quasi-governmental coalition and hire private security to maintain order.[35]

William Bratton explained the implementation of quality-of-life policing:

Think of malaria; for years and years the response to malaria was to swat at all those mosquitoes. But we are never going to kill all those mosquitoes. What was generating all of those mosquitoes? Swamps. Not until people went in and drained the swamps did they start dealing effectively with the problem.

New York City's crime rates did drop substantially. Between 1994 and 1996, serious crime rates decreased by 36% and murder rates decreased by 45%.[36]

Critics have argued that the crime rate reductions were also observed in other cities and were caused by several factors, including the decline of crack cocaine and the economic upturn. Some studies have noted that the reductions in crime only hold for the early years of New York City's quality-of-life policing. For instance, Harcourt (2001) found that between 1989 and 1998, violent crime increased as misdemeanor arrests increased.[37] One possible explanation is the focus by police on misdemeanor crimes rather than on serious crimes. Overall, there is little empirical evidence that the drop in crime rates in New York are due solely to the new arrest policies. Further, critics contend that quality-of-life policing in New York has widened the net of individuals who are drawn into the criminal justice system. The new arrest policies had especially negative consequences for Black people and Hispanic people. A study by Spitzer (1999) found that police stops, in which people were temporarily detained, questioned, and sometimes searched, disproportionately targeted Black people and Hispanic people. Even though Black people made up only 26% of New York's population, they accounted for 51% of all stops. Hispanic people who made up about 24% of the population accounted for 33% of all stops. In comparison, White people make up about 43% of the population but accounted only for 13% of all stops. Police often used quality-of-life infractions as probable cause for these stops. Quality-of-life policing resulted in a wider range of

Continued

CASE STUDY 9.1: QUALITY-OF-LIFE POLICING IN NEW YORK CITY—NET WIDENING AND RACIAL BIAS? (CONTINUED)

arrestible offenses and thereby widened the net of offenders drawn into the criminal justice system.[38]

In addition, critics also argue that homelessness and public disorder crimes have not been reduced but simply displaced to private spaces or marginal and remote public spaces.[39]

lifestyle, they would serve as a role model for others. By taking away the proceeds of their crime, it takes away the benefits and changes the cost–benefit analysis for offenders.

Third, another goal of asset forfeiture is incapacitation by taking away financial resources needed to continue the criminal enterprise. It costs money to engage in organized crime. For instance, drug traffickers need money to operate the planes that transport the drugs across borders. They also need substantial amounts of money to pay for the next load of drugs. Similarly, people who engage in financial crimes on the stock market need money to invest. Without these financial resources, the criminal is not able to continue their criminal enterprise or at least not able to continue it at the same level. For these reasons, there has been much effort to take away money used to finance terrorism. Terrorists needs a lot of money to plan and carry out attacks. If they have no financial resources, they become much less potent.

Fourth, asset forfeiture provides funds to compensate victims. Typically, victims of crimes can receive restitution after the offender has been convicted. Asset forfeiture allows the government to seize the property of the offender prior to a conviction in court. This ensures that the offender does not spend or give away their property to avoid restitution appointments. In addition, victims have to pursue restitution privately, which can be a heavy burden emotionally and financially. Forfeiture is carried out by the government and then distributed to the victims. Thus, it removes the burden from the victim.

Fifth, asset forfeiture is highly visible to the community and thus provides an opportunity for law enforcement to show that crime does not pay. For instance, it is very difficult for business owners, such as restaurants, who follow the law and work hard to make a profit when they are competing against businesses that have illegal funding sources.

Finally, asset forfeiture can contribute resources for socially desirable causes. For instance, property can be converted into a shelter for battered women. However, there is a fine line here that is easy to cross for police and the federal government. They may seize property of people who are not running a criminal enterprise in order to bring in money to the police department and to fund certain projects that otherwise could not be realized.[40]

Asset forfeiture has become part of the practice of widening the net of offenders drawn into the criminal justice system. Some police departments have been very aggressive in seizing property from people who do not have a criminal record or are part of organized crime cartels.

Stop and Seize: The Practice of Asset Forfeiture

After the 9/11 terrorist attacks, the police took on the role of assisting the Department of Homeland Security. Police officers were tasked to be more aggressive in their search for potential terrorists, drugs, and other contraband. The federal government invested millions of dollars to train officers. However, the hunt for terrorists and contraband also had other consequences that are rarely discussed in public. These consequences include the use of asset forfeiture laws and "highway interdiction" techniques taught by private companies to seize money from motorists and funnel them into the police department budget. One private company created a private intelligence network, which allows law enforcement agencies to share information about motorists, including social security numbers, addresses, and tattoos. The network is called Black Asphalt Electronic Networking and Notification System.

Under asset forfeiture laws, police officers can stop motorists and seize money and contraband without making an arrest. The motorist will only get their property back if they can prove that it was obtained legally. This can take several months, court days, and lawyer fees. *The Washington Post* gathered data on these stop-and-seize practices. They found that between September 11, 2001, and 2014, police seized $2.5 billion from 61,998 motorists. Of the $2.5 billion, $1.7 billion was kept by the state and local authorities and $800,000 went to federal agencies, including Homeland Security. About half of these seizures were for less than $8,800. Many motorists chose not to challenge the seizure. Specifically, only one sixth of all seizures were challenged in court. High legal fees and the long process of the appeal constitute substantial hurdles for motorists who want to challenge the seizure. But of those who did challenge the seizure, 41% received their property back. In 40% of the cases that were appealed by the motorists, the process took more than a year and motorists often had to sign a release form that they would not sue the police department or officers. The extent of the stop-and-seize practice has increased over the past years. Since 2008, 298 police departments and 210 police task forces have made seizures that equal 20% of their annual budget. There are many stories of innocent motorists who were stopped by police and had their property seized. For instance, in 2012, Manual Stuart from Staunton, Virginia, was stopped by police for a minor traffic infraction. The police took $17,550 from him, which he earned from his small barbeque business. He challenged the seizure and insisted on a jury trial. Eventually, he received his money back, but he lost his business because he was not able to pay the costs of running his business without the money that had been seized by the police.

There are also concerns that the data shared via the Black Asphalt Network violates privacy rights and civil rights. Critics contend that sharing private information of motorists who have not committed a crime violates their constitutional rights. Some state and federal agencies have also voiced their concerns.[41]

▶ **Photo 9.2** What would you do if you were stopped by the police and had assets seized?

In Defense of Asset Forfeiture

Former U.S. drug enforcement agent Steven Peterson defends the stop-and-seizure practices. He argues that the monies seized help police departments provide training and equipment for their officers. The practice is also important as it protects the public by aggressively pursuing criminals. He argues that police officers don't go hunting for money, they go hunting for the bad guys who bring drugs, weapons, and counterfeited goods into the community. The aggressive stop practices have also helped stop terrorists before they can kill innocent people.

In addition, police departments follow a compliance process with regard to the asset forfeiture program and the Black Asphalt data network has proven to provide correct and valuable information about individuals who engage in criminal enterprises.[42]

The practice of asset forfeiture is highly controversial and especially the practice of police seizing money and property without making an arrest. Several cases have been heard by the courts challenging this practice. The most important case to date was decided by the U.S. Supreme Court in 2019. *Timbs v. Indiana* (Case Study 9.2) challenged the legality of civil asset forfeiture from offenders who were arrested for dealing in drugs but were not running or part of an organized crime cartel.[43]

CASE STUDY 9.2: TIMBS V. INDIANA

On February 20, 2019, the U.S. Supreme Court decided the case of *Timbs v. Indiana*. The case dealt with the constitutionality of civil asset forfeiture under the Eighth Amendment. The facts of the case are as follows: Police arrested Tyson Timbs and he pleaded guilty to dealing in a controlled substance (i.e., heroin) and conspiracy to commit theft. He was sentenced to one year of home detention and five years of probation. Additionally, Timbs had to pay fees totaling $1,200. At the time of his arrest, the police seized his Land Rover SUV under the asset forfeiture laws. Timbs had bought the Land Rover for $42,000 with money he had received from a life insurance policy after the death of his father. The state claimed that the vehicle had been used to transport heroin with the intent to sell. The maximum fine against Timbs for his drug conviction was $10,000.

The trial court found that the forfeiture of the Land Rover was excessive under the Eighth Amendment and the police should have returned the vehicle. However, the Indiana Supreme Court overruled the finding of the trial court stating that the Eighth Amendment does not apply to Indiana's civil forfeiture statutes. Timbs appealed the forfeiture of his Land Rover and the case was eventually heard by the U.S. Supreme Court in 2018. Timbs argued that the forfeiture of the Land Rover violated his rights under the Eighth Amendment, which prohibits excessive fines. The Eighth Amendment reads "excessive bail shall not be required, not excessive fines imposed, nor cruel and unusual punishment inflicted." The Court held in favor of Timbs stating that the Eighth Amendment applies to the states.[44]

Discussion Questions:

1. So, the Supreme Court held that the Eighth Amendment applies to the states. What do you think will happen next? Should Timbs get his Land Rover SUV back?

2. Who do you think should get the seized money: law enforcement, the city, or somebody else? What should the seized money be used for?

3. Only one sixth of all motorists who had money seized by police file an appeal. What may be the reasons?

SUMMARY

This chapter discussed the concept of net widening, which is referred to as the increase in juveniles and adults who are drawn into the criminal justice system. This typically occurs due to the implementation of a new practice targeting juveniles and adults who have committed minor offenses or who are at risk of committing criminal acts. These programs attempt to intervene in the life of the person before they commit a serious crime. The intent of these programs may well be good, but they often draw in minor offenders and youth at risk who would otherwise have had not been in the criminal justice system. Several studies have shown that the main purpose of diverting offenders away from detention into community-based programs was not successful as detention rates have remained stable and instead people were added to fill up the community-based programs. Thus, more people have become drawn into the net of the criminal justice system. This has greatly negative consequences for their life and the lives of their family.

The chapter also discusses the net widening effects of civil asset forfeiture and the Supreme Court case *Timbs v. Indiana* (2019), where the Court decided that the prohibition of excessive fines of the Eighth Amendment applies to all states.

KEY TERMS

Civil asset forfeiture 178
Diversion programs 171
Human capital 172
Net widening 171
New nets 171
Quality-of-life policing 175
Stronger nets 171
The Cut 178
Wide nets 171

DISCUSSION QUESTIONS

1. Discuss the concept of net widening and which policies and programs have contributed to net widening.

2. Discuss how diversion programs can contribute to net widening.

3. Discuss how community policing can be the foundation for restorative justice.

4. Discuss how restorative policing can be implemented in police departments.

5. Debate the pros and cons of diversion programs.

6. Some people argue that drugs courts have done much harm to racial-ethnic minority drug offenders and communities. Debate the pros and cons of drug courts.

CHAPTER 10

COMMUNITY POLICING AND THE USE OF FORCE

The beating of Kelly Thomas was so brutal that his face was unrecognizable when he arrived at UC Irvine Medical Center in a coma. Thomas, a homeless man, who suffered from schizophrenia, died five days after an altercation with six Fullerton, California, police officers. Thomas had sustained major head trauma and brain injury from the beating. On July 5, 2011, Fullerton police responded to a call of someone vandalizing cars. Officers arrived at the scene to find a shirtless, dirty, disheveled man in the area. While attempting to search him, the man became uncooperative. Although witnesses confirm Thomas was uncooperative, they also witnessed the interaction between police and Thomas. At one point, Officer Manuel Ramos put on latex gloves and was heard to say, "Now you see my fists? They are getting ready to (expletive) you up."[1] Officers' voice activated recorders and nearby mounted surveillance cameras corroborated witness statements that detailed the brutal beating by officers as they tried to subdue Thomas. Witnesses also related that Thomas had begged for his life and called out for his father as the officers continued to beat him. Thomas, the son of an Orange County deputy, was known to police as a transient in the area. Initially, the beating of Kelly Thomas garnered some media attention but built momentum when Thomas was removed from life support and passed away.[2]

Two issues fueled public outrage. First, the amount of force against Thomas was incongruous to the seemingly minor offense; and second, police handling of a vulnerable mentally ill subject was inhumane. The Orange County District Attorney's office indicted Officer Manuel Ramos, Corporal Jay Cicinelli, and Officer Joseph Wolfe on murder charges. Three other officers were not charged. Nearly two and a half years later, Ramos and Cicinelli were found not guilty, and the district attorney dropped the charges against Wolfe. All three officers were fired from the Fullerton Police Department. Protests over the verdicts lasted several days.

Learning Objectives

1. Define use of force, abuse of force, and lethal force.
2. Identify how procedural justice relates to police legitimacy.
3. Explain the definition and use of force models: continuum of force, ladder of force, wheel of force, and force options model.
4. Discuss how community policing and use of force are compatible.

Although the officers were found not guilty, large financial settlements were awarded to the mother ($1 million) and to the father ($4.9 million) of Kelly Thomas.[3]

INTRODUCTION

In this chapter, we examine the compatibility of community policing with use of force. Hostile relations between police and the public stemming from police brutality were the impetus of police reform. It was expected that community policing would be the panacea to help communities heal and to improve relations between citizens and police. The conundrum is whether use of force and community policing can be reconciled. Can there ever be a trust if police continue to use coercive force on citizens? This chapter examines use of force, including deadly force, in the community policing era. The surprising conclusion is that community policing proponents are not against the use of force, but they are against excessive force. To get a deeper understanding of the issues, we will explore the definition of force, force options, prevalence rates, officer-involved shooting, legitimacy and procedural justice, departmental training and policies, and officer discretion. We will also take a closer look at the use of force against poor people and people of color, historically and today. In part, implementation of community policing hold promise to address past abuse by police against members of these communities.

▶ **Photo 10.1** Justice for Kelly Thomas protest.

COMMUNITY POLICING AND USE OF FORCE

Research has been scant in the area of community policing and use of force. Limitations on such research involve the extent to which departments truly engage in community policing as well as the heterogeneity of such programs. Some departments say they have community policing but do not, while other departments only employ specific aspects of community policing, for example, leaving out the problem-solving component.[4] Additionally, increasing police–citizen contacts through innovative but unwelcome community policing endeavors in already hostile communities is unlikely to achieve positive results. Community policing itself cannot erase a history of public distrust, especially in areas of high crime and social disorder. Regardless of whether a department has full engagement in community policing, it may be difficult to confirm a link between community policing and lowered rates of force by police. Researchers argue that community policing programs should be tailored to the community served and focus should be on building police legitimacy through procedural justice. In some communities, the task of rebuilding trust will be more challenging; however, without trust, effective policing is unattainable.

The underlying assumption of community policing adoption was that it would lessen the need for coercive means to deal with crime and criminals. It was assumed that community policing would lower the use of force rates in three ways. First, community policing would boost the trust citizens had in their police and, thereby, increase cooperation and compliance. Second, the police would engage in a new kind of problem solving that (mostly) did not involve physical force. Third, building relationships would enhance familiarity, counter negative stereotypes, and break down the social distance between police and the public.[5] Whether less force would lead to better relationships or better relationships would lead to less force is the question. Only recently have researchers taken a closer look at the link between community policing and use of force, especially deadly force.

Police Legitimacy and Procedural Justice

According to former Attorney General Loretta Lynch, speaking on the commitment to community policing, law enforcement officials and residents should view one another as allies rather than as adversaries.[6] Therein lies the key—a relationship between police and the community must be well established and fortified with mutual trust and respect; otherwise it is more likely than not that the public will perceive any use of force as an assault against them. The community must believe that police decisions and actions serve their interests, and not the interests of police. Nearly 200 years ago, according to Sir Robert Peel's principles, the extent to which the cooperation of the public can be secured diminished proportionately the necessity for physical force and compulsion for achieving police objectives.[7] Greater trust means more cooperation and less need for physical force. Through community policing endeavors, police hope to build strong relationships to promote cooperation with the public. Community policing and use of force, then, are not mutually exclusive, nor are they incompatible.

Police Legitimacy. Use of force, especially excessive force, negatively affects public perceptions of police. The harm of a single use of force incident can be long-lasting, and in some cases, irreparable to police legitimacy. Where does legitimacy come from? Is it earned or a given? The authority of police to enforce the laws, to keep the peace, and to uphold the rights of citizens to be free and secure is the extension of the legitimacy of government. **Police legitimacy** is the social contract between citizens and their government, an agreement where citizens give up certain rights in exchange for police services.[8] Enforcing the laws is not what puts the police in conflict with the public however. Conflicts arise from perceived or real bias in how police enforce the laws. Throughout their shift, police officers resolve minor disturbances and interact with disorderly people. The methods that police use to gain compliance and restore order may involve coercive force. Legitimacy results from the belief by the public that police possess not only the legal authority but also the moral authority to enforce the laws.[9]

Procedural Justice. There is an expectation that police decisions and actions will be fair and just. That the police are trustworthy, fair, and effective is referred to as **procedural justice**. Procedural justice has been identified as the cornerstone of police legitimacy and public trust.[10] Law enforcement agencies are instituting procedural justice concepts to forge better relationships. The four pillars of procedural justice are fairness in processes, transparency in actions, opportunities for voice, and impartiality in decision-making. How a person is treated by a police officer, even when the contact involves unlawful behavior, has lasting impressions on that individual and the public as a whole. This is especially true in incidents involving force. While not every encounter involves coercive force, even the mere presence of a police officer causes angst in most people. Think about how you react when a police car is directly behind you in traffic. Traffic stops are the most common types of encounters people experience, whereas other types of encounters are less prevalent. How often do police have to go hands-on to control a situation? What types of force do police employ to gain compliance?

Prevalence of Force

Actual data on prevalence rates of use of force are difficult to collect and assess for several reasons. First, the definition of use of force is not universal and is subject to interpretation. Second, the variety of types of force are numerous. Attempts have been made to estimate the numbers and types of force. For example, force options include verbal commands, and use of flashlight, batons, Taser, chemical spray, motor vehicle, chokeholds, and firearms. Third, the data come from several different sources. These include court records, arrest records, department use of force records, citizen complaints, injury reports, surveys of citizens and officers, and self-reports. Fourth, as with crime data, some incidents of force are unreported. Also, not all police departments require officers to self-report use of force, or the department does not keep records of such incidents. However, more may be known from injury reports due to documentation of rendered aid. Obviously, the most accurate data would come from cases involving use of deadly force.[11]

Many people are interested in accurate data about use of force, especially excessive and lethal force. Police agencies are interested in such data for policy and decision-making, training, supervision, and discipline. The public wants to know for transparency reasons, and specifically, if police officers have been held accountable for misbehavior. They have a right to know when members of the public have been killed at the hands of police and the circumstances surrounding incidents involving police use of force. Police agencies are not always forthcoming with information, often evoking concerns of privacy or its investigatory status. In some cases, victims or civil rights groups must take law enforcement agencies to court to force release of information, such as body cam footage or identity of officers involved.

Massive data collections, especially by state and federal government, are available on most topics; therefore, it is quite surprising that there is a dearth of comprehensive data on officer-involved shootings. Such a task has been taken over by public and news agencies. When looking at the data collected by these entities, one study found more accuracy on deaths by police officers and less accuracy on nonfatal shootings.[12] The media broadcast incidents as they occur, sensationalizing the occurrence. *The Washington Post* has been keeping a database on fatal police force by name of deceased, state, gender, race, age, mental illness, weapon used, if there was a body camera, and whether that person was fleeing the scene. According to *The Washington Post*, 1,004 people were shot by police in 2019,[13] while another site puts that number at 1,099.[14] The difference between these data reflects the problem in obtaining accurate information. Collection of data by news agencies allows them to publicize the information and frame it in such a way that it inflames public sentiment, especially about race. What is not reported is that there are significantly more assaults on police and most of the deaths of suspects by officers were as a result of those assaults.[15] Recently, in response to this method of collecting and broadcasting statistics, the FBI created the first national database of police involved shootings, which includes deaths and injuries. The data collection began in January 2019.[16]

As noted, prevalence rates for use of force are difficult to ascertain due to a number of variables, including force inflicted on individuals that do not involve shooting. Force may involve kicking, hitting, and pepper spraying. Law enforcement agencies generally have a reporting requirement by their officers anytime force is involved in an encounter with a member of the public. As many officers are taught, "when in doubt, write a report." Documenting force, especially discharging a gun, may be required by the officer's agency, but no such requirement for the agency to report use of force to state or federal authorities. Because there is a body, it is much easier to see why fatal police encounters are documented when compared to other force options. That being said, up until recently, such data were not collected. Police scholars and administrators are seeing the value of such data for training and policy decisions.

Race and Use of Force. Much of the data now being collected and analyzed are showing some troubling trends, especially where it concerns officer-involved shootings of racial minorities. There may be an assumption that jurisdictions with high crime rates have corresponding high fatality rates; however, some cities with low crime rates documented high rates of police violence.[17] Between January 2013 and December 2017, 27% of the police killings were by police in the largest U.S. cities, and although only 21% of the population in those cities were Black, 39% of the people killed were Black.[18] According to one source, a leading cause of death for young Black men in America is being killed by police.[19] In a recent survey, Black people reported being unfairly stopped five times more than White people.[20] In a nationwide poll conducted on 1,223 adults, it is noted that police are too quick to use deadly force against a Black person. Black people also thought that police were treated too leniently by prosecutors when they injured or killed civilians, which had an impact on perceptions of police by Black people during contact with police.[21]

Police Brutality. Police brutality is not a new issue, and some believe police are abusive most of the time, especially against people of color. The case illustrated above involved a White man; however, examples of controversial beatings or deaths of racial-ethnic minority individuals at the hands of police are numerous, for example, Freddie Gray, Eric Garner, Michael Brown, Amadou Diallo, Abner Louima, Malice Green, and Rodney King. In many of these cases, the beatings or deaths were discordant with their crime, and some were particularly gruesome and violent. In 1997, New York Police Department officers, responding to a fight at a club, singled out Abner Louima, a 31-year-old Haitian man, and arrested him for disorderly conduct, obstructing police, and resisting arrest. Louima was beaten on the way to the station, and once at the station, was taken into the bathroom where he was sodomized with a toilet plunger, which was then forced into his mouth, breaking his teeth, and he was handcuffed all the while with his hands behind him. Two officers were convicted and received 15 and 30 years in prison. Other officers were charged with the coverup. Many felt that the punishment of the officers was too light. Louima was paid $8.7 million in a civil lawsuit against the city.[22]

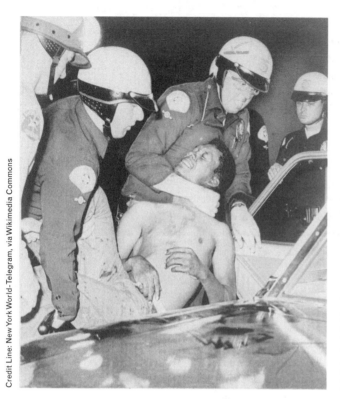

▶ **Photo 10.2** Police arrest a man during the Watts riots, August 21, 1965 (World-Telegram photo by Ed Palumbo).

The cases detailed above are vivid examples of brutality by police. Both fanned the flames of public disgust and outrage. As if the beatings were not bad enough, to add insult to injury, the public believes the police were not justly punished. In the case of Kelly Thomas, the officers walked away scot-free. Unfortunately for police, the public tends to paint all officers with one broad brush. One bad officer or, one high-profile incident can taint the reputation of all law enforcement. These views are especially prevalent in racial-ethnic minority communities where there is historical violence, excessive force, and abuse by police.

The President's Task Force Report in 1968 and the President's Final Report in 2015 recommended the adoption of community policing with the objective of improving community relations, especially in poor and racial-ethnic minority communities. In these communities, police brutality, corruption, and abuse were reportedly rampant, leading to hostility and animosity between citizens and police. High-crime communities had a greater demand for police services, more citizen police conflicts, and higher rates of physical force. Images from civil unrest in the 1960s, for example during the Watts riots, reveals the extent to which police and citizens clashed.[23]

In most cases, civil unrest is sparked by police actions, followed by an escalation of violence, death, and destruction, and spreads rapidly to other cities. Ironically, physical restraint used by police to invoke arrest and gain compliance resulted in more force being used to squelch the rebellion. It was a vicious cycle leading to a loss of police legitimacy. Community policing was the answer, according to the President's Task Force, to improve community relations and to stem the violence. That was the hope, but was it just wishful thinking?

Legacy of Police Brutality. Few issues garner more public outrage than police brutality. The public abhors acts of violence against its citizenry by police, even when warranted. Perhaps, from a law enforcement perspective, in many circumstances the force was warranted and appropriate, but regardless of that viewpoint, if the public perceives the force to be unfair, unwarranted, or overzealous, it creates a serious problem for police. Incidents involving police use of force, especially lethal force, have huge social costs. The loss of police legitimacy hinders the ability of police to do their job. A single incident can erode years of community relationship building. As illustrated earlier, quite often, police officers walk away, unscathed by charges of brutality and wrongful death, and continue working as police officers. The gravity of an excessive force incident is compounded when there is no justice for the victims.[24]

USE OF FORCE POLICIES AND LEGAL CASE PRECEDENT

Certainly, use of force, deadly force, and excessive force are very different issues; however, the line is often blurred. One might wonder how a police action is determined to be either acceptable or nonacceptable. There are guidelines, policies, and statutes that grant authority for use of force, even lethal force, if necessary and under certain circumstances. Although no universal definition of use of force exists, the International Association of Police Chiefs generally defines **use of force** as "the amount of effort required by police to compel compliance of an unwilling subject."[25] Such a simple statement belies the complexity of the use of force issue, however. Police officers must act in accordance with policy; however, all occurrences of force result in an investigation within or outside the department; and many cases, especially in the case of deadly force, will end up in court to be scrutinized further.

Use of Force Policies

Over the past century, as policing became more professional, clearer guidelines, policies, and laws covering use of force emerged. Officers have a great deal of autonomy and discretion when addressing situations in the field; and therefore, parameters for acceptable behavior must be communicated through training, laws, and policies. Additionally, because no two situations are the same, the definition is vague and overly simple. Use of force policies vary from department to department but have similar content, such as the words "reasonable force." It would be a long document if all circumstances and responses were included in the policy. Some departments do outline circumstances and appropriate responses, although there is a danger in lack of inclusion of unique or extraordinary situations. Therefore, use of force policies are meant to be guidelines and not instructive in every case. Mostly, according to the policies, officers are to rely on training and experience employing common sense and reasonableness in their decisions. Based on a database of large agencies across the United States, top content categories in use of force policies include:

- Force and deadly force defined
- When deadly force is authorized
- Reasonableness, probable cause, resistance defined
- Required equipment: what uniformed officers must carry
- What actions are prohibited; for example, using a Taser on a child, shooting at or from a moving vehicle, firing a warning shot, when action endangers bystanders, using oleoresin capsicum (OC) spray on individuals already under control
- Deadly force against an aggressive animal is permitted and notification procedures
- Notification and reporting duties of officer and supervisor
- Provide medical aid following the use of force

- Failure to report excessive force of other officers when witnessed
- Investigation process, often performed by another agency
- Review process by the district attorney
- Expectation of continued training as well as demonstrated knowledge of the policy

Use of force policies are updated and reviewed periodically. Often policies are updated to address new weaponry, additional limitations or restrictions, and new requirements. Modifications are based on legislative and legal actions. For example, in 2004, a car theft suspect was hit 11 times with a flashlight by Los Angeles Police Department (LAPD) officers who were attempting to capture him at the end of a pursuit. The beating was captured on video by two news helicopters. Following the release of the video, the public was outraged by the sight of officers using a flashlight in that manner, and quite frankly, so were police administrators and other officials. The mayor said, "I am demanding a complete investigation and explanation, and it better be good."[26] Due to the publicity of this case and others, several agencies reviewed and changed their use of force policies to restrict the use of a specific type of flashlight, the Maglite, which is a large heavy flashlight that can apparently do more damage than a baton.[27] Police officials often make prudent modifications of their policies as a preemptive strike against legal liability; however, quite often, use of force cases end up in the courts for juries and jurists to decide, and changes are mandated through legal or legislative means.

Legal Precedence and Liability. The standard in use of force prior to precedent-setting cases was the Fourteenth Amendment, which concerns the due process rights of detainees. Later the courts clarified the distinction in cases involving excessive force during arrest by applying the Fourth Amendment, applying the Fourteenth Amendment in cases of abuse during pretrial detention, and applying the Eighth Amendment in claims of abuse for those already convicted and in custody. Below are a few precedent-setting cases with which most law enforcement officers are familiar:

- *Graham v. Connor*, 490 U.S. 386 (1989): The officers' actions under the Fourth Amendment were deemed to be "objectively reasonable" during an investigatory stop.

- *Tennessee v. Garner*, 471 U.S. 1 (1985): Known as the **fleeing felon rule**, police were limited to nonlethal force when a felon is trying to escape, unless the suspect poses an immediate danger to the officers or others.

- *Terry v. Ohio*, 392 U.S. 1 (1968): Refers to **"Stop and frisk"** searches and establishes that officers can stop and frisk a subject based on reasonable suspicion and not the higher standard of probable case. Also known as the **Terry stop**.

- *Plakas v. Drinski*, 811 F. Supp. 1,356 (1993): There is no constitutional duty for an officer to use lesser force or resort to alternatives when deadly force is authorized.[28]

Off-Duty, Current, and Retired Law Enforcement Officers. Departments have the additional responsibility and legal liability concerning the actions of off-duty and retired law enforcement officers. Qualified current and retired law enforcement officers are granted authority to carry off-duty guns under the Law Enforcement Officers Safety Act (LEOSA) enacted in 2004. Such a law extends liability of the city, county, state, and federal jurisdictions when it comes to current and retired officer use of force.[29] Case law, legislative law, and departmental policies proscribe guidelines of what weapons and magazines off-duty or retired officers can carry, where they are permitted or prohibited from going, for example, on an airplane, across state lines, in no-gun zones, native lands, and federal buildings, and reporting procedures if force is used. Some policies and legal liability coverage may be void if the officer was drinking alcohol or violating a policy that either prohibited or required certain actions or behaviors. The liability burden is significant when you think of the numbers of retired and off-duty officers (see Case Study 10.1). There are many cases where an off-duty officer took action in a situation and killed someone. One such case was in Barstow, California, where a suspect attempting to rob a McDonald's restaurant was confronted by an off-duty officer who thwarted the robbery. When the suspect opened fire, the off-duty officer returned fire. In the end, the gunman and an innocent nine-year-old girl were dead.[30]

LAW ENFORCEMENT TACTICAL GEAR

Law enforcement is the only profession, aside from military service, where officers carry weapons, not just guns but an array of tactical equipment to use in the course of their work. Officers carry pepper spray, TASERs, batons, semiautomatic pistols (usually both duty and backup guns) and have access to shotguns and patrol rifles in their police vehicles. Officers also carry multiple high-capacity magazines, two pairs of handcuffs, and other tactical gear such as a knife. To the average citizen, this may seem extraordinary. A popular 1960s television character, Deputy Barney Fife from *The Andy Griffith Show*, carried one bullet in his pocket just in case he needed it. Perhaps this gave the public the naïve notion that police do not need a large arsenal. Most people are not knowledgeable about what police carry nor do they comprehend the dangers of police work. Certainly, with the large arsenal available to an officer, the potential that weapons will be deployed exists. Police officers are not jumping at the chance to shoot their gun; however, they cannot be reluctant to deploy their weapon if the situation warrants its use. Officers are guided by training and policy, and hopefully, common sense. Through the hiring process, they are carefully selected, screened by numerous hiring tests, subjected to thorough background investigations, and must graduate from the rigorous academy and field training to ensure their suitability for this stressful job, one that involves life-or-death situations.

TRAINING, FORCE OPTION MODELS, AND OFFICER DISCRETION

Law enforcement agencies must train officers and maintain their training. Because officers have a significant amount of discretion that includes making life-and-death decisions, it is imperative that their training be on par with the level of responsibility they have in society. We would not go to a doctor to have surgery if the doctor had never operated on

CASE STUDY 10.1: OFF-DUTY POLICE OFFICER KILLS MAN IN COSTCO

On June 14, 2019, three people were shot at a Southern California Costco. Corona police surrounded the building while shoppers fled. Police rendered the scene safe and took a man into custody for the shooting. The shooter was identified as an off-duty Los Angeles police officer. He killed a man and injured two but was acting in self-defense, according to his account of the incident. The deceased man, later identified as Kenneth French, 32 years old, who apparently suffered from mental illness, had shoved the off-duty police officer, Salvador Sanchez, who was holding his infant son. Sanchez was knocked to the floor and rendered momentarily unconscious. When he regained consciousness, he thought he was under attack, identified himself as a police officer, pulled out a gun, and began shooting, hitting French and his parents. The altercation was caught on surveillance cameras, although that video has not been released by the district attorney's office pending the decision of the grand jury. French died and his parents were injured.

On September 25, 2019, Riverside County prosecutors declined to file criminal charges against LAPD officer Salvador Sanchez, following the grand jury's decision not to indict the off-duty officer for the killing of Kenneth French. After the decision not to file criminal charges, the surveillance video was released to the public. Although Sanchez will not face criminal charges, he is not off the hook when it comes to his future career. LAPD is currently conducting an administrative investigation as to whether Sanchez violated the department's off-duty use of force policy.[31] Additionally, the family is calling for a federal investigation into the matter, and this could lead to federal civil and/or criminal charges against Sanchez.[32]

Public outrage following the announcement of the decision stems from the belief that Sanchez used excessive force by using a gun, being off-duty, not in uniform, and out of jurisdiction. Central to the investigation was whether Sanchez was under imminent threat for his life and that of his son's. The attack was unprovoked, according to the evidence, and the parents admitted that their adult, mentally disabled son had recently been put on new medication and had seemed combative prior to the shooting. Public expectations of what should have happened differed from the official determination. Such differences create animosity between the community and their police, and are fodder for debates about use of force.[33]

Discussion Questions:

1. How might the officer be justified in this shooting?

2. What are some of the reasons the officer might be guilty of excessive force?

3. Identify the short- and long-term implications of this incident.

---

someone or had not done so since medical school. There is a significant amount of assumed liability every time an officer straps on their gun, gets in a patrol car, and responds to calls for service. Today, there is more public scrutiny of law enforcement decisions and actions than ever before, mostly due to the wide dissemination of information through social media. The public is likely to ask why a certain action was taken and not another. Law enforcement officials have a strong incentive to ensure officers are provided with a high level of training throughout their career.

Basic academy provides recruits with both practical and classroom instructions where they learn about criminal laws and codes, investigations, arrest and control, self-defense, force options, firearms training, patrol procedures, and first aid. However, it takes years of training to develop mastery and muscle memory or automatic capability that does not involve an officer having to think about how to do something, for example, reloading a gun during a gunfight or communicating on a radio during a pursuit. If you have to think about it, it's probably too late. Second, crime trends change, so what an officer learned in the academy might not be sufficient to address new weapons or crimes. Responding to an active shooter was not something most recruits learned in the academy years ago. **Less-than-lethal weapons** are increasingly being introduced and officers need to learn how to use them. Less-than-lethal weapons are those weapons that can be deployed by officers in resistant situations to help officers gain control without going hands-on physically with the suspect. Keeping officers up-to-date and refreshed on their skills is not only necessary, it is legally mandated.

Training

Officers are trained how to use their equipment to control a situation, and hopefully, to bring about a peaceful solution. In the early days of policing, there was no requirement for officers to have continued training after they left the academy; for example, some officers never again fired their weapon. Currently, law enforcement officers are required to qualify or certify with their weapons annually, quarterly, or even monthly. In addition to firearms proficiency, tactical training in what are called **perishable skills** continues throughout the officer's career. Any skill that diminishes after a period of nonuse is called a perishable skill, such as pursuit driving, tactical firearms, force options, arrest and control, and verbal communications.[34] These requirements resulted from litigation of a wrongful death suit where an officer killed a fleeing man. In this case, the court proclaimed that the department and city should be held civilly liable because the training was "so severely deficient as to reach the level of gross negligence."[35] The officers in this department only went to a public range to shoot every six months and never received training, such as night shooting, moving target, or shooting in residential areas.[36] In this and other cases, the courts have concluded that firearms training needs to be relevant, realistic, and conducted regularly.[37]

Training not only includes learning the tactical skills necessary but also helps officers manage their physiological responses to stressful situations. Training scenarios are very realistic. Most recently, officers across the nation have ramped up training to address mass shooters. Many training opportunities are now available for law enforcement. An Orange County, California, sheriff academy, for example, offers both academy and advanced officer training in a realistic setting called Laser Village. Other agencies are invited to use Laser Village and participate in a full day or two of training. Laser Village was built to look like a real city, and during the training, officers wear uniforms, drive patrol cars, engage in radio traffic, and are permitted to use an array of specially designed equipment to simulate the same equipment officers carry. Mini scenarios are set up so that officers can play cops-and-robbers games, such as hostage situations, robberies in progress, and dealing with gun-wielding suspects.[38] However, these scenarios are not fun and games, they are tests of officers' split-second decisions and actions. In fact, they are so realistic, the officers who participate feel the same level of anxiety and stress that are experienced in actual situations in the field despite knowing it isn't real.

Not every agency has the luxury of having a whole village for training. Other options include use of simulators, such as, tactical defensive driving simulators. Officers are given the opportunity to learn driving techniques that include pursuits through city traffic with pedestrians, rainy weather, sudden turns, shots fired, all while talking on a radio. Many departments set up the Shoot–Don't Shoot Simulators. Today, these simulators are quite sophisticated, allowing officers to interact with players on screen. An example of a training scenario might be a domestic violence call where the officer is faced with a suspect who is holding a knife to a woman's neck. The officer participating in the simulation may shout commands at the subject, who will either comply or escalate. If the officer is forced to make a life-or-death decision, they can shoot at the screen and the suspect falls to the ground. An evaluation is conducted on the officer's actions, revealing how many shots were fired as well as which shots were fatal. Some equipment has gone quite high tech, using surround screens, physical structures, foliage, lighting, props, and sound.[39] While both simulator and mock scenario training are important, law enforcement officers must also have the foundational knowledge of force options.

Force Option Models

A use of force continuum, developed in the 1980s, outlined an escalation of force depending on the circumstances. The continuum evolved from a linear diagram, to a ladder, and then to a wheel. The levels included:

1. Officer presence, also known as command presence: Officer uses mere presence and symbols of the job, such as the badge, gun, and uniform, to gain compliance.

2. Verbalization: Officer issues calm, nonthreatening commands and may increase the volume or shorten the directives to words such as "Stop" or Don't Move."

3. Empty-hand control: Officers uses both soft and hard techniques, such as using a wristlock and then, if necessary, may use punches or kicks to restrain the individual.

4. Less-lethal: Officers use baton, projectile, chemical spray, or TASER to gain compliance when the subject has ignored previous commands.

5. Lethal force: Officers use lethal weapons to gain control of a situation when suspect poses a serious threat to officer or others.[40]

Original iterations of force models were designed for the officer to move from one level to the next level as needed, with the expectation that the officer would use the continuum as steps toward lethal force. Both the continuum and ladder force models demonstrated that each level permitted greater physical force. The wheel force model puts the officer in the middle, at the hub of force options, allowing them to use any level of force that was reasonable in the situation. In the wheel force model, it was not necessary for an officer to use the previous steps if the situation dictated immediate use of lethal force. In the early days of the force models, the courts often lambasted officers on cross-examination about their failure to adhere to the use of force escalation model. For example, the attorney would ask the officer on the stand, "Did you use a verbal

THINK ABOUT IT: LESS-THAN-LETHAL FORCE OPTIONS

Less-than-lethal weapons have been used by the police since the first police departments were established. Police carried billy clubs, saps, or wooden batons. Today, less-than-lethal weapons are far more sophisticated. There are several incentives for these less-lethal alternatives to the firearm. For one, police can stay safe from a distance as they attempt to gain compliance. Second, there are fewer injuries and less likelihood of death to both the officers and suspects. Third, the public is becoming less tolerant of fatal shootings. Pepper spray has become one of the most common forms of less-than-lethal weapons, followed by the conducted energy device (CED), such as the Taser. Both pepper spray and CEDs have been known to cause injury and even death. While nearly 11,000 law enforcement agencies currently use Tasers, some agencies are concerned about legal liability.[41] There is some indication that officers are quick to use the Taser in situations where they might have tried other methods to gain compliance. On the other hand, use of the Taser has been shown to be effective in reducing the likelihood of injury or death to both the officer and suspect. However, although the CED has been found to be safe in uses with healthy adults, cautions need to be taken in the case of children, older people, those with underlying heart disease, and pregnant women. Some subjects, who come to the attention of police through abnormal mental status, can pose a problem when deploying the CED as it may increase the delirium.[42] When comparing pepper spray and use of CEDs, officers were four to five times more likely to use the CED than pepper spray. Pepper spray has a couple of drawbacks in that a decontamination process has to be initiated following the deployment and, second, there is a likelihood that the officer will also be affected by the spray. CEDs have after-care measures as well, with some agencies requiring medical removal of the prongs and the possibility of the officer being shocked during the process of removal. However, agencies that have equipped their officers with Tasers find that officers deploy the Tasers far more often than use of pepper spray. The fact that Tasers are easier than pepper spray may confirm the concern that officers are using the Tasers too frequently, in situations that do not warrant this type of force.[43] Without these less-than-lethal options, officers may be more likely to reach for their firearm. Officers have many options, including the ability to retreat and re-engage if there is a known suspect. Officers can deploy nonlethal force and attempt to get the upper hand in a dire situation. And, in a worse case scenario, officers are permitted to use deadly force to stop the threat. Courts have ruled, however, that officers choosing deadly force over other options, when deadly force is warranted, should not be held legally or civilly responsible.[44]

Discussion Questions:

1. Do you think officers should carry Tasers? Why or why not?
2. What are the pros and cons of less-than-lethal force?
3. Should officers be held responsible if they use less-than-lethal force and the subject dies from that force?

command before firing your weapon?" If an armed suspect is threatening the officer with a gun, it is reasonable that the officer meet force with like force, not going through a series of levels before lethal force can be applied. Research and logic indicate that situations often escalate out of control well before application of lower levels are possible.[45]

Officer Discretion and Use of Force

Judicious use of force remains the standard in policing today. To be sure, officers, for the most part, are reluctant to go hands-on, to use force at all. They would rather have compliance and not risk their own safety or the safety of others. As we will discuss, there are some officers that readily use force while others do not. There is a tipping point for officers to engage in force, but no clear formula exists. Is that tipping point ego, danger, or risk? When making an arrest and placing handcuffs on an individual, the officer must be prepared for resistance where the force can escalate. As for prevalence rates, one must drill down to the kinds of force counted and reported. When we take into consideration the frequency of police-citizen interactions, the use of physical force is rare and the use of police force is even rarer.[46]

Officers possess a significant amount of discretion on the job. They work autonomously in the field with little supervision. Many variables influence officer decisions and actions. Decisions to use force involve the circumstances, the type of crime, the suspect's behavior, the officer's experience, and training. With the focus on the officer, research has examined the role of officers' background, job assignments, and arrest activity to explain use of force rates. One study found, for example, that a few officers are responsible for a large portion of use of force incidents, while other officers have less frequent uses or none at all.[47] That finding in itself would be enough to warrant investigation into why that is. Why are certain officers more prone to use force? Supporting the notion that police are biased, studies have found force was used more often and more vigorously against suspects who were male, non-White, poor, and younger.[48]

Characteristics of the Officer and Use of Force. Research findings have been mostly inconsistent when it comes to characteristics of officers and force. For example, with regard to the likelihood of a particular type of force used, some studies found that there are no gender differences.[49] However, other studies found male officers were significantly more likely to use force.[50] Research findings also show that racial-ethnic minority officers are more likely than Whites to use force.[51] That finding is also inconsistent with others that show no difference with regard to race.[52] Another interesting finding reveals that officers use less physical force if they have a four-year college degree and more experience.[53]

Job Assignments and Use of Force. Where officers work, the jurisdiction, the shift, the number of officers in a beat area, whether assigned to patrol or a specific detail or assignment, all figure prominently in the likelihood or unlikelihood of force being used. Police officers assigned to patrol, especially in high crime areas, are faced with a greater likelihood of conflict with subjects who are hostile. Officers assigned to a special detail, such as prostitution stings or drug buybacks, also may face greater danger than an officer assigned to traffic control. Time of shift assignments is an important factor as well. Similar to officers who are assigned to high crime areas, officers assigned to peak crime times (9:00 p.m. to 3:00 a.m.) are more likely to use force.[54]

Arrest Activity and Use of Force. Most use of force incidents occur during arrest.[55] Some officers are more productive than other officers, generating a higher volume of arrests. Productivity may be a characteristic of the officer—or as mentioned above, it may be a function of the area and shift the officer is working. Officers who make more arrests increase their chances of having to resort to physical force.[56] However, some researchers argue that officers who work in high crime areas are likely to avoid making arrests because they are familiar with the people, disorder, and behavior in the area and use other tactics to negotiate a solution.[57]

Misconceptions Regarding Use of Force

In general, the public have little or no knowledge of police use of force policies nor do most people comprehend the dynamics of police encounters involving use of force. There are numerous misconceptions about use of force. For example, a common misconception is that police shoot to kill; however, the reality is that officers are trained to "stop the threat." In the Wild West television shows, cowboys would shoot the gun out of the bad guy's hand or shoot at his feet to "make him dance." When a person is killed, people will ask, "Why did they have to kill him, why didn't they just injure him?" Trained police officers shoot to stop the threat, shoot at the center mass (the chest is largest target) to minimize harm to others, and continue shooting until the threat is eliminated.[58] When does an officer determine that force is necessary or reasonable, how much force to use, and under what circumstances? What we do know is that officers do not have much time to ponder these questions. Most of the time they react instantaneously to a threat by employing their quick reflexes, muscle memory, training, experience, and the like.

HOW CAN COMMUNITY POLICE REDUCE USE OF FORCE ISSUES?

As has been mentioned throughout the book, community policing is not the answer for everything; however, there are many ways that it can minimize the risks of abuse of force. First, by forging a relationship and partnership with the community, police and the community would work together to solve problems. In a true partnership, there is accountability by both citizens and police. The relationship building would enhance the trust and belief that police are working in the interest of the community. Second, the organizational structure of community policing would allow greater **officer discretion**. Police officers would be empowered to be innovative and creative in problem-solving strategies. Police officers, under a traditional police model, were generally discouraged from thinking outside the box. Under the community policing model, supervision would be coaching rather than directing. Third, often times, training new officers was about folding them into the police subculture and traditional methods of policing. Instead, field training officers would encourage rookies to think of long-term strategies to persistent problems, such as homelessness, drug dealing, car burglaries, and juvenile delinquency. This helps the new officer develop critical thinking skills. Fourth, officers would be rewarded for not using force, for coming up with creative solutions, and for working with the community, especially vulnerable or at-risk individuals. Finally, and most importantly, officers would be selected, hired, and trained with community policing in mind. Individuals would be selected because of their desire to work with the community.[59] Departments who do this have one-third fewer sustained complaints on use of force than those who do not.[60]

It appears there is no lack of research on police and use of force. The problematic inconsistencies and conflicting findings hinder clear policy direction. What we do know, however, is that there are implications for careful and intentional officer recruitment, selection, training, and supervision. The bigger question is whether community policing

THINK ABOUT IT: USE OF FORCE STANDARDS FOR POLICE

The standard for use of force, including deadly force, across the nation is that police may employ lethal force under reasonable circumstances. As of August 2019, California has adopted the toughest standards to date. The new language in the law changes "reasonable" to "necessary." The previous law stated that [police] "are authorized to use reasonable force to effect the arrest, to prevent escape, or to overcome resistance." The new law sets out that the use of physical force . . . is a serious responsibility that shall be exercised judiciously and with respect for human rights and dignity. . . ." And further, "use of deadly force [may] only [be used] when necessary (emphasis added) in defense of human life."[61] On its face, the change may seem innocuous; however, the ramifications are huge. Ultimately, the new law directs the judicial system—judges, juries, prosecutors—to examine the "totality of circumstances" including the actions of the officer and the suspect that led up to the physical force and/or deadly force. While this does not appear to be different to the layperson, as part of subsequent investigation of any use of force incident, an assessment will be made of the officers' decisions and whether the officers created the jeopardy in the first place. Certainly, policing is a dangerous profession and officers take risks; however, officers should not recklessly put themselves in danger. For example, if an officer steps in front of a car and then shoots the driver because the car moved toward him, the reviewers are required to examine whether the officer put himself in danger; and therefore, lethal force could have been avoided.[62] The new law has raised concerns with both citizens and police; it appears that there is support for and challenges to the changes by both sides. Groups such as Black Lives Matter argue that the law does not go far enough, while police officers claim that it will make them reluctant to get involved in most dangerous situations. Provisions for amendments to relevant penal codes that guide use of force are included in A.B. 395 and S.B 230. Additional changes include training requirements and guidelines for specific language in department policies. Although not widely publicized, the changes are making waves in other states and many legislative bodies are following suit. We probably will not know the impact of such a law for many years to come.[63]

Discussion Questions:

1. How does the new law differ from the previous standard of reasonableness?
2. Why should or shouldn't the officer's perception supersede civilian perception when force is used, for example, during a trial?

and problem solving reduce the frequency and severity of use of force by police. Does community policing hold the promise of better relationships with the community, greater trust and faith in the police by the public, and ultimately smarter, less coercive policing methods? There is some indication that indeed community consultation and partnerships do lead to lower rates of police use of force.[64]

SUMMARY

In this chapter, we have discussed the complexity of use of force by police. Throughout the decades, although essentially rare considering the hundreds of thousands of nonphysical police–citizen encounters that occur every day, use of coercive force remains a hot button issue, especially in racial-ethnic minority and poor communities. No other police action garners greater emotional and political consequences than use of force, abuse of force, and lethal force. Social costs include loss of trust and faith in police, lack of cooperation and compliance, outright hostility from the public and a lasting negative impact on members of the community. Abuse of force incidents can hurt police departments in terms of funding, resources, and political support, which in turn can undo or halt community policing and problem-solving efforts.

Community policing builds relationships and forges partnerships between citizens and police. Citizens and police must work together to solve community problems. Citizens must have trust and faith in the system so that they are willing to work with police to fight crime in their neighborhoods. When citizens and police work together, there is a commitment to help one another. Not only does community policing promote strong partnerships between citizens and police, community policing reduces the need for use of force by incorporating smarter policing strategies. Community policing encompasses a vast number of programs that include problem solving, intelligence-led policing, Compstat, and third-party policing. Innovative problem solving involves prevention and long-term solutions. Smarter policing means the public are partners and they work with police on problems that plague their community. With smarter strategies, arrests are used as a tool to solve larger problems, not a solution for every immediate problem.

KEY TERMS

Force Option Models 196
Less-than-lethal weapons 195
Officer discretion 199

Perishable skills 195
Police legitimacy 187
Procedural justice 187

Stop and frisk 192
Use of force 191

DISCUSSION QUESTIONS

1. What is the definition of use of force?
2. Describe the force option models?
3. Define police legitimacy and procedure justice. How are they related?
4. Explain why community policing is compatible (or not) with use of force?
5. What are some of the ways police departments can reduce use of force through community policing?

PART IV

DEPARTMENT ORGANIZATION AND CHALLENGES

CHAPTER 11

POLICE AND THE MEDIA

On Thursday, April 12, 2018, in a Philadelphia Starbucks, two Black men were arrested because they were sitting in the coffee shop, allegedly waiting for a friend to transact a real estate business deal. They had not ordered anything.[1] That is the story the media hyped following the now viral video posted on Twitter by a Philadelphia resident, Melissa DePino. The cell phone video shows two Black men being handcuffed and "perp walked" out of the Starbucks. The truth may vary a bit by point of view, however. The police asserted that they were doing their job and the police chief contended that the police at the scene did nothing wrong. The store manager called police after she had asked the two men to order something or leave, and they refused. Upon arrival, the police asked the men to leave and again the men refused. The men were then arrested for trespassing and creating a disturbance, handcuffed, and transported to the station, although the men were later released without being charged.

Most likely, no one would have concerned themselves with the lives of these two men, the police, or Starbucks but for the Twitter post. Newsfeed immediately picked it up on a Thursday and by Saturday, it was one of the biggest news stories. In the days following the incident, headlines were intentionally geared to evoke high emotion and spark rage; and they were successful, as evidenced by protesters invading the Philadelphia Starbucks with bullhorns and signs. Popular networks, such as CNN, used race to bait readers with headlines: "Two Black Men Were Arrested in a Philadelphia Starbucks for Doing Nothing" and "'Whites Only' spaces still exist."[2]

While this story illustrates the ability of social media to garner national and international attention, it also demonstrates media's power to interpret, spin, and broadcast their version of events. The media, and therefore, the public's focus of this incident changed from abusive police tactics to the perceived racial bias of a White

Learning Objectives

1. Explain the symbiotic relationship of media and police.

2. Explain media impact on public perceptions of crime and policing.

3. Describe media portrayals of police in news and entertainment.

4. Explain impact of media on fear of crime.

5. Explain how social media presents both opportunities for crime innovation and crime prevention and solving.

store manager and Starbucks, albeit the underlying premise for both views was racism—poking those embers into flames. And just like that, the central stakeholder in the game changed from police chief to the CEO of Starbucks. Within days, the CEO of Starbucks issued an apology to the men and offered them a settlement. Additionally, it was announced that 8,000 Starbucks would close for an entire day so that employees could attend antibias training, at a significant cost to the corporation. The media's initial "abusive police" narrative aligned with many people's personal viewpoints, and therefore, was not easily dismissed. What impact did this incident have on the relationship between the police and the community? Was the relationship harmed by the media attention or did the media attention bring much needed scrutiny to an already unhealthy relationship?

INTRODUCTION

In this chapter, we will learn how the media shapes our perceptions about police and policing rather than through firsthand experience. Additionally, we will explore the pros and cons of the media's impact on justice. We will see that technologies, such as the cell phone, and the proliferation of social media have both positive and negative implications for community policing. The strength of police and community relationships is based on trust, respect, and legitimacy of police authority. Many factors can influence, facilitate, nurture—or conversely, interfere or destroy the rapport. What role do the media have in fostering a positive community/police partnership? Alternatively, what role does media spin play in public outrage, civil disobedience, riots, and mass violence? To what extent do the media construct our reality of police, crime, and justice? News and entertainment media not only contribute to public perceptions of police but may have greater influence on our reality than does actual direct personal experience. Today, social media contributes an additional layer of complexity to the police and community relationship, because of the pros and cons of social media.

POLICE AND MEDIA: A SYMBIOTIC RELATIONSHIP

The police and the media make strange bedfellows, but it's not a marriage made in heaven.[3] Neither like each other, but both need each other. "The police are sought, quoted, and catered to on one hand, but marginalized and criticized on the other" (p. 114).[4] A majority of police officers believe media treat the police unfairly and those officers who hold this view are more likely to feel frustrated in their jobs and see a disconnect between themselves and the public.[5] Attitudes of police toward the public as well as the public's attitudes toward police are the subject of much research. Less is known about how the media influences or interferes with police–community relations.

Some may argue that the public's negative view of law enforcement is tied directly to the heavy-handed tactics and racism demonstrated by police officers against citizens; however, the real blame for public sentiment may be tied to media portrayals of these events. This is not to suggest that the blame for police abuses should be shifted to the media; however, the media is quick to cover such occurrences.[6] For example, indicating that stories about police in general and police brutality in particular are increasing, one reporter admits, "We're not seeing more police shootings, just more news coverage."[7] Although historic police brutality against people of color is well documented, present-day mainstream media coverage continues to focus on the negative history rather than any positive strides police have made. Journalists and reporters have the ability to push a narrative to the forefront and that narrative sells more papers and gets more viewership when the news is sensational, such as a police shooting. Public perceptions about police are more about what the media chooses to share and how the story is framed, than about what the police actually do. However, that doesn't mean bad actions by the police should be framed differently or not covered at all, only that the balance of coverage and media perspective may not reflect reality.

IMPACT OF THE MEDIA

The media has a powerful impact on shaping citizen attitudes about crime, justice, and specifically, the police. The media shapes and molds people's perceptions and may ignite their emotions. When an incident occurs, such as an officer-involved shooting, the media is not merely covering the public's reaction but fanning into flames where sparks exist. For example, focusing on the officer's and suspect's race is likely to stoke the flames in racial-ethnic minority communities, where deep-seated beliefs of "the racist cop" persist. This is not meant to suggest that race of the officer or suspect is never an issue, only that the media is quick to inject that narrative. Negative attitudes of citizens toward police undermine community policing efforts. Those negative perceptions destroy belief in the legitimacy of police to act in a fair and unbiased manner in the ways in which they do their job and treat citizens.[8]

Social Construction Versus Direct Experience

The public perception of law enforcement is mostly derived from what we read, hear, and see through media sources and not necessarily direct knowledge. Few of us have been with police when they chase down a suspect, get into a vehicle pursuit, wrestle a criminal into handcuffs, or walk the beat in a high crime area; but we do know what that is like from watching a police drama on television, for example. What people believe about law enforcement, whether they support police or despise police, may have more to do with what they have learned about police through the media and less about what they have personally experienced.

News and entertainment—via print, radio, television, film, and now social media—are the main sources of our understanding about police and policing. Regardless of whether the content is news or entertainment, perceptions about police are socially constructed. **Social constructionism** is the theory that people develop knowledge, perceptions, and beliefs of reality through social interaction and the media plays a large role in that construction. In the process of understanding the universe through media consumption, people are exposed to and influenced by what is presented. The media filters out or emphasizes aspects that might support or oppose viewpoints. In the interaction with others, more so than our direct experience, people's perspectives are shaped and molded.[9] Viewers take away information from media, such as whom and what to fear, whom to glorify, and how to punish those who violate the law. The more we watch (television), the more we believe the real world to resemble that of the pixel world; as a result, society becomes more homogenized and differences regress.[10] Despite the argument some might make that they never watch television or consume any media, they are influenced by their social network of people who do consume media.

POLICE PORTRAYALS IN THE MEDIA

Crime and justice are popular topics for public consumption and the media has capitalized on that interest. Crime sells. The images, ideas, and narratives that permeate the media influence how people think about crime and justice. This includes the behaviors we think should be criminalized, who we feel should be punished, what the punishments should be, and how we think the police, judges, attorneys, correctional officers, criminals, and victims should act. All these behaviors are influenced by media portrayals of crime and justice (p. 416).[11] In all types of media—print, radio, film, television—crime and justice have been popular and profitable and have influenced our lives. Just think about the numerous TV shows that have been on the air for many years, such as *Law & Order*, *CSI*, *Bosch*, *Jack Ryan*, *Homeland*, *Chicago PD*, *Sopranos*, and so on. The success of these shows strongly indicates the popularity of crime and justice among the population. Shows such as *Law & Order* regularly pick up on real events and use them in their storyline.[12]

Entertainment Media

The most featured component of the justice system in entertainment media has undoubtably been law enforcement. Other than traffic violations, most people do not have direct experience or contact with police; therefore, it is inevitable that perceptions about police and crime-fighting come from media and especially entertainment media through books, magazines, television, and films.[13] How accurate those depictions of police and police work are is not clear, although most findings suggest that prime-time portrayals are uneven or front-end loaded (i.e., more about police and crime-solving than punishment); overdramatized or full of action; and overwhelming pro-police (able to solve crimes in an hour).[14] There are some exceptions, of course, when police officers are shown to cover for each other or act violently toward suspects, such as Officer Stabler in *Law & Order*. However, the vast majority of depictions show the police as crime-fighters and problem solvers.

A detailed overview of police programming cannot be accomplished here, because there are too many shows to cover. Media portrayals of police have been slapstick, comedic, serious, realistic, and actual. Media police officers have been bumbling, incompetent, heroic, clever, funny, charming, brutal, wise, corrupt, and tough. In whatever way they are portrayed, the public cannot get enough, and what better way to get viewership than to capitalize on popular content. Criminal justice programming continues to enjoy high viewership. Early silent films introduced us to the bumbling and funny; Keystone Cops; and today we have reality-based shows with actual footage showing the excitement, danger, and action of policework.

Even a detailed overview of police television programming would likely miss mentioning many popular programs. Older generations may remember their favorite police dramas, which included Westerns, G-men, hard-boiled detectives, and the like.

Although it is easy to find many of these older shows on platforms such as YouTube or cable television to view again, because of our more discerning, modern, sophisticated tastes, we find most to be laughable and corny, including television shows that were meant to be serious, such as *Highway Patrol* (1955–1959).

Popular police television programs included: *Dragnet* (1951–1959); *The Untouchables* (1959–1963); *Car 54, Where Are You?* (1961–1963); *Adam-12* (1968–1975); *CHiPs* (1977–1983); *Hill Street Blues* (1981–1987); *COPS* (1989–2020); and *Law & Order* (1990–2010).[15] Many of these shows have enjoyed second and third iterations with new actors. *Law & Order* has had several spin-offs with equal success (*Special Victims Unit, Criminal Intent, LA*). *The Shield* (2002–2008), a show about a police officer who doesn't like to follow the rules, a police officer who was the hero, and an antihero all rolled into one, had a record 5 million viewers for its premier in 2002. The public loves a good crime show and a gritty, no-holds-barred police officer to save the day. In 2013, five of the top ten most watched shows were crime dramas and police procedurals (*NCIS, NCIS: Los Angeles, Blacklist, Persons of Interest,* and *Blue Bloods*).[16]

Entertainment media does play a role in public perceptions about police. How the police are portrayed in crime and police dramas is one of the factors that influences attitudes about police. Viewers of police dramas portraying police as corrupt or abusive are more likely to believe that police are brutal, corrupt, and racist. However, most police dramas on television and film show police in a positive light. "Good guys wear blue" is a common theme and the fact that police solve the crimes and catch the bad guys every night on TV is helpful to promote positive attitudes toward police.[17] Many police and crime shows are on cable, Netflix, and Hulu; and therefore, most people are learning about police, crime, and justice through entertainment platforms. Even though news media is not as popular as entertainment media, especially among young people, the news media has a substantial influence on the citizens' perceptions of police as well.

Some of the programs mentioned are still being produced, such as *COPS*. Reality-based crime shows, such as *COPS*, have become more popular than ever. Infotainment are shows that both inform and entertain. Reality and infotainment shows are more popular now than are fictional shows. Created in an entertaining style, these docudramas include *20/20, 48 Hours, Dateline, Cold Justice, First 48,* and *Dr. G. Medical Examiner.* News content is the source of many of the most popular fictional shows' storylines, such as *Law & Order* and *CSI. Law & Order*'s slogan is "ripped from the headlines," borrowing storylines and perhaps embellishing details to increase excitement, danger, and violence.[18]

Two recent additions to the reality-based television programming have taken police docudramas to a whole new level with the advent of the body cam. One show is called *Body Cam* and the other is *Live PD*. Both shows use an officer's body camera or the dashboard camera to highlight various encounters with police. These snippets of body cam footage are carefully selected, and at times highly edited to depict the excitement and danger in the life of law enforcement officers. A particular noteworthy episode shows an officer shooting through his windshield as he is driving after an armed suspect who is shooting at him. Another episode shows two officers immediately ambushed and shot as they respond to a domestic violence call. The officers recovered but the pregnant wife, son, and suspect were found shot to death after a 17-hour stand-off. While these shows are obviously meant for entertainment purposes, and the use of body cam footage is the basis for the shows, the release of body cam footage to the public in controversial police shootings has been the subject of recent litigation. This further shows the blurring of lines between entertainment and the news.

News Media

The lines between news and entertainment have been blurred so much so that people no longer watch the news for the sake of being informed but for the entertainment value. Most people rely on the news for information about crime. Studies have found that citizens, especially people of color, hold negative attitudes about police when police misconduct is highly publicized.[19] Much of the attention-grabbing headlines are those involving crime and police, especially if it hints at scandalous, gruesome, or bizarre events. The incident at Starbucks in Philadelphia discussed earlier has the elements of racial bias, which makes it particularly newsworthy. News media's obsession with violence and the police is based on ratings. "If it bleeds, it leads" is a phrase often used to describe the priority given to attention-getting headlines. Media takes little responsibility for "body bag" journalism, claiming it is the public who desires this content. People tend to gravitate to content that either entertains them or supports their view of the world. Crime, especially murder, is an attention getter. It has a high entertainment value and many people believe that crime is rampant and murder rates are high. These beliefs are certainly reinforced by the news media. Car chases are also one of the favorites on TV, for both the news media and viewers.

Crime and justice news comprise a significant portion of news media.[20] Incidents featured in the news depends on three components: newsworthiness, timeliness to publication or broadcast, and whether it fits into established themes.[21] However, people assume that what they see is what is happening in the world and do not question the reason it is featured. The more harmful and unique or rare the event is the more likely it is to be in the news. Additionally, there is a somewhat new aspect of reporting that has taken over the news media: investigative reporting.

Investigative Reporting. In contrast to reporting, the new trend is investigative reporting. Instead of just reporting a story, an **investigative reporter** goes deeper, with the goal of unearthing evidence, usually of wrongdoing. Investigative reporters see themselves as detectives. Investigative reporting has a ring of status and authority to the title and therefore no longer is the news just being reported, it has greater value. Investigative reporters believe that the public has a right to know; and certainly, the motivation is to produce something that will lead to more readership or viewership. Investigative reporters are not beholding to the police or anyone except the media organization they work for and—beholden—for the public interest.[22]

The public believes that crime, especially violent crime, is rampant and imminent. Much research has examined media content, extent of media consumption, and impact on fear of crime. The link between media content and fear of crime is multidimensional, in that it may depend on saturation and type of content; the characteristics of the person viewing, such as race, gender, level of education, and age; prior victimization; and the environment.[23]

THINK ABOUT IT: WHY ARE POLICE PURSUITS POPULAR TO WATCH ON TELEVISION?

Televised police pursuits became popular more than two decades ago when 90 million viewers tuned in to watch former football great and star O. J. Simpson, in the now infamous white Ford Bronco slow-speed chase through California freeways. On television, people could be seen standing on the freeway overpasses cheering Simpson on as the Bronco inches down the freeway followed by a cadre of police units. Special camera equipment has been developed just to cover chases. Televised police pursuits, especially in California, are often the only reason regular television programing is interrupted. Not only are pursuits likely to cut into regular programing, they are usually carried on multiple channels. Viewers can switch channels to see which one has the best coverage. Network helicopters vie for air space with law enforcement air units to cover the chase and the competition for the best air space has led to several deadly accidents. In one horrific incident, two news helicopters crashed in Phoenix, Arizona, while covering a police chase, killing all four people on board.[24]

High-speed police chases have been called a spectator sport, with the news media competing for coverage of the most spectacular chase. On February 17, 1999, two simultaneous police chases occurred freeways apart on the same day, ended up on the same freeway, and television news stations had to use split screen to show both pursuits. It was exciting for those

▶ **Photo 11.1** What makes police pursuits and other "breaking news" events so captivating for audiences?
Credit Line: iStock.com/Anson_iStock

Continued

(Continued)

who enjoy a good pursuit.[25] Some chases have a comical edge to them.[26] For example, when the police were in a vehicle pursuit with a naked driver on the freeway and broadcast live on television, it got a lot of laughs from the news anchors as they were covering the chase.

Police chases are not without controversy and many believe that police should not pursue criminals this way. Some even question as to why the media thinks they are newsworthy. There is an inherent risk of imminent tragedy. Over 5,000 bystanders and passengers have been killed since 1979, and thousands have been injured.[27] Following a tragic on-air ending of one pursuit, television stations now have an automatic delay to prevent viewers from witnessing bloody outcomes. In 1998, news stations broke into children's cartoons and afternoon talk shows to broadcast a police chase. At the end of the pursuit, a very agitated man, Daniel Jones, a maintenance worker, could be seen facing off with police. Suddenly, in dramatic fashion, he picked up a shotgun, pulled the trigger, and killed himself on live TV.[28] News stations are much more cautious and have implemented the seconds delay; however, pursuits, fatal or otherwise, end up on YouTube, or are used as footage in entertainment programing.

Discussion Questions:

1. Should police pursuits be televised?
2. What purpose do media have for showing police chases?
3. Why are police pursuits popular?
4. Does television coverage of pursuits increase the danger or promote public safety?
5. Does the popularity of police pursuits increase the likelihood that criminals will seek their 15 minutes of fame by fleeing, knowing that the chase will be televised?

FEAR OF CRIME

Fear of crime has become an important area of research, and better understanding includes an examination of the genesis of fear of crime. Fear of crime has little to do with actual victimization, and as discussed previously, social constructed perceptions about crime, police, and the justice system originate from mass media.[29] Because murder and violent crime are prevalent on prime-time news and entertainment, the public believes crime, and particularly violent crime, to be rampant. Fear of victimization has little to do with statistics presented in the news and more to do with the images of criminality. Actual visual imagery, choice of crime, and victims to be featured transmit information of the likelihood being victimized, especially if the person viewing it is similar to the victim in the story. Over time, these media images are enforced into collective feelings.[30] The purpose of broadcasting real crime and crime statistics in the news is less about informing the public and solving crimes and more about entertaining and gaining viewership.[31]

Reduction of crime has been touted as an important assessment of police effectiveness. However, early studies found that fear of crime, more than crime itself, has a significant impact on quality of life issues for most people. Citizens are more likely to experience fear of crime than actual victimization of crime. Police officials were hard-pressed as to how to respond to fear of crime. "If someone breaks in, call us." There is nothing much police can respond to if the person is just afraid of being victimized. While findings were mixed with regard to crime reduction, early foot patrol studies in Newark, New Jersey, and Flint, Michigan, found that increased foot patrol did have an impact on citizen satisfaction with police services, feelings of safety, and reduction of fear of crime (p. 126).[32]

Therefore, fear of crime was the impetus for community policing efforts and remains a central focus. In part, assessment of police services and public satisfaction of police is based on whether the police have substantially reduced fear of crime. In 1994, a Bureau of Justice Assistance report said, "An effective community policing strategy will reduce neighborhood crime, decrease citizens' fear of crime, and improve quality of life in the community" (p. 45).[33] Addressing fear of crime is an important aspect of community policing.[34] If people feel safe, their satisfaction of police goes up.[35] Researchers have used citizen satisfaction surveys to evaluate the effectiveness of police services, especially community policing; however, most agree that other aspects must be taken into consideration.[36]

Fear of crime affects the way people live their lives. When people feel unsafe, they are likely to change the way they live by becoming isolated or not engaged with the community. Many people do not have the means to move away from a high crime area and continue to feel vulnerable and victimized. They are unlikely to interact with the police even if it means helping them feel safer. A recent incident illustrates not only real violence but also how people are fearful to come forward to help police solve crime. Eight people were wounded, three critically, including a teen boy, in a dice-game shooting at a San Bernardino apartment complex. Handguns and rifles appear to have been involved in the shooting, however, no weapons were recovered. Victims and witnesses refused to cooperate with police, making it difficult to determine what happened. According to police officials, when police responded, they were confronted with a hostile crowd and had to request help from additional agencies. Investigators were asking the public to help them solve the crime, but there was little cooperation. San Bernardino police spokesperson Captain Richard Lawhead said, "I think some of it is fear. There are really good people who live here because this is where they can afford to live. And they're held captive by some of the violent, illegal activity that occurs here. But unless we know it's occurring and know who the people are involved in that, we can't stop it."[37]

Fear of crime and fear of victimization are compounded by beliefs that police are unable to do anything about crime. In addition, media content about police and how they do their jobs impacts public perceptions. Incidents of police use of force, abuse of force, and corruption covered in the news—while truly rare—feed into a narrative of police as racist, violent, and out of control. Disdain or dislike for police, a belief that they are unfair and biased, and a belief that they are unable to solve crime or make the environment safe undermine community and police relationships. In recent times, negative aspects of policing are often featured in the news. If we look back at our Philadelphia Starbucks story, we can see that the media was quick to jump on the racist, abusive police scenario and it had a lasting impact even if it turned out the police did nothing wrong. Are the police deserving of public scrutiny? Probably so. However, what if everything they do is framed in a manner that makes them seem like public enemy number one? Police–citizen relations are unlikely to improve as long as citizens only see racist and abusive officers featured in the news. Police officers and departments have the distinct disadvantage that they very rarely receive positive feedback, that is, that the news media reports about officers who helped a victim or a citizen. Unfortunately, these acts are not typically newsworthy. Police accomplish many positive results every day and help many people, but citizens are not aware. The picture many citizens have is greatly distorted and does not resemble the everyday work of police officers.

MEDIA ROLE IN THE "WAR ON POLICE"[38]

Addressing fear of crime is a valid concern of police, especially important to community policing efforts. Some evidence reveals that media do play a role in elevating the level of fear through their focus on violent crime. Another way in which the media can influence public perceptions and behaviors is by portraying law enforcement as biased, racist, brutal, and corrupt, thus eroding the legitimacy of police. During the Obama administration, a strong media-enhanced antipolice sentiment resounded throughout the United States and continued into the Trump administration. Although it was taken out of context, the comment from President Obama was widely published. In July 2009, following the arrest of Harvard professor Henry Louis Gates by campus police for disorderly conduct, Obama was quoted as saying, "...the police acted stupidly," among other things. Although law enforcement officials objected to Obama's comments and criticized his handling of the issue, Obama's apology the next day did nothing to counter the antipolice sentiment. The seeds were planted and took root. The media began promoting any incident involving police and minorities and framed it with racial overtones. It appears mainstream media had an obsession with portraying police officers as hypermilitarized racist occupiers on the hunt for Black men.[39]

The 2012 Trevon Martin shooting and acquittal of George Zimmerman and 2014 police killings of two Black men, Michael Brown and Eric Garner, gave birth to Black Lives Matter (BLM), through the social media hashtag #BlackLivesMatter. The BLM Movement, while decentralized, has built its strength and power through protests, riots, rallies, violence, and law-breaking. Well-known Black leaders, such as Reverend Al Sharpton, were front and center of any media attention. There were many impacts on police and their ability to do their jobs. In terms of federal and state support, community-oriented programs were not funded. Additionally, police agencies had considerable problems recruiting officer candidates and fewer people were attracted to a career in police. Being a police officer seemed like a thankless and an especially dangerous job. Between 2013 and 2014, the number of officers killed doubled from 27 to 51, the largest increase since 2011. Many of the officers killed in that time period were ambushed. It has even become popular to say that policing should be abolished. The killing of police officers continues under the Trump administration, despite a more pro-law enforcement stance. An official from the National Law Enforcement Officers Memorial Fund says, "Too often, their service and sacrifice are taken for granted."[40] A more supportive federal administration translates into more support and funding for law enforcement programs; however, this does not necessarily mean the police are treated any better in the media.

CONTROLLING MEDIA INFORMATION

In order to maintain organizational legitimacy, police administrators need to control what and how messages about police actions and crime occurrences are disseminated to the public. Most police departments employ public information officers (PIOs) who deal directly with reporters and the public. PIOs are the only members of the police force, in addition to the chief or other administrators, who are authorized to speak for the department. PIOs must respond proactively to scandals, find political and public support for police actions, and provide information and data to satisfy media inquiries.[41] When incidents happen, such as an officer-involved shooting, messaging is carefully crafted to minimize harm to the police image and to maximize its positive image. Ideally, the message must contain just enough information to satisfy the public's need to know, curiosity, and concern for personal and public safety.

The police goals to ensure legitimacy, promote, and instill public trust are important tenets of community policing. Without such trust and legitimacy, police would not be able to enforce the laws.[42] Legitimacy is also a product of how effective police are at fighting crime and whether the public believes they are safe from harm. The media plays a large role in citizen perceptions both positively and negatively. To bolster a strong community–police partnership, the police must be transparent in their communications. For example, when an officer is involved in a shooting, law enforcement officials need to be forthcoming with information about the shooting without hampering investigative efforts. However, it is not unusual for the public to demand the release of officer body cam video footage aside from the police narrative provided. To deny this request could spark unrest; however, releasing the footage could have a similar outcome (see Case Study 11.1).

Police officials must walk a fine line between balancing the need to know and the right to know. The media has considerable power and influence over how something is reported. Perhaps, a decade ago, police had the upper hand in how the message played out. In the last decade or so, social media or a cell phone video footage may make any individual a "reporter" and police have little or no control over that messaging.[43] Often times, cell phone or other video footage is the only evidence individuals have to counter a police narrative that is meant to minimize harm or cover up wrong doing. Police have even less control over the media's decision to edit and release only specific snippets of video. Before cell phones, citizen videos have made the news and had serious consequences, for example Abraham Zapruder's famous tape of the assassination of President John F. Kennedy and George Holliday's video of the beating of Rodney King (Case Study 11.2).[44]

SOCIAL MEDIA

The mainstream media is partially responsible for negative coverage; however, much of what makes the news comes from the public through social media. The younger generations rely heavily on social media for their news and entertainment. In 2018, approximately 85% of U.S. adults, including older adults, get their news from social media, which is up from 65% in 2016.[45] A story once tweeted out on Twitter or Instagram is difficult to control. Police may record encounters with citizens with their body cams and dash cams, however people are also recording their encounters with police.

Any discussion of media would not be complete without including and highlighting social media because it is more used than other forms of media. Not only does the use of social media contribute to the social construction of crime and justice, it presents new concerns. Social media, with its speed, broad dissemination, and instantaneous communication across geographic distances creates a venue for innovative crime ideation and methodology. Police, throughout the United States and internationally, are now realizing that social media has taken on greater significance in law enforcement efforts. It is quite possible that the tech-savvy and tech-dependent generation will change the landscape of policing completely. The biggest implications are yet to be discovered; however, some are

CASE STUDY 11.1: SHOULD POLICE BODY CAM AND DASH CAM FOOTAGE ALWAYS GO PUBLIC?

On July 19, 2015, a White University of Cincinnati police officer, Ray Tensing, stopped Samuel DuBose, an unarmed Black man, for a missing front license plate and a suspended driver's license. When DuBose suddenly started the car and pulled away, Tensing fatally shot DuBose in the head. Tensing and two fellow officers claimed that DuBose dragged the officer by the arm, which was caught in the vehicle's window. Fearing that he would be killed, Tensing shot him. The body cam showed that the officer was not dragged at all. A grand jury indicted Tensing on murder charges. Two trials resulted in hung juries. Although charges were dismissed, Tensing was fired from the police department. This was little consolation for DuBose's family, however.

In another case, dating back to 2014, Laquan McDonald, a Black teen who was 17 years old, was fatally shot by White Chicago police officer Jason Van Dyke. The officer was charged with murder. The trial took several years to be decided and relied heavily on the dash cam video, which disputed the officer's statement that McDonald threatened him with a knife. The dash cam footage shows McDonald walking down the street away from the officers who were called to the scene of someone breaking into vehicles.

Van Dyke was shot 16 times. Witnesses said that McDonald had fallen to the ground after the first few shots and the majority of shots came after he was on the ground. As in the earlier DuBose case, release of the dash cam video and body cam video was critical in the outcome. In both cases, the release of the footage was demanded by family, the community, and the media. Van Dyke was convicted October 5, 2018, of second-degree murder and taken into custody immediately following the verdict. Protesters were waiting for the verdict, ready to move forward with anger if the officer was acquitted. Following the verdict, many were angered at the second-degree murder conviction and believed the officer should have received a first-degree murder conviction. Afterward, peaceful protests and celebration were covered in the news.[46]

A public outcry for the body cam video after the shooting has led to a debate over whether the footage is public record. Some argue that law enforcement should release any body cam footage as a matter of transparency and accountability, especially in the event of a shooting. Courts across the country are dealing with this issue and there is no lack of cases for the courts to consider. Currently, no standard has been established. In the earlier discussed case of the shooting death of 43-year-old Samuel DuBose, the footage was released 10 days after the killing despite DuBose's family and supporters demanding to see Tensing's body cam video. The media joined in on that demand. Eventually the video was released reluctantly. City officials said they had planned on releasing the video after the investigation was complete. Prior to the release, the police chief said, "The video is not good."[47]

Discussion Questions:

1. Is body cam video footage public record?
2. Who should get to see the footage before it is released?
3. Why should the media have a role in what happens to the footage?

already apparent, such as increases in on-demand consumption (Netflix, Hulu); decline in both television viewing and print media consumption (newspapers); an uptick in digital sources (smart phones, tablets); and reliance on social media (Instagram, Twitter, Facebook) for news and entertainment. Police are several steps behind the curve; however, the more sophisticated departments are hiring social media specialists and looking for officer candidates who have computer and social media skills, not just brawn. Fortunately, social

CASE STUDY 11.2: THE BEATING OF RODNEY KING, FIRST VIDEO TO GO VIRAL BEFORE SOCIAL MEDIA

Shortly after midnight, on March 3, 1991, following a short vehicle pursuit, police initiated a traffic stop on a Black man, later identified as Rodney King. The beating of Rodney King, though shocking and horrific to all who saw it, was not what made the event so important for the police–citizen relationship and community policing. It was the fact that it was videotaped and broadcast on national news that changed the course of history, not only for police but for the media as well. Many officers viewed the actions against Rodney King as normal and within the scope of their duties. That view was not shared by the majority of people across the country as apparent by the outrage over the police's actions. From then on, the behavior of police became a major point of focus and controversy in the news media and everyday life.

George Holliday, a private citizen, was awakened by the sounds and commotion of sirens and a helicopter outside his window. He grabbed a camcorder and from his balcony, he videotaped 12 minutes of the police encounter with Rodney King. Holliday was so concerned with what he had witnessed, he took the video to the police; however, they were not interested. Knowing his tape was important, he decided to share it

▶ **Photo 11.2** Police and rioters clash at a shopping center in widespread riots that erupted after the acquittal of four Los Angeles Police Department (LAPD) officers in the videotaped arrest and beating of Rodney King on April 29, 1992, in Los Angeles, California.

Credit Line: Donaldson Collection/Contributor/Getty Images

Continued

CASE STUDY 11.2: THE BEATING OF RODNEY KING, FIRST VIDEO TO GO VIRAL BEFORE SOCIAL MEDIA (CONTINUED)

with a local news station. By the end of day, the footage went "viral" and suddenly it was national and international news. The public was shown a 90-second clip of four police officers savagely beating a Black man with batons while numerous officers looked on, further attesting to its normalcy for officers. It was probably the first view of police brutality for mainstream Americans, and it became the visible proof of police brutality against people of color. That 90-second clip shown over and over, in the news and in the courts, is arguably the most viewed video of all time.

Surely, everyone saw it the same way and believed that the four officers would be held responsible. However, over a year later, the four officers were acquitted of all charges. One of the most destructive riots in history followed the announcement of the verdict. After six days and six nights of violence, over 1,000 buildings destroyed, 53 people killed, and thousands injured, the assessment of damages came close to a billion dollars, making it one of the costliest riots in U.S. history.[48]

Discussion Questions:

1. If social media platforms had been available at the time of the Rodney King beating, and the acquittal verdict of the four officers a year later, would there have been a different outcome? What would have changed?

2. If social media platforms were not available, would it have changed the outcomes of the shooting of Michael Brown in Ferguson, Missouri? What would have changed?

3. Why is the Rodney King video important today?

4. If the police had collected the video from George Holliday, would America have seen it?

media also has many positive uses, especially for police. The pro-social, proactive uses of social media by police include tools for enhanced public relations, crime prevention, and for eliciting citizen participation in crime solving.

News Media and Social Media Merge

Much of what has been presented here has shown the negative side of the media and its impact on police and the public—for example, creating misperceptions of what police do or can do; sensationalizing controversial police actions through intensive, often biased coverage; and through promulgation and unbalanced focus on violent or bizarre crime. Police and the media also have a pro-social relationship, whereby good things police do are featured and celebrated. A police officer in New York was featured in the news for buying a homeless man a pair of socks and boots when he came across the barefoot man in the frigid cold. By the time the police officer went to dinner that night, a tourist's snapshot of the New York Police Department officer and the homeless man had gone viral.[49] The tourist who took the picture got nothing but the satisfaction of having her photo go viral.[50]

In another poignant story, a San Diego police officer's last act of kindness was caught on surveillance video. The officer bought lunch for a child who had asked him for 10 cents to buy a cookie. The officer was killed while sitting in his patrol car shortly after that act was captured on video. The surveillance video and the story were widely broadcast.⁵¹

These stories generate positive feelings about the police but are not as frequent as the negative coverage the police get. As far as the kind acts police do, Patrick Lynch, president of the Patrolmen's Benevolent Association of the City of New York, said, "It happens all of the time and it is usually unseen."⁵² The media does not cover many positive stories about police unless it happens to fall near the Christmas holiday season when the police are featured giving gifts to needy children. Police invite media to cover charitable events, so they can be featured in a positive caring light to the community. In one illustrative case, in December 2013, more than 3,000 children from low-income families lined up outside the Eastside Library to receive toys and books from the Riverside Police Department and a charity group. Proclaiming that this event provides an opportunity to meet the public under positive circumstances, Lt. Guy Toussaint said, "Community policing, the only way it works, is if there's that bond. I'd much rather do this than take people to jail."⁵³

These examples illustrate how media can be used for public relations purposes. While some may be happenstance, such as the image of the officer giving socks and boots to a homeless man, other events or incidents are well planned to highlight police and bolster their image within the community. Though few and far between, positive stories are certainly welcomed by police when they can get that kind of exposure.

▶ **Photo 11.3** In one community program, officers take students on a back-to-school shopping spree, helping build positive connections between the students, the police, and the community.

Police are also proactively using social media and related technology to prevent crime, solve crime, and in general, to better communicate with the community they serve. As of 2013, of the local police agencies serving 10,000 or more residents, more than 90% have their own website and more than 80% used social media.⁵⁴ Departments have their own Facebook, Twitter, and Instagram accounts. Most departments serving 50,000 or more residents posted crime statistics on their websites. Larger departments, those serving 250,000 or more residents, also posted crime statistics at the district, beat, neighborhood, or street level during 2013 (Figures 11.1 and 11.2).

Investigation of Social Media Crimes

Not only do police use social media to disseminate information, they investigate crimes committed using social media, such as cyberbullying and stalking, hacking and fraud, identity theft, cat-fishing, sex trafficking, and gang activity, to name a few. In today's star-struck society, people are the producers and distributors of their own performance.⁵⁵ Ray Surette, an expert in justice and the media, argues that crime used to be clandestine operations, but now criminals boast about their crimes through texts and social media and shockingly record their crimes and broadcast them through their social media accounts.

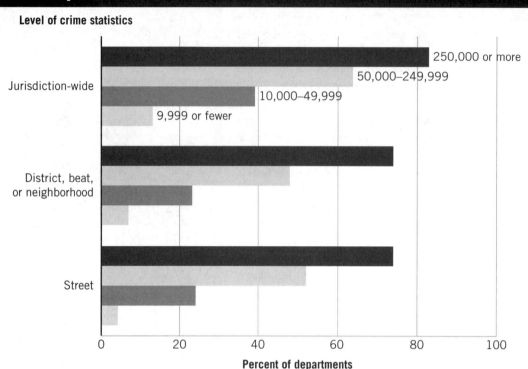

Figure 11.1 Percentage of Local Police Departments Using Websites and Social Media, by Size of Population Served: 2013

Source: Figure 6, page 7 of that publication: Local police departments using websites and social media, by size of population served, 2013. Page 6 Local Police departments, 2013 Equipment and Technology July 2015, NCJ248767 (https://www.bjs.gov/content/pub/pdf/lpd13et.pdf)

Notes: Most recent data available.

One horrific example was in April 2017, when a man hanged his 11-month-old daughter before killing himself on Facebook Live because he was upset with his wife.[56] In some circumstances, prosecution of these criminals is easier because evidence is available for juries to consider. However, the potential negative impact is that juries expect physical visual evidence. This phenomenon is like the documented "CSI effect" resulting from viewing a fictional yet popular CSI television series. From viewing that television program, juries hold unrealistic expectations about forensic evidence available in trials.[57]

Public Use of Social Media. People use social media in many ways, connecting and staying connected with friends and family, research and information gathering, conducting personal and professional business, and as entertainment. Especially true with the millennial generation, social media connects like-minded people for the purposes of social, civic, and political engagement.[58] Civic engagement denotes a sense of noble citizenship; however, it may involve less altruistic ideals. People connect on social media to arrange protests, with mostly peaceful intentions; however, riots and looting can break out spontaneously. There may be some planning to participate in law-breaking as well. Police may obtain such intelligence

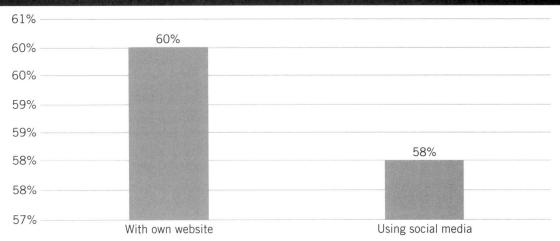

Figure 11.2 Level of Crime Statistics Provided on Local Police Department Websites, by Size of Population Served 2013

Source: Table 7, page 6: Level of crime statistics provided on local police department websites, by size of population served 2013. Page 7. Local Police departments, 2013 Equipment and Technology July 2015, NCJ248767 (https://www.bjs.gov/content/pub/pdf/lpd13et.pdf)

Notes: Most recent data available.

prior to a riot in order to monitor, prepare, and prevent harm of people and property. While protesting is a constitutionally protected form of expression, blocking traffic, damaging property, looting, and assaulting opposing parties are not protected activities. Today, people have the technological advantage of cell phones and social media platforms of Instagram and Facebook to instantaneously crowdsource. Spontaneous uprisings, riots, and protests spread rapidly, build intensity and passions, and multiply possible harm. By monitoring social media communications, law enforcement can better assess and plan for threats against public safety (see Case Study 11.3).

SUMMARY

Certainly, traditional boots-on-the-ground type of policing is not going to be abandoned. Foot patrol, vehicle, alternative forms of patrol, and face-to-face citizen–police encounters are all integral to any policing strategy. However, law enforcement must keep pace with the people they police and that includes not only using social media but moving it up in priority. Community policing relies on communicating with the public to prevent crime, to solve crime, and to build relationships. Police can communicate day-to-day, hour-by-hour, or minute–by–minute with the community members. Communication can be about upcoming events such as Coffee with a Cop at a local restaurant, or when it is something serious such as an active shooter in a mall or at a school.

Most recently, the law enforcement community has been using social media to assist people in the COVID-19 stay-at-home order. The stay-at-home order issued by the president and the governors requires the assistance of law enforcement to enforce such an

Table 11.1 Local Police Departments Using Electronic Methods to Exchange Information with Citizens, by Size of Population Service: 2013

Population served	Receiving crime reports or other crime-related information from citizens			Providing crime statistics or other crime-related information to citizens		
	Total electronic	Agency website	Other electronic means	Total electronics	Agency website	Other electronic means
All sizes	66%	44%	49%	60%	37%	49%
1,000,000 or more	100	100	43	100	100	64
500,000–999,999	97	97	48	100	100	90
250,000–499,999	98	96	56	98	96	82
100,000–249,999	90	85	53	91	83	76
50,000–99,999	92	88	52	93	84	77
25,000–49,999	81	75	45	89	71	72
10,000–24,999	83	73	54	79	59	62
2,500–9,999	71	49	53	65	41	50
2,499 or fewer	47	15	44	37	11	33

Sources: Table 8 Local police departments using electronic methods to exchange information with citizens, by size of population service, 2013 p. 7. Local Police departments, 2013 Equipment and Technology July 2015, NCJ248767 (https://www.bjs.gov/content/pub/pdf/lpd13et.pdf)

CASE STUDY 11.3: HOW POLICE USE SOCIAL MEDIA

Brevard County, Florida, sheriff Wayne Ivey created an entertaining, yet productive way of encouraging citizen participation in solving crimes using the agency's Facebook and YouTube pages. In Ivey's "Wheel of Fugitive" broadcast, he spins a wheel of fugitives with warrants to garner the public's help in locating the individuals. According to Ivey, these efforts reach between 300,000 and 700,000 people weekly. Ivey knew that many people use social media and that this was a way to reach far more people than traditional methods. Members of the community are entertained, engaged, and at the same time, informed about potential dangers and criminal activity in the area.

One example illustrated the power of Facebook to solve a crime and save costly detective work. After a local restaurant's countertop charity bank was taken, Ivey's social media team put a photo online. Within five to 10 minutes, via private message, the thief was identified.

One fugitive said that he watches it every week to see if he is on the wheel. Another fugitive, Alicia Pack, posted on Facebook that she saw herself on the show but added she was going to the beach where police tracked her down. Ivey said, "They ended up

▶ **Photo 11.4** Brevard County sheriff Wayne Ivey spins the "Wheel of Fugitive" board inside the media studio at the Brevard County Sheriff's Office in Titusville, Florida. Sheriff Ivey's Wheel of Fugitive highlights local fugitives in Brevard County and yields an 88% capture rate for each of the featured fugitives.

Credit Line: The Washington Post/Contributor/Getty Images

Continued

CASE STUDY 11.3: HOW POLICE USE SOCIAL MEDIA (CONTINUED)

having to Taser her. So, we put up on our Facebook page, tanned, Tased, and arrested all on the same day." Some people have criticized the idea saying that it is public shaming. "If you don't want to be on the 'Wheel of Fugitive,' don't commit a crime in Brevard County. Don't be a fugitive in Brevard County, because our team is going to come after you, our community is going to come after you," Ivey said.[59]

order. In some communities, officers are issuing tickets to people who are not abiding by social distancing orders by gathering in social spots such as the beach or parks, while other communities have directed officers not to enforce such orders. Social media has been particularly helpful to the police and the public during the quarantine circumstances caused by COVID-19.

Police do not have to rely on having a local news station to connect with the community, although that may be part of the strategic communications. Departments can control their own message on social media and be their own advocate (p. 37).[60] Police can use social media to get a message out about a lost child, a missing older person with dementia, a shooting, or a crime that needs to be solved. Citizen participation is an important part of community policing, bringing police and citizens together to improve the quality of life for all who live there, by facilitating problem solving and encouraging an open dialogue. The media, especially new types of media, are essential to the quality of communication and to relationship building with the citizens and their police.

KEY TERM

Investigative reporter 208   Social constructionism 205

DISCUSSION QUESTIONS

1. Describe the relationship between the media and the police? What are the pros and cons of the media relationship?
2. How are police depicted in entertainment and news media?
3. What is the War on Police and what is the media role?
4. What is meant by blurring the lines between entertainment and news with regard to police?
5. What are the pros and cons of social media?
6. What are the ways police use social media?

CHAPTER 12

COMMUNITY POLICING: THE MEN AND WOMEN IN UNIFORM

On March 16, 2020, 32-year-old officer Christopher Walsh, a four-year veteran of the Springfield (MO) Police Department, was shot and killed by a mass shooter. After reports of shootings across the city, officers responded to the location where the shooting suspect had crashed his vehicle into a convenience store and began shooting. Neither of the responding officers hesitated to enter the store to save people and stop the shooter despite knowing the danger. When they arrived on the scene, they were fired upon by the suspect. Officer Josiah Overton, 25, received non-life-threatening injuries and Officer Walsh was killed. After it was all over, five people were dead, including the shooter and Officer Walsh. Officer Walsh was the first Springfield officer to die in the line of duty since 1932, when six law enforcement officers were massacred.

Nearly 90 years apart, both events demonstrate the dangers of law enforcement but also the rarity when we think of the hundreds of thousands of police encounters that occur every day without incident. The unpredictability of law enforcement makes it one of the most dangerous occupations. Although there are other occupations with a higher fatality rate, on-the-job fatality rate increased in 2018, and unlike other occupations, it is mostly due to homicide. As noted in the previous chapter, traffic fatalities also figure prominently in the law enforcement death rate. One would think that the danger is enough to make a career in law enforcement less attractive; however, it is precisely the reason people are attracted to the profession. In the community policing era, law enforcement agencies must attract uniquely talented individuals who hold community service, rather than excitement and danger, as their motivation for seeking a law enforcement career.

Learning Objectives

1. Explain the role of police officers in the successes and failures of early community policing programs and lessons learned.

2. Discuss the subculture of policing: attributes, values, and resistance against community policing.

3. Describe the police officer recruitment, hiring, and training for community policing and problem solving.

4. Explain the need for diversity and education in police recruitment for community policing and problem solving.

INTRODUCTION

In this chapter, we will take an in-depth look at the men and women in uniform, including the attraction to the profession, the career journey from recruitment to retirement, and the ubiquity and influence of the police subculture. We will explore proposed changes in how police officers are recruited, selected, trained, and retained for community policing. To understand current personnel challenges, it is important to review lessons learned from earlier implementations of community policing programs. In the early years, there was little or no knowledge that community policing might require a different kind of police officer; and therefore, police leaders focused on altering organizational structure, policies, and practices in the adoption of community policing and problem-oriented policing. Although there were some successes, it was the failures that revealed what and where changes needed to be made. One of the biggest lessons was the finding that traditional values of police officers did not closely align with community policing ideals. The success of community policing, or its failure, ultimately depends on the purveyors of police services, the police officer.

Much research has focused on the uniformed patrol officer, including issues of job stress, job satisfaction, police personality, career milestones, decision-making, police subculture, and job performance, however, not necessarily within the context of community policing. It would be expected that within the context of community policing, there would be significant differences when comparing community police officers to traditionally selected and trained officers. Such research might inform police administrators which attributes and behaviors are favorable to community policing. Additionally, it would assist in the development of specific community policing and problem-solving training for veterans and new officers. Local police agencies have received help through federal funding and research grants; however, there has been little consistency in community policing adoption.

Renewed interest in community policing had to do with occurrences of high-profile, controversial police actions that put police at odds with the community, as evidenced by protests and riots. Altruistically, local police departments wanted to rebuild trust with their communities and forge productive relationships, but they also wanted to insulate themselves as best they could from legal liability by reducing the hostility and confrontations of police and citizens through smarter policing. In general, police administrators have been increasingly concerned about who they hire and how they are trained due to legal liability. Both legal costs and **vicarious liability** cause concern for cities and departments. Vicarious liability is a legal concept whereby the city, the department, and supervisors are held accountable and legally liable for their officers' decisions and actions. Lawsuits resulting from police actions can cost taxpayers millions in punitive damages. For example, three years after the now infamous beating of Rodney King, King was awarded $3.8 million and $1.7 million in attorney fees.[1] In 1991, the Los Angeles riots sparked by the acquittal of four officers in the Rodney King beating trial left 50 people dead, 2,000 or more injured, and an estimated one billion dollars in damages.[2] Public outrage and rioting following officer-involved shooting incidents take a toll on public safety and property. Officers and departments do have legal liability for their actions, so it is justified that departments are motivated to find ways to hire and train officers to solve problems without coercive, invasive, and violent means.

In 2015, President Barack Obama's Task Force on 21st Century Policing presented its final report proscribing six pillars of recommendations, one specifically recommending a commitment to community policing. Additionally, a focus group of experts held a forum on recruitment and hiring. Community Oriented Policing Services (COPS) and the Police Executive Research Foundation (PERF) challenged police leaders to revamp their recruitment and hiring process with the objective to select attributes that will forward community policing ideals.[3] Experts at the forum identified three concerns. First, police departments would benefit from hiring candidates who share the values and vision of the community. Second, the hiring process needs to be more efficient, with the goal of preventing qualified candidates from dropping out of the lengthy process. Finally, police departments should advance objectives that increase diversity and inclusiveness by hiring officers who reflect the communities they serve. The challenge for police leaders is to accomplish these objectives. What needs to be done is dependent upon officer acceptance or resistance to change. In order to bring about positive change, police leaders have to counter the resistance from police officers.

President Obama's report was not the first one that called for community policing. In 1965, President Lyndon Johnson formed a task force of criminal justice experts, in part, to examine the problems between police and the community. In 1967, the task force report recommended community policing to improve community relations. What followed was a rush to implement community policing, most of which were experimental, stand-alone programs. Over the next few decades, some of the programs were successful, while most were viewed as costly failures. Let's take a look at the lessons learned from early implementation of community policing.

LESSONS LEARNED FROM EARLY COMMUNITY POLICING IMPLEMENTATIONS

President Johnson's Commission on Law Enforcement and Administration of Justice called for sweeping changes in police personnel practices to achieve better community relations, especially in poor racial-ethnic minority communities and with juveniles. Although many of the suggestions were more wishful thinking than actionable, positive changes began in earnest. For example, two federal agencies were developed to support police reform, the formation of the office of Community Oriented Policing Services (COPS) and the Law Enforcement Assistance Administration (LEAA). The latter was the funding arm of the government to finance innovations in policing.[4] Many departments claimed to have community policing but did not. Certainly, going forward, there have been greater incentives and advantages to embrace community and problem-solving policing. Haphazard, albeit well-meaning, implementations of community-oriented policing followed this directive over the next few years; but it took many years of both successes and failures to realize inherent organizational and philosophical conflicts between community policing and professional or traditional policing.

It was not long before many researchers, administrators, police experts, and police officers viewed community policing as a dismal failure and costly experiment. Several problems threatened to derail sincere efforts to bring about positive change in policing. The sources of problems were both outside and inside the department. The crack epidemic, for one, hit communities in the 1980s and 1990s, putting pressure on law enforcement officials to return to traditional law and order, with its tough, military-like response to combat the violence. Second, the high turnover of police leadership resulted in a lack of continuity and commitment to new programs. Additionally, these programs were expensive to maintain. Many programs, created and implemented on seed money, contained no long-term funding plans, leaving departments with limited resources to continue innovative but costly programs. At the time, many departments did return to traditional policing. Finally, the biggest hurdle to the success of community policing was the officers themselves. Police officers had never bought into community policing, in part because the concept lacked clear definition, an explanation of practices and policies, and specific training. In other words, police officers were unclear as to what to do and how to do it. Community policing officers received no structure, training, or support to accomplish goals of building community relations and addressing quality of life issues. Supervisors, not trained in community policing principles, were at a loss as to how to provide support and supervision to these officers. Additionally, tensions between community policing and traditional officers caused serious internal conflicts in the department.

Split Force

Original iterations of community policing programs consisted of a few officers assigned to a specialized unit within the department, separate from the traditional motorized patrol unit. Officers assigned to the patrol unit continued their traditional law enforcement duties. Community policing officers, on the other hand, were expected to forge relationships and partnerships with community members. The department was thus divided into two types of responses and two types of officers, often with conflicting tactics, ideals, and outcomes. An illustrative story involved community police officers who talked to and shared ice cream cones with a couple of juveniles, with the goal of chipping away at the wall of distrust. Later that same day, patrol officers spotted the juveniles, stopped them, made them sit on the curb, took their names, searched them, and in the end, let them go with a warning to keep out of trouble. What impact do encounters such as these have on the juveniles and on the community? Which encounter left more of an impression?

The organizational division of traditional and community policing units created a split force with two differing and often conflicting mandates. The split force concept was used in many of the foot patrol experiments, for example, Newark and Flint. Community policing expanded the officer's role to include forging relationships and partnerships and to address signs of disorder. The community policing mandate not only expanded the role of officers, it put them at cross-purposes with the arrest and control function of professional policing. Patrol units had contempt and little respect for the community policing officers assigned to the specialized unit. The patrol officers viewed the community policing units as "not doing real police work" when tough enforcement was necessary.[5] Departments also made the mistake of assigning officers to the community policing unit whether they wanted to be assigned or not, or assigning recruits right out of the academy to the specialized role. Officers assigned against their will were unlikely to be committed to doing a good job. Additionally, new officers, fresh from the academy, were chomping at the bit to make arrests, to fight crime, and to "kick ass and take names"—a common theme among new recruits.

▶ **Photo 12.1** Positive interactions between the police and the community are important. How can split force policing both help and hinder that goal?

Multiple issues contributed to the unraveling of community policing in the early years; however, reform efforts were renewed in the wake of high-profile and controversial police incidents and practices. Community relations, police legitimacy, and public trust continued to be central to police reform in the 21st century. Many of the lessons learned from early implementations helped make current efforts more intentional and informed. For example, the problematic split force revealed the need for full engagement by the department. Attaining full engagement means getting all employees to buy into the reasons for the process as well as the desired outcomes of community policing. In any business, making organizational changes is difficult for employees, and police officers are no different. Police officers are resistant to changes that threaten their value system.[6] To encourage a shared vision and enthusiasm for community policing, police administrators had to counter officer resistance with positive information. In addition to providing clearer directives and training, they also needed to confront the strong subculture by challenging traditional police values with concepts of smarter policing.[7]

OFFICER-CENTERED CHALLENGES TO COMMUNITY POLICING

Challenges to successful implementation of community policing centers around the police officer. The example of community policing officers trying to build trust with juveniles while patrol officers treated them like criminals illustrates the problem of specialized community policing units rather than department-wide adoptions. The structure of separate units was problematic because it brought to light that community policing ideals ran counter to how police viewed themselves as crime fighters and law enforcers. That value system carries from one generation of officers to the next. It is intrinsic in how they are hired and trained. Newly hired officers, motivated by excitement and crime-fighting, are quickly socialized into the normative values of traditional policing. Such values are institutionalized and difficult to change.

Police Subculture

Community policing will only work if police officers buy into it.[8] The task of transforming policing is not an easy one. Administrators with lofty objectives to revamp, transform, and improve their departments must first address the resistance of the **police subculture**. The police subculture is defined as members of a close-knit group who share values and informal norms and differentiate themselves from the general public. This separation from the public breeds mistrust and suspicion of citizens and creates the us versus them mentality, a notion that fosters the belief that police officers are the good guys and everyone else is potentially bad. Fifty years ago, police scholars wrote about the police subculture, citing its focus on crime-fighting and its hostility toward the public.[9] This ethos influences police behavior and ensures absolute loyalty among police officers to one another, manifesting in a **code of silence**, often referred to as the blue wall of silence, meaning that police officers cover for each other in cases of wrongdoing. In addition to the shared suspicion, officers are focused on risk, danger, bravery, and law enforcement solutions to problems. The subculture suppresses individualistic identity and behavior. Interestingly, the recommendation to bring diversity into the force—for example, the hiring of women and people of color, with the hope that it will bring police and community closer—is likely to disappoint. Newcomers adapt and adopt the values of the subculture, or they are ostracized.[10] One researcher of police subculture instructs police reformers to have a solid understanding of the subculture before launching into transformative ventures (p. 364).[11]

The success or failure of community policing comes down to the purveyors of police services. A police department's investment in recruitment, hiring, and training is extremely important because a department's service is no better than those who perform the day-to-day tasks (p. 40).[12] Who is hired and how they are trained are key. The challenge, then, is to find individuals who have the capacity and desire to serve the community and to build relationships with its citizenry. However, despite the desire to incorporate community policing ideals, police officers also must be selected and trained for the critical aspects of their job. Police are not being asked to give up traditional police tactics in exchange for community policing but to enhance their service and capabilities. Tactical readiness and courage to head into dangerous situations remain a top priority. At the same time, there is a continued emphasis for police to work closely with the community, to form proactive productive partnerships, and to work smarter. Without public trust, legitimacy, and confidence, police will not be effective in any aspect of their job.[13]

THINK ABOUT IT: IS POLICE WORK REALLY THAT DANGEROUS?

The same thing that attracts some people to policing may be the exact thing that deters others, and this is why the recruitment, hiring, and selection process is critical. Historically, one of the reasons that police work was considered a man's job was the notion that it required physical prowess. Women were hired as policewomen, not to do the same job but to take less dangerous aspects, such as contact with juveniles. Today, men and women do the same job and therefore must be equally suited for its rigors. However, understanding what constitutes police work becomes very important for recruitment, hiring, and training. Whether policing is truly dangerous is a critical determinant to attracting the right kind of people. It would be disingenuous to say that policing is safe. In fact, law enforcement officers die every year and some years are worse than others. For example, according to the FBI, 55 law enforcement officers were feloniously killed in 2018, nine more than in 2017.[14] In 2020, 16 officers have died compared to 12 officers this time last year.[15] In 2018, 51 officers were killed accidentally, 34 of those in motor vehicle crashes. According to the Bureau of Labor Statistics for the same time period, loggers, fishers, pilots, and roofers topped this list of work-related fatalities.[16] However, those workers were not murdered. Police officers were murdered in the line of duty, with 51 of the 55 killed by a firearm, and four were killed with their own gun. Which shows that anytime a police officer is involved in an altercation, there is potential that the firearm will get into the wrong hands. Officers died while attempting to arrest, responding to robberies in progress, confronting mass shooters, pursuing a fleeing suspect, and in rare cases, being ambushed.

Shockingly, while 55 officers were killed in 2018, 159 officers took their own lives. Law enforcement officers are exposed to human tragedy, horrific accidents, and killings throughout their careers. According to one source, the rate of depression is five times higher for police and firefighters than for the civilian population.[17] Officers are unlikely to seek psychological support for fear of being thought of as weak. Police are more likely to kill themselves than be killed by someone else, yet more time is spent on supporting the belief that the danger they face comes from the public.

The law enforcement career is dangerous in many ways and takes unique people to fulfill the law enforcement mission of public safety. People are attracted to the danger, to the excitement, and to putting their life on the line. If recruiters play up the concept of community policing, will it change who is attracted to this profession?

Discussion Questions:

1. Is police work dangerous, and what makes it dangerous or not?
2. Can police officers be trained for the dangers and the community service?
3. What can law enforcement officials do to address the suicide rate?

THE HIRING PROCESS

A police department's investment in recruitment, hiring, and training is a priority because a department's service is no better than those who perform the day-to-day tasks (p. 40).[18] The selection and training of officers is key to organizational integrity and effectiveness. Although there are no universal standards for the hiring and training of police officers, departments across the United States have similar practices. Law enforcement agencies rely on local or state-level governing boards to regulate and set standards for the hiring and training of public safety professionals. For example, California utilizes a Commission on Peace Officer Standards and Training (POST) to assist with regulations and resources.[19] Existing recruitment, selection, hiring, and training processes have been in place for decades. Over the years, legal challenges to the equitability and job relevancy of prescreening tests for employment have been the impetus for change and improvement.

Some of these legal challenges have been based on discrimination of women and people of color that violate Title VII of the 1964 Civil Rights Act. For example, the claim is that physical agility tests have an adverse impact on female applicants and that written tests have an adverse impact on racial minorities. The prescreening tests must have job relatedness, the courts have ruled. Many of the early legal challenges have resulted in greater prescreening validity, meaning the tests do represent what officers do on the job. However, a lack of gender and racial-ethnic diversity in police departments to be an issue. Therefore, one objective is to attract a diverse applicant pool and design tests that do not eliminate people of color and women. Another objective is to develop prescreening tests that are able to identify personality traits and abilities that align with community policing objectives.

The hiring process is lengthy, often four to six months from application to appointment. During that period, potential candidates are subjected to rigorous prescreening. Although some variations exist, preemployment screening comprises the application, physical agility test, written examination, interview, background investigation, psychological assessment, polygraph, medical examination, and final interview. Each step of the process weeds out individuals who fail to meet certain criteria. Successful candidates must be physically and mentally capable of the job, have no criminal record, possess high standards of integrity and honesty, have a good work history, and be psychologically sound. Prescreening is designed with the least costly aspects of the process in the front end, with the goal of weeding out large blocks of unqualified individuals (e.g., the physical and written tests), and ending with the more costly tests, such as the psychological, polygraph, medical examination, and the time-consuming the background investigation. The first step of the hiring process is the recruitment phase. Recruitment is key to attracting the right people to the profession.

Police Recruitment

The hiring process begins with recruitment efforts. In some cases, an interested person seeks the career on their own. Most departments cannot rely on self-selection to accomplish their hiring need, however. Large departments have continuous recruitment and testing events scheduled throughout the year, whereas a smaller agency only recruits and hires as needed. Hiring advertisements attract thousands of applicants for only a few openings. The "weeding out" or "selecting in" process begins at the recruitment stage. Recruitment may be the single most important way of selecting a new type of officer. Transformation is a slow process, and the first step begins with intentional and focused recruitment strategies.

Although police department administrators find recruitment to be very challenging, they are making it their top priority. A major recommendation of the Final Report of the President's Task Force on 21st Century Policing is the need to diversify police forces by race, gender, language, life experience, and cultural background.[20] In order to promote the principles and practices of community policing and problem solving, police administrators select police officer candidates who reflect the values of the community and possess innovative problem-solving ability.[21] Research supports the notion that increasing the number of racial-ethnic minority officers may go a long way to improving community relations.[22] Additionally, the report recommends placing value on educational achievements when making hiring decisions and providing officers with financial incentives to continue their education. Recruitment campaigns need to be designed with these objectives in mind.

Law enforcement is not a field that people wander into; it is one where people are attracted to the career for specific reasons. Self-selection has been the primary path to a law enforcement career. In the past, departments could rely on a continuous pool of interested and qualified people from local communities. However, the number of qualified, interested people has declined. First, qualified candidates have been difficult to find. Casual drug use, lack of positive work history or education, excessive tattoos and piercings, illegal downloading of music and other media as well as inappropriate social media usage—all common among young people today—are automatic disqualifiers for many departments. Second, people are less attracted to police work because of negative publicity about police and the job. Some attribute this decline to the **Ferguson effect**, which argues that a loss of legitimacy in policing has soured many on the career itself, especially among racial minorities.[23] The Ferguson effect is the term used to address the impact on racial-ethnic minority communities that followed the shooting death of Michael Brown, an African American man, by a White police officer Darren Wilson in Ferguson, Missouri, in 2014. Following the death, protesters took to the streets for several days. These protests re-erupted in November 2014 after the police officers responsible for the death of Brown were acquitted of all charges. While some research has debunked the validity of the Ferguson effect, confidence in police among people of color has declined.[24] The career does appear to be getting more dangerous. For example, the ambush killing of five police officers in Dallas, Texas, and three in Baton Rouge, Louisiana, in 2016, made people disinclined to seek a police career.[25] Good pay and benefits, often a draw to police work, are hardly compensatory to the danger of policing. In addition to a drop of interested persons, current positive economic conditions, such as low unemployment rates, mean police departments are in competition with other employers for qualified individuals. The same qualities that police management look for in candidates are the same ones that other employers want as well. Because the hiring process for police positions is invasive and lengthy, many highly qualified people are inclined to take a swifter path to a different career or are quickly snapped up by other agencies or businesses.

Ultimately, recruitment campaigns must make the job look appealing, spark an interest, or feed an interest that already exists. Today, recruiters need to focus on attracting a new breed of officer by using messaging that highlights community relations but avoids slogans that reinforce traditional views of policing. Recruitment efforts have to be well informed and have greater outreach. First, recruiters must design the message, and second, they have to know where, to whom, and how to deliver the message.

The Recruitment Message

If departments do want to diversify their forces, it would be important to find out how to attract a diverse applicant pool. The message or slogan has to resonate with people and attract them. For example, the first step in designing an effective message is learning what motivations people have for entering policing, thus making a connection between the interest and the career. Recruitment ads from earlier decades promoted excitement, using slogans such as "are you looking for an exciting career" or egocentric messaging such as "something for whatever you're into" or "it's not just a job, it's a career." Most research on motivations, outdated and based on White male perspectives, had found that recruits were influenced by helping others, job security, enforcing the laws, and fighting crime.[26] A 2008

▶ **Photo 12.2** Old police pamphlet with an African-American.

study of San Diego's Police Department police officer applicants found that a majority cited wanting to help others and/or the community, certainly a positive sign for police departments with community policing programs.[27] Another study examining motivations of New York Police Department recruits, with a focus on gender, race, and ethnicity, found similar results, that helping people was a top reason. This was true for both people of color and women but was also true for White males. Perhaps fluctuations in career motivations are generational rather than attributed by demographics such as race and gender.

Who, What, Where of Recruitment. Police recruiters are proactively targeting women, people of color, and educated officers. As we discussed, the messaging to attract these groups is critical but so too is the location of where to find them. Many recruitment strategies include a focus on targeted groups. Recruiters have become creative in their efforts by attending sporting events, job fairs, parks, beaches, clubs, places of worship, colleges, and high schools. They are also forming partnerships with other agencies to cut costs and to share efforts. Recruiters also use social media as a way to advertise. Additionally, recruiters are traveling or advertising in other states. Recruiters receive training on how to attract individuals, how to get them in the door, and how to keep applicants committed to the process ahead. Many recruiters are also trained to administer the written tests to expedite the process.

Recruitment of Women and People of Color

A police recruiter is unlikely to entice people who do not already have an interest in policing. Tailored and targeted recruitment efforts are needed to entice those who never contemplated a police career, especially under-represented groups, for example, Black Americans, Hispanic Americans, Asian Americans, and women. A few problems hinder the effort to target racial-ethnic minorities and women, however. First, some Black and Hispanic people tend to view police and policing negatively, perceptions that range from mild dislike to hatred. Race continues to be a major predictor of attitudes toward the police.[28] Distrust in police is a significant deterrent from desiring to be one of them. Second, because the police profession is most often associated with male traits of physical prowess and bravery, women doubt their fitness for the job. Women, and some men, for that matter, lack the self-efficacy and social support for pursuing a police career.[29] Third, others may not be attracted to the excitement of policing, unaware that other aspects would interest them. Asian Americans, for example, are underrepresented in policing, with studies showing that they tend to be more interested in entrepreneurial endeavors, science, business, and information systems rather than policing.[30]

To maximize a diverse pool of applicants, the objective is not only about finding the correct message, it is also about tailoring the message to specific individuals or groups, by countering negative beliefs, self-doubts, and unfamiliarity with the job. The message should reflect the variety of duties performed by police and highlight the rewards of a law enforcement career. Ironically, and pertinent to this chapter's focus on recruitment for the community policing model, the very reasons individuals assume that they do not have traits of an ideal police officer—e.g., bravery, toughness, aggressiveness, machoism—are exactly the reasons they should apply.

Recruitment of Educated Officers

According to the 2015 task force on recruitment and hiring, one of the objectives was to diversify the police force and to increase education incentives for officers. With that in mind, college settings are ideal for recruitment campaigns because potential candidates are older, diverse, are earning a college degree. However, only about 15% of agencies across the country require some college and only 1% require a four-year degree.[31] There has been much debate over whether officers need a college education to perform well. Others argue that because community policing, problem-oriented policing, and intelligence-based policing require police to engage in complex data-driven decision-making, education is a necessity. One would assume that college education would enhance decision-making and critical thinking skills and improve the professional status of police. A few studies have found that, when compared to their less educated peers, college-educated officers have more positive attitudes toward citizens and community policing policies.[32] Although some research findings are mixed as to the relation of education to performance, professional attitude, reliance on use of force, propensity to embrace cultural, ethnic, and racial differences, and community awareness, every indication is that college-educated officers do perform better in the community policing era. Therefore, requiring higher educational levels upon entry into the profession is desirable for community policing and problem-oriented policing objectives.

While a college education is a desirable qualification, especially for the professional image and status, some fear that such a requirement could have an adverse impact on recruitment of women and people of color.[33] Regarding female representation, agencies with a higher education requirement had a higher percentage of women.[34] There is some evidence that requiring a four-year degree, in fact, does lower the number of racial-ethnic minority applicants.[35] Some departments are returning to a lower education requirement to ensure a more diverse pool. For example, the New Orleans Police Department lowered its requirement to a high school education equivalent in order to increase its pool of applicants, while other cities have raised the level to a four-year degree and were able to meet their recruitment goals.[36] Women and people of color have greater access to college than in years past; however, some disparities continue to exist. In terms of recruitment, college settings are ideal for fulfilling goals of diversity and education. The expectation is that smarter policing demands smarter police, or at least, college-educated ones.

Hiring a new type of officer with a predilection for working with the community is a good start; however, most departments hire a few new officers but have a large force of seasoned officers. The question is how to reshape the mindsets of traditionally hired and trained officers. Draconian methods such as starting afresh with new officers have been tried in recent times. For example, a police leader in Camden, New Jersey, replaced nearly the entire force so that, enthusiastic, community-minded officers could pave the way for improved community relations. Replacing veteran officers may not be realistic for most cities and agencies to do, however. Training existing officers in the principles of community policing and hiring new officers who are already motivated to do that kind of work may be the only pathway available to agencies.

POLICE TRAINING

Recruitment and hiring with community policing objectives in mind move the needle in that direction; however, basic, field, and in-service training must include those same objectives or the socialization process will undo well-intentioned practices. Once applicants successfully move through the hiring process, they attend an academy for four to six months. After successfully graduating from the academy, police recruits are assigned to a training officer for field training until the new officer is ready to be on their own. The third kind of training is in-service training, where all police officers learn new skills or practice old skills. Throughout their career, officers engage in training for their perishable skills, those skills that need to be practiced in order for officers to remain proficient (e.g., firearms qualifications, arrest and control tactics, cardiopulmonary resuscitation CPR, and first aid).

During the basic and field training, the focus is on acquiring technical skills as well as gaining legal, policy, and procedural knowledge. During field training, the rookie is placed with a seasoned officer in the field to observe and learn the job. Field training is mostly a continuation of the "boot camp" mentality of the police academy—a cross between learning police tactics and being yelled at. Training, therefore, becomes more a rite of passage into the police world and less about preparation for problem solving, communicating, building relationships and partnerships, and thinking creatively. In the paramilitary structure of policing, such freethinking is highly discouraged. Similar to military boot camp, officers experience extremes of conformity and leadership, submission to and respect for authority, physical intimidation, and domination. Much of the process is reinforcement of masculinity, bravery, and resiliency. What is not part of that experience is learning how to use communication, trust, compassion, creativity, and problem solving. These types of skills are referred to as soft skills and they would work well in a policing model where officers must collaborate, communicate, and facilitate innovative problem

THINK ABOUT IT: INTEGRITY TESTING AND POLICE CORRUPTION

Police departments have a responsibility to protect the communities they serve, even from their own police officers. Police misconduct is a betrayal of public trust and can have lasting negative impact on community relations. Misconduct includes bribery, fabrication of evidence, excessive force, biased enforcement, and other unacceptable behaviors. In part, departments must ensure that their police officers demonstrate integrity and continue to be well suited for the job throughout the career, not just at the beginning. The intensive prescreening of police officer candidates somewhat guarantees that the best of the best are hired; however, it does not eliminate the possibility of misconduct in the future. Are these preemployment screening tests adequate to determine the likelihood of future misconduct by officers?

Police departments throughout the country utilize similar prescreening tools to assess an applicant's appropriateness for police officer positions. As part of the battery of prescreening examinations, candidates are subjected to personality tests. Most agencies use the Minnesota Multiphasic Personality Inventory (MMPI). Other popular tests are the California Personality Inventory (CPI), the 16 Personality Factor Questionnaire (16PF), and the Inwald Personality Inventory (IPI).[37] The personality tests have proven to be predictive and legally defensible in selecting-in traits that will identify successful candidates. These survey tests are conducted in conjunction with a psychological interview to assist with interpretation of the results. Selecting-out or weeding-out candidates is done through assessment of risk, maladjustment, or psychopathology that may surface in this high-stress environment. Because the results must be interpreted by a clinician, they are costly.[38]

While these tools are used to assess fitness for the position, they are not specifically used to identify propensity toward future misconduct. Because misconduct is an ongoing issue throughout the career, one might wonder if the tests could, in fact, predict an officer's likelihood of engaging in misconduct in the future. One study examined personality scales conducted during the preemployment phase against officer's performance post-hire. Examining current records of police officers, both violators and nonviolators, it was found that violator's item responses in their earlier personality tests did indicate they were more likely to have engaged in delinquent acts in other settings (e.g., high school). Nonviolators were more concerned, according to their original item responses, about behaving conscientiously and maintaining stable interpersonal relationships with friends, family, coworkers, and supervisors (p. 43).[39] However, agencies base their hiring decisions on pass or fail recommendations by the psychologist's professional judgment. Most of these test results are not retained by the hiring agency and therefore not relevant to prevent or respond to current misconduct. Additionally, the preemployment testing does not account for changes to officer's personalities over time.

Many agencies are looking for means to better evaluate police officers throughout their career, such as integrity testing. Although integrity testing is not new, postemployment assessments have not been subjected to the same validity measures applied to preemployment measures. One early case of misconduct resulted in the use of integrity undercover investigations.[40] The Knapp Commission uncovered department-wide corruption in the New York City Police Department. Simulated corruption opportunities led to numerous prosecutions.[41] While this type of corruption warranted such tactics, what police departments are seeking are screening tests, similar to the survey methods of preemployment screening, with the objective of prevention. Today's integrity testing is used to assess "counterproductive work habits" (CWH), for example, absenteeism, terminations, drug abuse, arrests, violent behaviors, thefts, driving under influence (DUIs), and other behaviors. In general, it does measure some of the same behaviors that the preemployment personality tests and psychological interviews measured but are less costly without the psychological interview.[42] Integrity testing may eventually become commonplace, but part of the problem for administrators concerns the legality of such measures.

Continued

(Continued)

Whether police officers should be subjected to regular and routine testing throughout their career has been hotly debated. Police unions have fiercely protected police officers from such testing, citing rights of privacy, for example. Random drug and alcohol screening has been implemented successfully by some agencies but usually as a means to support officers by providing resources. In regard to integrity testing, how such testing and results will be applied to personnel decisions (e.g., suspension or termination) is the concern.

Discussion Questions:

1. Should officers be subjected to routine psychological screening or integrating testing throughout their career? Why or why not?

2. Why would integrity testing of police officers be good for community policing?

3. Why would an officer object to being tested later in their career?

solving in partnership with the community. When incorporating community policing into their departments, police administrators should focus on the importance of training and hiring with soft skills in mind.

Soft Skills

Research examining community police officers has identified the need for officers to possess **soft skills** (e.g., communication, problem solving, leadership, flexibility, collaboration, and creativity).[43] Much has been written about the importance of soft skills and emotional intelligence in business environments and, only recently, law enforcement.[44] **Emotional intelligence (EI)**, sometimes referred to as emotional quotient (EQ), is defined as an individual's ability to recognize and regulate the emotions of self and others toward successful environmental adaption.[45] People with high EQ usually make great leaders and team players because of their ability to understand, empathize, and connect with the people around them, such as in police work.[46] Ironically, it was the U.S. Army Command that first used the term at a Soft Skills Training Conference in 1972, speaking of skills that were not technical in nature but apparently critical to workplace productivity.[47] Police officers are technically trained and equipped to deal with situations in the field; however, most of their work involves interacting with people through communication and not necessarily using physical means, although they must be prepared for that as well. Domestic abuse situations, for example, ripe with emotion and potential for violence, require careful handling by police. Officers engage soft skills and EI by controlling and managing the emotions of parties involved, including their own.[48] Police often learn soft skills on the job, after years of doing the job; they hone their communication and negotiation skills, for example. Some individuals inherently have more soft skills, while others acquire such skills on the job. Women, once thought to be too soft and emotional for police work, often possess traits more conducive to community policing than do their male counterparts. Their attributes include trust, compassion, empathy, and cooperation.[49] Although both male and female police officers use these skills every day, it is not valued as much as a good arrest. What officers

CASE STUDY 12.1: MAN LANDS JOB AFTER OFFICER GIVES HIM A RIDE TO INTERVIEW INSTEAD OF A TICKET

Ka'Shawn Baldwin has a new job, thanks to a Cahokia, Illinois, police officer who gave him a ride to the interview. Police officer Roger Gemoules made a traffic stop on Baldwin for an expired license plate. The situation got worse because Baldwin did not have a valid driver's license. Later Baldwin told the NBC reporter, "I was nervous. I was thinking, I was just going to get some more tickets and have some more fines that I really can't afford to pay." According to Officer Gemoules, the routine procedure would be to issue a ticket, tow and impound the vehicle, and take the person to jail. Baldwin told the officer he had to get to a job interview, and that he had borrowed the car but did not know it had expired plates. The officer informed him that he could not drive the car but agreed to follow him back to his driveway to park it and then to Baldwin's surprise, the officer gave him a ride to the job interview.

Officer Gemoules thought Baldwin was a nice young man who really needed the job. The officer later told the reporter, "I knew if I gave him a bunch of tickets and towed his car, it would be tough to recover from." Baldwin posted the story on Facebook and messaged the officer to thank him later and to tell him he got the job. The story went viral and Officer Gemoules was lauded by both the police department and the city's mayor. The assistant to the mayor stated, "On behalf of Mayor Curtis McCall Jr., I would like to thank Officer Gemoules for showing compassion and being a great example of how community-oriented policing actually works." Baldwin, who had little interaction with police before, said, "I never looked at police in a negative manner or in a positive manner. I never looked at them in any manner. But this just made me give them more respect than I did before."[50]

believe to be important about their jobs and what they actually do in the field are not the same thing. The police value system celebrates catching the bad guy and not necessarily positively impacting the life of an at-risk youth. In the community policing era, police officers need to have, use, and value soft skills.

Whether soft skills or EQ are teachable has been debated. In fact, some research indicates that one must have such traits in order to be amenable to training as well as achieving successful outcomes of the training.[51] One scholar noted that soft skills are a misnomer because they are the hardest skills to learn, which means that police administrators need to select individuals who possess such skills coming into the occupation rather than attempting to train them afterward.[52] Some employers, including military and police agencies, are using assessment tools during the preemployment screen process to identify candidates who have soft skills.[53] Police hiring committees and administrators also have a hand in selecting certain traits over others. Law enforcement officials, for example, are looking for able-bodied, clean-cut men and women, who embody mental and physical prowess, especially physical competency. In general, they are not looking for, nor are they able to identify "soft skills" during the hiring process. The hiring process does include interviews; however, the results are pass/fail and not necessarily rating competencies in communication or other soft skills (see Case Study 12.1).

SUMMARY

The transformation process from traditional to community policing calls for changes in organizational structure, officer attitudes and behavior, and culture. In part, these things need to change because traditional values would not be conducive to community policing principles. Occupational values in the traditional policing model include emphasis on distrust of citizens (us versus them mentality), crime control over service, and reliance on aggressive police tactics. One study found that officers with traditional views use higher levels of force than do community police officers.[54] While it is not clear that an objective of community policing is to reduce use of force, when force is used, police legitimacy is at risk. Therefore, the incentive for police leaders and city officials is to ensure a level of commitment to the citizenry and not to the strong cultural values of police officers.

Clearly, police departments have legitimate concerns over whether new and veteran police officers will be amenable, trainable, and successful in a community-policing model. On a recent television show called *Live PD*—which follows officers in the field of action—officers were asked what they liked about police work. One officer responded, "Driving fast and catching the bad guys." The other officers nodded and laughed. This is a simple answer that has many connotations, and perhaps, invokes fear in the hearts of police administrators. The reality is that people are attracted to aspects of policing that are about excitement and catching the bad guy. And, unfortunately, the inherent power and authority of the badge attract all types of people, but only certain people should wear the badge. When it comes to implementing changes, top-down approaches are rarely well received by officers. Successful adoption of community policing requires intentional recruitment, hiring, and training practices that select-in traits and weed out others. Alternatively, police departments should not eliminate those individuals who are motivated by excitement and bravery but elevate a desire to work with the community over other reasons.

Transition from the traditional or professional era of policing to the community-oriented policing has been, and continues to be, a long slow process, through many decades and trials. The most important aspect of police reform is the police officer. Community policing and problem-solving policing demand a new breed of police officer. In 2015, President Obama spoke about community policing and about the wonderful things that were going on in Camden, New Jersey, with its community policing program. He recognized that officers needed to be brave and compassionate at the same time. "Now, to be a police officer takes a special kind of courage. . . . It takes a special kind of courage to run towards danger, to be a person that residents turn to when they're most desperate. And when you match courage with compassion, with care and understanding of the community—like we've seen here in Camden—some really outstanding things can begin to happen."[55]

KEY TERMS

Code of silence 228
Emotional intelligence (EI) 236
Ferguson effect 231

Police subculture 228
Soft skills 236
Us versus them mentality 228

Vicarious liability 224

DISCUSSION QUESTIONS

1. What are traits of the police subculture?
2. What were some of the problems and lessons learned from early implementations of community policing?
3. How would you design a recruitment campaign to attract individuals for community policing and problem-solving policing?
4. How would you target certain groups to apply for police officer careers?
5. How would you train officers for community policing and problem solving?

CHAPTER 13

COMMUNITY POLICING AND POLICE ADMINISTRATION

Learning Objectives

1. Explain the leadership role of the top cop.
2. Debate the use of incentives to adopt community policing.
3. Discuss the impediments to the adoption of community policing.
4. Explain why crime rates and budgets are difficult for innovation and why departments resort to traditional policing in the face of economic downturn.

William Bratton has had a long, illustrious law enforcement career in several large police agencies. To recount his career here would fill the pages of this book. He was recruited specifically for his leadership and **change management** ability. Although there is a lot more to it, a simple definition of change management is the ability of leadership to prepare and support employees through organizational change. In New York City, Mayor Rudy Giuliani selected Bratton to serve as police commissioner to address high crime, fear of crime, and social disorganization. Following his first term as commissioner, he became the chief of police for the Los Angeles Police Department, serving in that position for seven years. In Los Angeles, officials looked to Bratton to restore a loss of police legitimacy following the Rodney King beating, the O. J. Simpson trial, and other negative events that had plagued the city and department. Bratton was a proponent of data-driven policing instead of traditional reactive policing. He employed the broken windows theory as the underlying foundation for addressing quality of life concerns in the community, looking at reducing signs of disorder, with the goal of decreasing crime. In both New York and Los Angeles, Bratton's goals were twofold: promoting community policing and "putting cops on the dots."[1] Putting cops on the dots referred to using statistics to identify high crime areas and subsequently deploying police to those areas.

Bratton employed data-driven policing with aspects of community policing. According to his unique view of community policing, accountability should be placed more on precinct commanders who are educated, older, and with more years of service, and less on the beat officers, who are young, less educated, and with fewer years of service. The precinct commanders should meet with the citizenry and discuss their concerns about the community. Using this model, police discovered that citizens were not interested in statistics but about what they witnessed happening in their neighborhoods, specifically signs of disorder. This knowledge, combined with the use of **Compstat**, determined where line officers had the most potential to improve the quality of life by addressing signs of disorder and deterring crime in the community. Compstat, which literally means computer statistics, is an analytical managerial accountability system developed by Bratton and his command staff during his time as commissioner for New York City to better understand crime patterns and to apply this knowledge to solutions.[2]

During the years that he served in New York and California, his strategies significantly reduced crime. Bratton believed that much of that success can be attributed to policing, especially the application of Compstat, hot spot policing, broken windows, and community policing. Academic researchers, on the other hand, debate the ability of police to affect crime rates, citing other factors that contribute to the increase or decrease of crime.[3] Nonetheless, crime rates did decrease and relationships with the community improved in two major cities under Bratton's leadership.[4] Visionary leadership is key to reform in any organization, especially when that organization has a long history of institutionalized values and norms.

INTRODUCTION

In this chapter, we examine the role of the police administration in program innovation, the motivations to adopt community policing, impediments to successful reform, and next steps for police leaders. Beginning with a description of the duties and responsibilities of the top cop, we examine the decision-maker's role in the implementation of innovative programming, including community policing, problem-solving policing, and intelligence-led policing. The career path of the top cop (chief of police or sheriff) is fraught with potholes, peaks, and valleys. A police chief is a political and public figure who represents the men and women in uniform who serve the community. The top cop is subject to the whims of other elected officials who put forth their own vision regarding the needs of the community. Additionally, the public looks to the top cop to provide a safe environment. The top cop also is subject to the political power of the police union, who at times make it hard for changes to occur. The top cop also has to ensure welfare of their troops. They must balance the department's needs against all the other entities who demand equal attention.

We will discuss the motivations of police leaders to adopt community policing. Three inducement categories are identified: coercive pressure, crime and social disorganization, and funding opportunities. Coercive pressure comes from outside forces such as political pressures, court decisions, and legislative mandates that require specific or general actions to be taken by law enforcement officials. Mandated changes, for example, may stem from cases involving officer misconduct, such as incidents of brutality against people of color. The second category of inducements has to do with crime and social disorganization. In some jurisdictions, high crime rates and social disorganization mean that something is not working. Historically, police have marketed their services to the public as the default crime experts, and therefore, the public looks to the police to resolve criminal matters. The public expects the police to do something about rising crime, fear of crime, or signs of disorder. Furthermore, the public believe that they have no role in solving crime. The last category of inducements to adopt community policing is the enticement of money. Most police agencies have limited budgets and are stretched thin when it comes to operational costs. Smart leaders look for sources of funding to supplement their budgets. Millions of dollars are available through state and federal grants allocated for startup programs and for the hiring and training of police officers for community policing.

Next, we will look at obstacles to reform. The impediments to successful implementation include the organizational structure and culture of policing, officer resistance, crime, and budgetary constraints. The paramilitary organizational structure, which is the foundation of American policing, is not conducive to a flattening of the command structure, a requirement for creative problem solving at the line officer level. Although officer resistance has been discussed in other chapters, it is worth revisiting here because it is often cited as the most common barrier to change management. Police leaders may not be well versed in change management strategies that deal with officer resistance and helping personnel be ready for change. Leading an organization is difficult but leading an organization through change has significant challenges.

Ironically, two additional impediments that will be discussed were also incentives for new program implementation. For example, high crime, drugs, gangs, and violence inspired police leaders to look toward community policing for help, but they were quick to abandon such programs in favor of traditional methods of policing during tough times. Budgetary constraints can severely hamper innovation. Many agencies, especially in low-income, high-crime communities, are financially unable to meet routine operating costs, let alone the costs and energy of innovation. Although special state and federal funding was mentioned as an inducement for adopting community policing, the costs of innovation also an impediment to successful implementation. Even when departments are awarded grants, police leaders often fail to plan for long-term costs when grant money runs out. See Case Study 13.1 to understand the firing of officers for misconduct.

TOP COP: WHO'S THE BOSS?

The job of chief is very complex and multifaceted. The police chief plays an important role in police reform. Primarily, the police chief's desire for change is the initial motivation to make changes. The police chief may see the benefits of working with the community to resolve community issues and to improve the quality of life for the residents. Although not directly responsible for the development of community relationships, the leadership promotes such partnerships through policies and procedures that promote positive interactions between police and the communities they serve. Effective leadership balances the needs of the community with those of the department. As in the example of the Phoenix, Arizona, a police chief, the top cop must walk a fine line to appease both the community and the officers. Ineffective leadership can result in a **vote of no confidence**, whereby the chief can be removed from office, which is a very bad thing for their career.[5] A vote of no confidence is a vote by a majority, in this case through the union, that reflects nonsupport for the actions or decisions of a leader. Such a vote is an indication that the officers feel the chief is not qualified or fit to continue as chief. While it looks bad, the vote may not succeed in removing the as chief, but it puts him or her on notice that the troops are not supportive.

CASE STUDY 13.1: PHOENIX, ARIZONA, POLICE CHIEF FIRES OFFICERS

Phoenix, Arizona, police chief Jeri Williams spoke during a press conference regarding her decision to fire officers for misconduct. One officer was fired following a cell phone video that went viral, which depicted the officer pointing a gun at a shoplifting suspect, the suspect's pregnant girlfriend, and two small children. Unrelated to that incident, a few officers were either fired or reassigned after discovery of the officers' discriminatory and hateful Facebook posts. The officers' posts were uncovered during an internal investigation using data from The Plain View Project. Started by a group of attorneys in Philadelphia, The Plain View Project is a database of current and former police officers' Facebook posts that appear to endorse violence, racism, and bigotry. The database is available to the public.[6] Chief Williams was horrified when she read the posts. Williams stated, "No chief ever wants to discuss discipline like this in a public format. I expect my officers to be respectful to be professional, to be courteous. And that is not what happened in (these) cases."[7] She noted that the Facebook posts were deeply offensive. Chief Williams is concerned about the impact of those posts on the public. "As a public servant, we wear this badge as a symbol of our commitment to a higher standard, one that won't erode the trust of those who serve or tarnish the pride that is involved with being a Phoenix police officer."[8] As the top cop of a large city police department with over 2,900 sworn officers serving 1.6 million people, what Chief Williams fears is the erosion of the community relationships they have worked so hard to build.[9] The story is not over yet, as the firings are creating turmoil among the officers and city officials. While some support the chief's decision, others say that she is caving to political pressures. Police union leaders vow to call for a no-confidence vote against the chief, after receiving nearly 300 phone calls and emails about Williams's decision to fire the officers. The president of the Phoenix Law Enforcement Association (PLEA) Britt London asserts that Chief Williams has lost the department.[10]

Discussion Questions:

1. Should officers' Facebook posts be grounds for discipline or termination by administration?

2. Should a police chief make public statements in the press about personnel matters? Why or why not?

Top Cop Responsibilities

The top cop has numerous responsibilities, including oversight of personnel matters, such as hiring and training; managing fiscal and resource allocation; directing operations, including communications and deployment; writing and executing policies, practices, and procedures; and ultimately determining the course of the agency in terms of its mission and vision. The chief delegates much of the operational decision-making to the command staff who provide supervisory oversight to various units within the organization. Police agencies can be quite large, requiring numerous midlevel commanders, while others can be small, with very few higher-ranking personnel. Managing a complex organization is difficult and full of challenges, especially when implementing change. As William Bratton noted, "You have to motivate other leaders in the organization to share your vision and sense of urgency."[11]

Top Cop and Elected Officials. In addition to the command staff within the organization, the top administrator works closely with elected officials outside of the organization. Police chiefs are in highly political positions, meaning that they must answer to the community they serve, elected officials, personnel within the agency, partner agencies (especially in jurisdictions where county and city boundaries are tangential), and the media.[12] Former New York police commissioner William Bratton discovered the hard way that even at the top, decisions must be collaborative with other officials. Referring to his falling out with Mayor Rudy Giuliani, Bratton said, "The mistake I made with the mayor was I didn't stay in his headlights—I didn't stay close enough to him and to his vision."[13] There are political pressures on police leaders to do something about crime. Elected officials, for example, campaign on a strong law and order platform. The push to take law enforcement agencies in the direction of community policing is popular and has had strong bipartisan appeal at the local and national levels.[14] Even if there was not a strong political incentive for change, without the support of elected officials and other partners, transforming a department from traditional to community policing would be difficult to manage. The police chief must work within and outside the agency to bring about reform. Elected officials and top law enforcement administrators alike must respond to public demands for change.

Top Cop as a Public Figure and Community Leader. The top cop is both a political and public figure who represents the department, and by extension, the jurisdiction the department serves, whether it be city, state, federal, or tribal. It is not unusual to see police leadership, either the chief or spokesperson, appear in the news media, especially when something of a critical nature occurs. Despite strong leadership, well-trained officers, and solid policies, one poor decision can overwhelm and undermine a department's standing in the community.[15] Police leaders must get out in front of the story, and control and manage the information, before the story gets legs of its own. An important part of the equation is whether police departments have a reservoir of community trust they can count on during difficult situations. The time to have a first meeting with the community is not when citizens are already angry about a controversial incident. Police officials cannot afford to be silent. The days of hiding behind a wall of silence are long gone. In order to maintain legitimacy with the public, officials must be strategically transparent with their communications. In their public relations role, they may defend officers' actions or provide justification for decisions that were made at the time. In some cases, they apologize to the public, with the goal of promoting positive community relations, or averting disastrous consequences. An apology may not appease the public, especially if contentious relations already exist.

Top Cop and the Police Union. One aspect that has received little attention when it comes to the management of a police organization is the influence of the police union on the ability of the top cop to execute decisions and make changes. The relationship between leadership and the union is often contentious. At one point, the International Association of Chiefs of Police (IACP) declared unions to be adversaries of police leadership.[16] The police union wields substantial political and legal power in protecting working conditions of officers, which can include creating roadblocks for administrators in their reform efforts. Following the turbulent time of the 1960s, the call for police

Photo 13.1 Seattle police chief Carmen Best takes tough questions from reporters at the scene of a shooting that left one person dead and seven injured, including a child, in downtown Seattle, Washington, in 2020.

reform triggered the growth and power of the police unions to protect officers from fervent reformers who rode roughshod over police officer rights. The political strength and power of the union to provide advocacy for officers also impeded reform objectives. While unions protect officers from arbitrary discipline, they also prevent police managers from implementing change in working conditions without bargaining agreements. Police unions have interfered with implementation of civilian review boards, officer name badges, and use of dash cams and body cameras—all[17] in the name of protecting police officers from harsh administrative oversight and public scrutiny.[18]

Police Officer Rights. Police managers must be cautious when talking to the public about officers' alleged wrongdoings, even after the matter has been investigated. As members of collective bargaining units, police officers are protected from overzealous administrators. According to the Police Officer Bill of Rights, officers are afforded certain legal protections. Like the Miranda warning, accused officers have the right to be silent, right to representation, and cannot be compelled to answer questions under threat of being fired.[19] For example, an officer accused of misconduct may have their identity shielded from the public, or in some cases, disciplinary records erased.[20] Unions have been blamed for shielding officers from accountability and discipline even in the most egregious cases. In general, pending the results of an investigation, involved officers are placed on paid or unpaid administrative leave. For police leadership, these cases present unique challenges, especially where it concerns threats to police legitimacy. Of equal concern for leaders, is the morale of the department. When personnel matters are mishandled, there can be irreparable damage to

THINK ABOUT IT: POLICE SAFETY AND THE CORONAVIRUS (COVID-19) CRISIS

Seattle police chief Carmen Best, who was one of the first police chiefs to face the Covid-19 crisis, and several other police leaders laid out best practices to other police leaders about how to keep officers safe during a time of crisis within their communities. The most important point pertained to communication and the effective sharing of information. The police leaders identified focused on best practices.

First, establish a consistent line of communication with police officers using an emergency operations center and the police operations center. In addition, the police leaders should use videos and a central website. The website must be consistently updated. Police officers should also receive daily emails with updates.

Second, reduce the number of people in the facility. For example, the Seattle Police Department split its command staff—six are on and six are off. The medical specialists of the federal government predicted that 30% to 60% of all people would get infected. That, of course, would also hold true for police officers. Thus, there is a great risk of being exposed to the virus and getting sick. That would greatly reduce the number of police officers available for a department. By reducing the number of officers and staff who have contact with one another, the risk and rate of infection could be reduced.

Third, it is important to the safety of officers to flag potentially positive Covid-19 calls for service. The 911 operators should ask questions with all callers that pertain to whether the caller or anyone at the location is sick with a cough or a fever and whether the caller or anyone at the location has tested positive for Covid-19. Officers who respond to a call for service where it is likely that they will be exposed to the virus should wear full protective gear and an N95 mask.

Fourth, police departments should set up testing sites for their officers and families. Every department has emergency medical technicians (EMTs) who could complete the training to be qualified to do the testing. The department also needs to secure housing for officers who test positive for the virus to be able to self-isolate. For instance, the Seattle Police Department secured a hotel for its officers. In addition, it is critical to provide childcare for officers as schools are closed and all children are home. Officers have to be confident that their children are well taken care of while they are at work.[21]

Discussion Questions:

1. There has been much discussion about providing hazard pay for police officers and other people who are working on the frontlines because they have a much higher exposure to the virus and they are risking their life and the life of their family during time of crisis. Make an argument for or against hazard pay. Explain your answer.

2. If you were the police chief, how would you ensure that your officers stay healthy and are able to do their job?

the morale, which in turn, can negatively impact officer support for administrative decisions and foster resistance.

Top Cop as a Visionary. It is one thing for a top cop to be a visionary and another for that person to be a leader of change. There are a number of challenges to implementing change, not the least of which is getting everyone to share in that vision. The easiest thing to do would be to do nothing. So, why would a police leader want to transform their department from traditional to community policing? We will explore the motivations to embrace community policing and the challenges that come with transformation.

INCENTIVES TO ADOPT COMMUNITY POLICING

One might argue that modern, well-informed police administrators perceived the virtues of community policing and bought into it; however, realistically, incentives to adopt community policing were probably less altruistic and more individualistic. In one study on community policing, respondents who were transforming their organization from traditional to community policing gave insight as to their initial motivations. In that study, researchers identified a number of incentives for adopting community policing. The most frequently cited reasons were:

- Chief desire to switch to community policing (62.8%)
- Reduce crime and fear of crime (62.8%)
- Become more pro-active as a department (58.7%)
- Establish partnerships with community groups (50.8%)
- Focus on specific neighborhoods with high crime levels (48.9%)[22]

When the researchers conducted focus groups and case studies, concerns about crime, race relations, and economics came to the forefront of the discussions. Various city departments were part of the focus group case studies referring to their crime issues in the 1980s. St. Petersburg, Florida, and Portland, Oregon, police leaders said that drug dealing was their major concern. The Portland chief also cited prostitution, transience, and increased homelessness as incentives for changing their current tactics. San Diego, California, and Tempe, Arizona, police chiefs cited their motivations for change included problems with gang violence, drug dealing, and growing fear of crime in the community. All expressed that there was a sense of urgency in the need for change.[23]

The research mentioned above was conducted 30 years ago, just at a time when a plethora of experimental programs emerged designed to address high crime and fear of crime. Since that time, there has been an evolution of community policing programming and a better understanding of what it constitutes. Today, scholars, elected officials, and police leaders understand that community policing is an umbrella term to encompass many innovations, all with community partnerships at the center. Similar to what researchers found 30 years ago, we can identify three areas of motivations to adopt community policing. First, police and other government officials have grave concerns about legal entanglements, costly lawsuits, and criminal and civil liability. Much of these civil liability issues revolve around race relations and incidents of police brutality, abuse, and corruption. Historical tensions between African American communities have been, and continue to be, a spark for public outrage, riots, and civil unrest. These uprisings trigger social justice and police reform through court mandates, legislative actions, and policy decisions. The second source involved social disorganization, including signs of disorder and high crime. Traditional methods are reactive to crime, whereas problem-solving methods offer long-term solutions to old problems. Traditionally, police have undervalued fear of crime and social disorder as a legitimate law enforcement concern. Police believed they had no purview over the citizens' quality of life issues. The third motivation for police leaders to adopt community policing is a financial one. Because most police agencies are on limited budgets, innovation grants provide an incentive to extend and supplement financial resources. Throughout the past few decades, large state and federal grants were made available to fund community policing and other programs, such as gang units, driving under influence (DUI) enforcement, domestic violence task force, and other specialized services. For financially strapped cities, the lure of money attracted even the most change-resistant departments. One downfall of getting one-time funds is that departments become dependent on such money and cannot sustain innovation without more money.

Coercive Change Motivations

Policing has had many decades of normalcy, where few changes have been made in philosophy, policies, or practices. Traditional policing, with its focus on crime-fighting and law and order, was what the public, and the police for that matter, expected. If everything was working, why make changes? Unfortunately, two serious problems emerged. One concerned the apparent abuse by police against people of color, creating hostile relations between the police and these communities. In response to that abuse, public outrage culminated in protests, riots and civil unrest to protest established brutal and discriminatory practices by police. Second, police seemed ineffective at addressing the spike in crime. It became apparent that change was necessary. Impetus to change came about from forces outside the law enforcement community, prompted by lawsuits, court decisions, and public outrage. Except for outside pressure, community policing would not have been necessary, at least not according to police or its leadership.

Riots and Civil Unrest. Perhaps the greatest source of pressure was not to improve community relations but to stop the bleeding, literally and figuratively. Much of the civil unrest was sparked by police actions, abuses, and corruption. Civil unrest, and riots against police left destruction of life and property in their wake. Civil and criminal lawsuits against police cost departments and cities millions of dollars. Following civil unrest of the 1960s, the President's Task Force report called for sweeping changes in policing, courts, and corrections, specifically calling for improvement in relations between police and low-income, urban, and racial-ethnic minority communities. The report spelled out new expectations for law enforcement that could only be met with radical reform. Progressive police leaders collaborated with police experts and academic scholars to transform policing.

Race Relations. In many communities, police–citizen encounters are mostly negative, and the police are something to be feared. Strained, and often volatile, relations between the police and racial-ethnic minorties top the list of troubles that community policing is expected to address and resolve. Police interactions with people of color were, and continue to be, the impetus for many court decisions that mandate changes of police enforcement practices. Although research has shown that, in general, the public are supportive of police, this is not true in poor and racial-ethnic minority communities. Much of the dissatisfaction with police has to do with their ineffectiveness at solving crime, endemic corruption, incidents of brutality, overreliance or misuse of lethal force, and general unfairness in their enforcement practices. In 2015, the public confidence in police hit a 22-year low of 52%, with a slight gain in 2017 to 57%. However, the rebound in overall confidence masks the drops in other categories, such as Hispanic people, Black people, younger adults, and liberals.[24]

Crime and Social Disorganization Inducements

Two crime-related inducements led police leaders to adopt community policing. The first one resulted from research that showed that police were ineffective at addressing crime. Second, police scoffed at the idea that fear of crime, signs of disorder, and quality of life issues were something about which they should or could do anything.

Crime. High crime rates, especially violent crime rates, are of concern to the public and to the police. In the decades from the 1960s to the 1990s, there was a continual increase in violent crime.[25] This increase coincided with the call for police reform. Interestingly, public perceptions about crime do not necessarily align with actual crime data.[26] For example, although in 2016 the public thought that crime had gotten worse since 2008, violent and property crime rates had in fact dropped by double-digit percentages.[27] Nonetheless, crime is a national concern and has been for many decades.

Earlier in this text, we identified three traditional methods police use to address crime: motorized patrol, rapid response, and investigations. Citizens were encouraged to call police through 911 and to mobilize the police to deal with the crime. Citizens were not part of the crime resolution but part of the activation to get crime-fighters to the scene. Researchers discovered that these methods failed to reduce endemic crime problems. Findings that police were ineffective at preventing, intervening, or solving crime served as incentive for police leaders to look for innovation in policing.

Fear of Crime. Public perceptions about crime, whether accurate or not, fuel their perceived likelihood of victimization. As was discussed earlier in the text, fears may be enhanced by what they see and hear in the media about crime.[28] Second, fears of victimization likely increase when there are signs of social disorder, such as graffiti, transients, public drunkenness, and homelessness. Both fear and social disorganization affected the quality of life for residents living in these communities. People would change their lifestyle and become isolated, fearful of going out, fearful of strangers, and even fearful of police who did not do anything to help them. Signs of disorder and fear of crime were outside the purview of traditional policing. Police could not address the intangibleness of fear of crime, nor did they believe minor signs of disorder were important enough to address through the enforcement of laws. The Flint Neighborhood Foot Patrol Experiment in Flint, Michigan, was one of the first studies to focus on fear of crime. Although the Flint experiment was considered by most to be a failure and as we will see, abandoned for budgetary concerns, the one good thing that came out of it was the elevation of fear of crime as a valid police priority. Likewise, broken windows policing, applied in New York City and further popularized by William Bratton, did a lot to further the importance of social decay, disorder, and disorganization as a valid police concern. Despite some misgivings and failures of both programs, today's police leaders are incentivized to create community policing programs that address fear of crime and social disorganization.

Financial Enticement for Change

Not all pressure was coercive however; some involved financial incentives. The promise of funding through grants and other sources spurred interest and innovation of community policing programming. These programmatic implementations are costly, requiring restructuring, reassignments, hiring and training new officers, and acquiring new technologies and equipment. Agencies on limited budgets are unable to make changes without help from outside funding sources. At the beginning of reform efforts, following the release of President Lyndon Johnson's Task Force report on crime, the Law Enforcement Assistance Administration (LEAA) was established pursuant to the Omnibus Crime Control and Safe Streets Act of 1968 under the jurisdiction of the Department of Justice. In part, the LEAA was responsible for providing funding to state and local law enforcement agencies in their fight against crime and disorder. Although the LEAA has since been disbanded, new sources of funding have been established. Since 1995, the Office of Community Oriented Policing Services (COPS) has awarded some $14 billion to state, local, and tribal law enforcement agencies for hiring, training, and technical assistance. As recently as October 2016, the Department of Justice announced $119 million in grant funding toward the hiring of community policing officers.[29]

IMPEDIMENTS TO CHANGE

The success of community policing depends upon the extent to which it is integrated within the traditional structure and operations.[30] In order to understand the challenges of bringing about innovation in policing, one must first understand how the complexity and culture of the paramilitary structure makes change difficult. Researchers have found that a necessary aspect of successful community policing implementation is a relaxing or flattening of the structure to allow creative and innovative problem-solving at the patrol level. Police administrators are reluctant or unable to let go of that control, and hence many police scholars argue that community policing will never be fully realized successful.

Despite the advent of community policing, traditional operations and structure of policing have not changed much over the past decades. Reform efforts have had to fit into the existing organization. If anything, the structure of policing has become more formalized, more hierarchical, highly centralized, and specialized over the past decades.[31] Many early innovations, such as team policing, failed largely because police leaders did not understand the need for organizational change to support reforms.[32] However, changes to any aspects of the police culture have the potential to threaten the legitimacy of policing. Even though some of what police do may not be popular, there is comfort in its normalcy. Therefore, change, while it may be warranted or wanted, is challenging. One of the biggest hurdles for police managers is countering officer resistance.

Officer Resistance

With all the other negative variables police leaders have to deal with when planning for community policing implementation, officer resistance tops the list. Change is difficult for most employees regardless of the job.[33] For police officers, change resistance is compounded by the police subculture, peer dependency, cynicism, distrust in management, and distrust in the public. Traditionally, officers view the public as the enemy, part of that "us versus them" mentality. Anecdotally, officers have had little desire to work with the public on problem solving. A classic work by a police scholar provided descriptive terms of citizens by police officers in the field: the suspicious person, the asshole, and the

▶ **Photo 13.2** In order for community policing to be successful, there must be trust between police and the community.

know-nothing. These categories were used as a kind of shorthand as to what actions should be applied by the officer.[34]

Police officer resistance also comes from their preconceived ideas about community policing. Negative perceptions of what it is and what it involves are likely to jade their acceptance of it before they even are exposed to it. Most officers are reluctant to work closely with people that they police. In Portland, Oregon, the police department was transforming to community policing, and administrators spent considerable time imparting community policing's theoretical aspects to the troops; however, they failed to relay information about how their roles would change and what the new expectations would be.[35] Ultimately, there was little support for the change. Mandating that officers accept community policing has not been effective. Short of replacing all officers who do not comply, leadership must deal with the resistance strategically and not as an order. Earlier we discussed William Bratton and his change management ability. Change management focuses on the problems associated with resistance from the employees. Part of the solution is empowering the officers and having them participate in the planning.

Challenges of Police Culture

The organizational culture, characterized by the shared values, beliefs, and norms, make reform efforts difficult. Those values and beliefs create a culture within the organization that is in some ways more powerful than the organization itself. Additionally, there are specific institutionalized components that are particularly challenging. One has to do with the rigid paramilitary command structure of the organization, which dictates the proper behaviors and expectations of its members. Second, the personnel share occupational background, training, and work experience. The police see themselves as crime-fighters. For the most part, the public shares that view as well. Third, the nature of the organization is a closed system, which inhibits outside interference.

Institutionalized Values, Norms, and Beliefs. American policing is a bureaucratic paramilitary organization, with a rigid formal structure, authoritarian hierarchical command and control, standardized training, and uniformed and armed personnel. It has been argued that this type of structure is not conducive to community policing.[36] The people who work in the paramilitary environment may not be prepared to work in a community policing environment. For example, those attracted to policing are usually attracted to the symbols of policing, what they believe the job to be (e.g., fighting crime). The values that support that objective are engrained through selection and training and what is learned on the job. The tools of the trade—the gun, handcuffs, ASP batons, pepper spray, Taser—lend credence to the fact that officers should be prepared for battle. The paramilitary structure, deployment of officers, use of military tactics, and training reinforce the expectations of the people who work in the organization.

Background, Training, and Work Experience. As was discussed in earlier chapters, men and women are attracted to policing for similar reasons. The selection of the officers leads to a specific profile of people that work as officers. Training reinforces the crime-fighting aspects, including placing the priority and focus on physical strength. Police subculture is developed through close-knit camaraderie and shared experiences. Top administrators themselves come from the ranks and are part of the organizational culture and subculture. Just because they become administrators does not mean they move away from that value system.

Quite often, law enforcement personnel have military backgrounds that reinforce the values of discipline, uniformity, and authoritarianism. Law enforcement has always been a pipeline for former military officers. According the U.S. Census data, 19% of police officers are veterans, while only 6% of the population have served in the military. Law enforcement is the third most common occupation for vets. In part, the high percentage of veterans in policing was due to federal and state laws requiring agencies to give preference to veterans over others without such background.[37] However, military background or not, police officers are trained utilizing the military command and control structure, a structure that limits discretion of the individual in favor of the uniformity.

Closed System. There is a belief that much of what police do should be secret for public safety reasons, for example, investigations, criminal activity, or practices and tactics used by police. There has been consensus among researchers that the traditional structure of policing is antithetical to community policing with its inherent attribute of transparency. According to systems theory, the criminal justice system as a whole and law enforcement in particular, is considered to be a letting the public know little of the inner workings. In the past, the **closed system** was a characteristic of classic police management with very little input from the environment as well as careful consideration of what is shared with outsiders. Much of that secrecy is no longer possible; for example, agencies are ordered by the courts to release information such as body cam footage. Police departments will never completely be an . An **open system** allows information to flow both ways, exchanging feedback from the environment or in this case the community. While police agencies have moved toward sharing more information with the public, they only release information on a need-to-know basis. The closed system of policing does negatively impact an ideal partnership with the community.

All or Nothing. Some might believe that it is an all-or-nothing approach, or contrarily, that it is like a buffet whereby you pick and choose programs, components, and aspects without buying the entire meal. It is inevitable that some aspects of the traditional organization structure do have to change in order to incorporate community policing. Police administrators may believe that the organization cannot be changed because of the nature of the work police do. Many traditional tactics and methods are needed for public safety, especially during critical incidents, such as in the case of active shooters and mass casualities. Police managers are unlikely to give up on traditional policing altogether, and no one is asserting that they have to. However, one must ask whether community policing can exist in such a traditionally rigid structure.

Crime and Budgetary Restraints

Impediments to change also involve a lack of financial resources. Financial challenges to community policing implementation are twofold: the grant awards are not large enough to address the endemic social conditions about which the police can do little, and one-time money cannot sustain long-term efforts.

Crime and Disorder Challenges to Implementation. A recent BBC reporter remarked that Baltimore, Maryland, was what poverty looked like in the United States. He also mentioned that the only jobs available for people were selling your body or drugs.[38] Unfortunately, some cities are so strapped for operating funds that even with extra money, innovation is unlikely. The cities that can afford community policing are the ones that may have little use for it. Where the need is greatest (e.g., in high crime, severely economically disadvantaged racial-ethnic minority and urban areas where there are little or no resources), community policing is a luxury that a city cannot afford.[39] These communities are often over-policed and the use of aggressive tactics like pretextual stops are all too common. **Pretextual stops** are when police use a minor infraction or violation to dig deeper and find unrelated crimes. Such policing strategies create public animosity and distrust. The distrust goes both ways. The disadvantaged cities also suffer from an understaffed police department, low morale among officers, poor pay and benefits, and a number of other issues that make working as a police officer untenable.

Budget Allocations for Policing. The number of police officers, of course, is driven by the budget.[40] The per capita police spending varies greatly across U.S. cities. A study by The Center for Popular Democracy surveyed several cities and found significant per capita funding differences. The funding in the surveyed cities ranged from $381 in Los Angeles to $772 in Baltimore. The funding of the police departments depends greatly on the money available to the city and county. Cities typically fund the local police department and the county funds the Sheriff's Office. The police department communicates its funding needs to the mayor of the city who then negotiates the budget with the city council. Similarly, the Sheriff's Office communicates its budget needs, which is then determined by negotiations between the county administrator or county executive, the board of supervisors, the city council, or another appropriate body. Cities and counties spend a sizable amount of their budget on policing. Overall, the United States spends about $100 billion on policing per year. For instance, the city of Oakland, California, spends 40% of its budget on policing. Cities raise money for their budget via taxes, charges for services, and user fees. The state and federal government also transfers money to cities and counties. In addition, police departments also receive grants from a variety of organizations and agencies, such as the Department of Justice, the Bureau of Justice Assistance, and the Center for Problem-Oriented Policing.[41]

Economic Downturns Create Hardships. Resource allocations are dependent on the economy. During an economic downturn, for example, police departments receive less money because cities and counties receive less money from taxes. Cities also spend more money on social services and other services that help residents who have lost their job, become homeless, or need other services. A survey of police agencies showed that in 2 010, police departments received less money than in 2009 due to the economic recession. On average, departments that reported a budget cut lost 7% of their budget. Many of these departments also stated that they anticipated another budget cut in 2011. The budget cuts also led to a decrease of 3% in the number of police officers across departments. Departments also experienced other consequences due to the budget cuts. Sixty-six percent of departments reported cuts on overtime spending, 58% reported reduced salary increases or no salary increases, 43% imposed a hiring freeze for sworn officers, 22% laid off employees, and 16% implemented furloughs. Furthermore, budget cuts impacted services provided to citizens. Overall, 47% of police chiefs reported that they were providing fewer services, including reductions in responding to motor vehicle thefts, burglar alarms, and noninjury motor vehicle accidents.[42]

Unintended Consequences for Departments. Many police chiefs have voiced their concern about the budget cuts and the unintended consequences. Garry McCarthy, chief of the Newark, New Jersey, police, reported that they had to lay off 167 officers, which negatively impacted their ability to respond to crimes and reduce crimes. He said: "It wasn't fat we had to cut; it was effective crime reduction strategies."[43] Police departments in Illinois were also hit very hard by the budget cuts. Jonathon Monken, the Illinois state police director, stated that police departments across the state had to cut a total of $450 million, which led to a reduction of sworn officers from 2,200 in 2009 to 1,750 in 2010. Police departments basically did not replace sworn officers who were retiring. The impact has been felt mostly with regard to traffic-related deaths because the main cuts were made to the motorcycle enforcement unit. During the three months following the personnel cuts, the state police agency reported 78 more fatal car crashes than in the same time period during the prior year. Thus, budget cuts have very serious consequences for civilian death rates.[44] Budget cuts also have serious consequences for the police officers patrolling the streets. They may have to drive and handle incidents by themselves, the wait for backup may be very long, and they may not be able to respond to a call for service in a timely manner or at all. This puts great strain on police officers. For instance, in Flint, Michigan, officers were laid off and called back to work repeatedly because the city did not have enough money to pay officers. Sergeant Robert Frost, one of the Flint police officers, stated: "We've got, like, eight people working at any given time for a hundred thousand people, and there's no way to be proactive. You get one call, you handle that call, you do the best you can, because there is nothing you can do about the other fifty calls that are sitting there. We're just scraping the bottom of the barrel, just trying to keep up."[45]

Impact of Budget Cuts on the Community. Residents also feel the impact of budget cuts to the police department. In Flint, residents stated that they felt abandoned by the police because calls for service went unanswered or the response time for emergencies was very slow. Many serious crimes, like murder, remain unsolved because of a lack of officers and investigatory power. For instance, one of Flint's female residents called 911 to report that men were shooting at several kids in her neighborhood. The dispatcher had to tell her that this was the third shooting of the day and police were very backed up and would not arrive anytime soon. The woman then commented: "They want shit like this to happen in Flint—they want all of us to kill each other so there won't be no more shit they have to come to. That's why all of our young black boys keep getting killed." Residents feel helpless toward the high number of crimes and the lack of police officers.[46]

As we can see, the ability of police departments to implement community policing is subject to many factors, many of which are out of the control of the agency itself, especially where it concerns a lack of funding to support transformation. Once an agency is ready to make the transition to policing, there are a number of resources and funding opportunities to assist them. One of the problems not discussed here is the turnover of police leadership. According to the IACP, the average tenure for a chief of police is 2.5 years.[47] The ramifications of high turnover are a lack of consistency and follow-through. Officers usually outlast the leadership and are likely to wait for the next police leader and whatever innovations or changes come with the new leadership. With all the issues, one might wonder what the success has been for transforming an agency to community policing. How many agencies have community policing?

COMMUNITY POLICING REALITIES

In the past few decades, the term *community policing* has been thrown about so indiscriminately that it has lost its true meaning. Police and other government officials use it loosely to discuss ways of improving police and community relations. In police management circles, community policing is one of the most abused and misunderstood concepts in the past 30 years. It became so fashionable that most law enforcement administrators jumped on the bandwagon with high expectations but without a clear concept of what to do; they just knew it was necessary.[48]

Most law enforcement agencies have, or claim to have, or adhere to, community policing principles. Whether they do or not is the question. We may have to look deeper into what it is that police do differently, what aspects of community policing they have, and whether it is articulated as part of their mission statement. In 2013, about 7 in 10 local police departments, including about 9 in 10 departments serving a population of 25,000 or

THINK ABOUT IT: RESOURCE ALLOCATION—RATIONAL CHOICE VERSUS CONFLICT THEORY

A main concern of law enforcement is crime control, and police departments' effectiveness is judged by their ability to reduce crime and arrest criminals. In order to control crime, police departments need resources, including enough officers, money, patrol cars, equipment, computer software, and so on. Some police departments appear to be underfunded whereas other police departments have many more resources. Academics have explored the question of why resources are allocated disproportionately. What are the political influences on resource allocations?

There are two main theories on how resources are allocated to police departments: (1) rational choice theory and (2) conflict theory. Rational choice theory states that resources are allocated congruent with crime rates, that is, cities with higher crime rates receive more resources. This would be a "rational" decision by politicians because public safety is one of their main concerns and high crime rates are not only a threat to the safety of the community residents, but they also hinder economic prosperity. Big companies do not want to open offices in unsafe neighborhoods and small business owners fear robberies and other crimes. Residents are afraid to walk around the neighborhood and thus avoid going shopping or dining in high crime neighborhoods. Thus, crime-ridden areas often have very few stores and businesses.

Rational choice theory also argues that resource allocation to the police department depends on the budget of the city. Cities have a certain budget, which heavily depends on revenue from taxes. Cities with a high poverty rate take in very little tax revenue and they often have to spend significant amount of money on services. As a result, the money that can be dedicated to the police department may not be sufficient to control crime.

In contrast, conflict theory posits that resources are allocated with the intent to suppress the minority population. This is also referred to as minority threat theory. The powerful class purposefully allocates resources to regulate threats posed by the minority group. Minority threat theory states that the powerful class is concerned about minority groups gaining power as their percentage of the population grows. In order to keep the minority group weak, the powerful political groups use law enforcement to overpolice minority neighborhoods and arrest and incarcerate high numbers of minorities. As a result, they cannot build power because they are economically disadvantaged, and they are incarcerated. In other words, the police protect the dominant group by focusing on crime in impoverished minority neighborhoods. Simultaneously, the dominant group shapes the policies that guide law enforcement practices, such as punishments for drug use. As the minority population increases, they are perceived as a greater threat to the power of the dominant group. Thus, resources for police departments also increase at a similar rate to ensure that the police can continue to police the minority group and protect the dominant group. However, if the minority group grows enough to outnumber the members of the dominant group, the political power of the minority group members increases and resources to police departments may be reduced because the police is perceived as the protector of the dominant group.[49]

Discussion Questions:

1. What do you think? Which theory do you believe is correct? Explain your answer.

2. Thinking back to Flint Town, which theory fits better with regard to the budget cuts?

more, had a mission statement that included a community policing component (p. 8).[50] It is obvious that having community policing is important to most law enforcement agencies.

To be sure, there were numerous honest and well-meaning efforts, often assisted by knowledgeable academic researchers and experts. However, as noted in a previous chapter, early programs were poorly planned, primarily experimental, and not sustainable over time without continued funding. Moreover, money was not awarded without evidence that a department used it for the said purpose of implementing community policing. Therefore,

THINK ABOUT IT: FUNDING AND RESOURCE HELP FOR LAW ENFORCEMENT DEPARTMENTS

Law enforcement officials no longer must go it alone when trying to set up community policing programs. In the past, police administrators attended IACP conferences and workshops to acquire innovation and technical information and how to transform and improve their services. Today, individualized support is available from the federal government through COPS and the U.S. Department of Justice. In December 2017, the COPS Office, the Department of Justice, the IACP, and several organizations formed the Collaborative Reform Initiative for Technical Assistance Center (CRI-TAC) with the objective to provide assistance and resources to state, local, territorial, and tribal law enforcement agencies. Assistance is available on a range of topics including crime reduction and community policing. For example, law enforcement agencies can receive help, resources, training, and information on active shooter response, gun violence, gangs, at-risk youth, domestic violence reduction, and community engagement. The alliance uses several methods to assist agencies, such as peer-to-peer consultations, analysis, coaching, and strategic planning. Law enforcement agencies can learn how to make major or minor changes to enhance their services, including implementation of community policing programs.[51]

Interestingly, the program is not remedial in nature. For example, if a police department is engaged in state, local, or federal investigations or where there is potential litigation, CRI-TAC can also withdraw assistance. This can happen if the agency is not making progress toward its objectives or has new allegations. In a recent example, Adele Fresé, Salinas, California, police chief, had asked for CRI-TAC assistance in implementing community policing. The collaboration was halted, however, due to controversy surrounding four fatal police shootings. Although the officers were cleared, the citizens of Salinas voiced great distrust for the police. As a result, the Salinas Police Department had their community policing program funding withdrawn. Police Chief Adele Fresé was disappointed because she indicated that at least 45% of the Department of Justice's recommendations had been completed. Chief Fresé vowed to continue the efforts they were making in community engagement despite the setback.[52]

Discussion Questions:

1. What are some of the first steps leadership should take to adopt community policing?

2. Why would the Department of Justice COPS office withdraw funding and support?

3. Thinking outside the box, how might the community become involved in CRI-TAC programming?

claims that an agency has incorporated community policing into its operations was not enough. Since 1997, the Law Enforcement Management and Administrative Statistics (LEMAS) Survey has been tracking agencies and their community policing efforts.[53]

SUMMARY

We have learned that there are many incentives and challenges for police leadership to adopt community policing and its progeny. Today, most police leaders are committed to building trust and forging productive community partnerships, while juggling hot button issues that threaten to derail these efforts.[54] There has been no shortage of controversial incidents involving policing over the decades. The advent of cellphones, surveillance technologies, and social media issues.

One might say that, since the concept was first introduced back in the early 1970s, traditional policing no longer exists. We cannot return to the olden days when the public had little input and choice in how they were policed. Anything short of involving the community would be akin to an occupying army. Community policing is about working with the public, including involvement in setting priorities and planning. Police leadership has to stay on top of innovations and technologies, such as evidence-based policing, less-than-lethal weapons, face recognition, and infrared imaging. However, no innovations in the world can take the place of good human relations and procedural justice. Community policing emerged as a panacea for all that was wrong with policing. While it may have addressed certain issues, community policing was not the panacea hoped for; however, it was exactly what policing needed at the time, and still needs now and in the future.

In the next chapter, we will look at progressive strategies for leadership now and in the future. We will discuss change management, strategic planning, officer training, technological advancements, community policing, Compstat, crime mapping, and other data-driven and intelligence-based policing. A large part of successful transformation is bringing on board the community and other stakeholders into planning and setting goals and priorities for the community. How can police leaders successfully implement change and overcome the obstacles? What are some of the next steps for leaders?

KEY TERMS

Change management 240
Closed system 252

Compstat 241
Open system 253

Pretextual stops 253
Vote of no confidence 242

DISCUSSION QUESTIONS

1. What role do police leaders play in the adoption of community policing?

2. Identify and discuss the incentives to adopt community policing.

3. Identify and discuss some of the challenges to community policing implementation.

4. Discuss why crime and budgets make implementation of community policing difficult.

PART V

WHERE DO WE GO FROM HERE?

CHAPTER 14

COMMUNITY POLICING, ENGAGEMENT, AND OUTREACH

Learning Objectives

1. Explain and discuss what community police officers do differently from traditional officers.

2. Explain and define community policing, community engagement, and community outreach.

3. Explain and discuss real world examples of community policing in action.

4. Discuss challenges and solutions police have for addressing specific populations' issues.

In the news, we often hear of mentally ill individuals being shot by police during a confrontation. These incidents have two common elements: police officers and people with mental illnesses. Take one example of a 52-year-old man in El Cerrito, California, who was shot and killed by police. On August 24, 2019, police responded to a man throwing a brick at his 70-year-old aunt. His aunt, who admitted that her nephew suffered from mental illness, claimed he was never violent. A short time after police arrived at the scene, her nephew had been shot dead by the officers. When they arrived, the mentally ill man had advanced on them swinging a shovel. The officers deployed the Taser, but when it failed to stop him, they were forced to shoot. Although both officers had years of experience on the force, they were unable to deescalate the situation, and it ended tragically.[1] In another incident, in San Jose, California, on Halloween night 2019, a man who had mental health struggles, aimed a replica gun at police and was shot to death. The man, three years previously, had tried to get the police to kill him when he refused to give up a knife.[2] Every police–citizen encounter has a potential to be dangerous, especially with instances involving mentally ill persons, as in the two cases here. What could the police have done differently? In *Graham v. Connor* the court ruled that the Constitution simply does not require police to gamble with their lives in the face of a serious threat of harm and that a reviewing court may not employ "the 20/20 vision of hindsight."[3] This ruling is little consolation for the families who have lost loved ones in police shootings.

Police know from experience that certain people, specific populations, and circumstances could be more dangerous and have a higher level of unpredictability than other situations. For example, calls involving people with mental illnesses, domestic violence, gangs, and drugs—all have potential for danger. Police must rely on their training and experience to handle all calls effectively. However, despite their training and experience, when it comes to mentally ill individuals, officers are not trained to handle the mental health needs. Such care is usually the venue of mental health clinicians and not police. Specific to this population, there are few alternatives for police. Unfortunately, arresting them does not connect them with the services they need. The criminal justice system is becoming the default depository for this population. Although criminal justice experts claim that jails are the biggest mental health facilities, there is little or no treatment provided.[4] After incarceration, these individuals are back on the street where the police encounter them again.

INTRODUCTION

In the past few decades, police experts and leaders have pondered what could be done differently and more effectively, and not just with people with mental illnesses. In the era of community policing, police leaders are exploring smarter policing through community engagement with specific populations and other key stakeholders to offer resources and find solutions to critical problems before they turn lethal. Later, in this chapter, we will discuss a comprehensive program for handling mentally ill individuals in the field.

What the police do differently is the focus of this chapter. We will examine community policing in action and explore how the police, the public, and various stakeholders are working together to resolve community problems through community engagement and outreach. We will see how a department moves from a community policing philosophy, building community partnerships to make that philosophy actionable. The Seattle (WA) Police Department is a good example of a department that is doing just that. The Seattle Police Department has developed Micro Community Policing Plans (MCPP) to address distinctive needs of each community. Unique and individualized programs are designed to address specific problems using a three-pronged approach of engagement among citizens and stakeholders, crime data, and police services. The goal is to approach each population with its unique issues and problems and apply appropriate resources and solutions.[5]

Second, we will look at community engagement and its role in community policing endeavors. Clearly, there is no point in the police changing what they do if the public does not want to participate. The key to program success may be the level of commitment and duration of the participation. Community engagement denotes a level of trust and relationship building. For example, it is suggested that in order for community policing programs to be successful, citizens should be empowered to share in the decision-making such as identification of problems, setting priorities, planning an intervention strategy as well as participating in the proposed program. The police cannot force community policing on citizens. Nevertheless, the police are not responsible for the well-being of the community. Residents and other stakeholders have a responsibility for their own community, and therefore, they should have a voice in how it is policed.

Lastly, we will highlight current police and community programs. Although there are too many to cover here, many successful programs include unique methods of delivery, such as bicycle, skate, and foot patrol, or fun activities like basketball, Shop with a Cop, Coffee with a Cop, and Sing Karaoke with a Cop. The primary objective of these types of programs is relationship and trust building. We will also look at programs that address things of a critical nature such as encounters with people with mental illnesses and homeless camps. Objectives of these programs include relationship building; however, they may also involve special tactical components that educate and train police to better handle victims, suspects, witnesses, and the community in general.

FROM COMMUNITY POLICING TO COMMUNITY ENGAGEMENT

According to the COPS definition, community policing is the philosophy that promotes organizational strategies that support partnerships with the community and problem solving to address public safety issues.[6] In review, community policing comprise three key elements: community partnerships, organizational transformation, and problem solving. Community partnerships involves the collaboration of community members and groups, nonprofits and service providers, other government agencies, private business, and the media. Organizational transformation happens within the department to align management, structure, personnel, and information systems. The third component is proactive problem solving versus the reactive response of traditional policing. Problem solving may use the SARA method: scanning, analysis, response, and assessment.[7] These three cornerstones of community policing help with the design, creation, and implementation of programs that target specific populations and problems. The Office of Community Oriented Policing assists agencies with implementing community policing, advising best practices, addressing specific problems, or providing comprehensive hands-on collaboration.[8] They offer assistance with various programs and training for police, sheriff, and tribal agencies. Topics include active shooter response, community engagement, crisis intervention, drug-related crimes, gangs, homeless population, and many other topics. An online portal is available for training officers.[9]

Law enforcement agencies have been creative when designing programs that target various populations with the goal of improving the quality of life, reducing crime, and building trust. Some community policing programs come about through months of collaborative work and are thoughtfully executed, while others flow organically through happenstance. Either way, police leaders must identify the purpose of the program, benefits, and outcomes before an investment of time, effort, and money is extended. Two aspects are critical to that decision. One is identifying the community to be targeted, and second, to engage that community and get buy-in from its stakeholders.

Community Defined

When designing a specific community policing program, police must take into consideration the community. Earlier in this text, we spoke about different groups and the characteristics, needs, and challenges they presented for police. It might be easy to identify a community by its race/ethnicity, socioeconomic status, and age, for example. However, the meaning of community is complex. As the definition of community engagement points out, geographic proximity, similar interests, and similar situations are elements of a community. In society, people belong to many communities; the communities are nested within each other, are organized in different ways, and have formal and informal institutions.[10] They may have symbolic or actual boundaries, such as in a neighborhood. People are bound by faith, gender, sexual orientation, political affiliation, race, ethnicity, and culture, and now social media creates virtual communities. You might live in a neighborhood but not associate with anyone in that neighborhood. You might drive to a church outside your geographic area, but you interact with people in the church and call that your community. People may interact with many different communities over the course of a day.[11] As police seek participation from the community, they must do so with careful consideration and understanding. They must identify the community and then consider the needs of the community through the perspectives of the residents and not their own. Research has shown that police see a neighborhood and its problems differently than do the residents. However, officers' experiences and perspectives are important and should not be left out of the planning.[12]

Community Engagement Defined

Although other disciplines contribute to the literature on community engagement as a theoretical concept, much of the literature is generated from the health professions. A definition of community engagement was established in 1997 by the Centers for Disease Control and Prevention (CDC) when discussing community well-being. The CDC defines **community engagement** as:

> [T]he process of working collaboratively with and through groups of people affiliated by geographic proximity, special interest, or similar situations to address issues affecting the wellbeing of those people. It often involves partnerships and coalitions that help mobilize resources and influence systems, change relationships among partners, and serve as catalysts for changing policies, programs, and practices. (p. 7)[13]

This definition is easily applied to community policing endeavors. Critical to the success of community policing programs is community engagement. Full community engagement involves the participation of police, residents, and other stakeholders. Community engagement gives community members and stakeholders a voice in setting policing priorities and activities, something they did not have under the traditional policing model. Community **stakeholders** include elected officials, business owners, nonprofit groups, other agencies, and the media. Often, individual citizens are not motivated to participate directly but may elect or nominate leaders that represent their views. Police are encouraged to facilitate and attend neighborhood meetings to learn about community concerns as well as to share ideas and strategies.

FROM COMMUNITY ENGAGEMENT TO COMMUNITY OUTREACH

Stakeholders, elected officials, other agencies, and services are essential in community policing endeavors. However, the two most crucial pieces of the puzzle are the patrol officer and the resident of the community the program is attempting to impact. Planning should be an inside-out job and not a top-down approach. For example, we know that homelessness is sometimes a choice; therefore, forcing a person into a shelter or into a program is likely to be resisted. Getting people involved during the program development is more important than after a program is ready to be implemented. It must be a collaborative effort.

Program Development: The Who, What, Where, How, Why?

Prior to implementation, police administrators should determine who would benefit from an outreach effort. **Community outreach** is the effort to engage members of the public through proactive strategies with a goal of building partnerships and forging positive relationships. What are the reasons for the program? Is the program geared toward relationship building, addressing fear of crime or quality of life issues, or crime prevention? Is the program focused on new strategies aimed at changing the way calls involving certain populations are conducted, such as those involving mentally ill persons? What are the steps agencies should take to start a program? Where should a program be geographically located, for example, in a neighborhood, at the police station, out in the field, or at a school? How should they go about getting stakeholders on board and engaged? How do the police build trust, strong relationships, and partnerships? Why are the police implementing a particular outreach program? Did it show promise? Did it work elsewhere? These are important questions to consider when maximizing the effective use of resources.

Scale and Focus of Outreach. Some programs are aimed at a specific neighborhood, a business district, a park, or a school. For example, a program may be located in a neighborhood substation where residents can report crime, or a large community center where citizens gather for events. The program may be an afterschool program where the police and kids meet and talk. The program may be focused in an area where signs of disorder are increasing, where drug dealing is occurring, or homeless people are camping. Some programs are large-scale efforts and focus on building trust with immigrant or Black American communities. In this chapter, we will illustrate a large-scale program in Aurora, Colorado, addressing immigrant community concerns during contentious political times.

Planning a Program. Police administrators must take the lead on program development and bring others into the process. Any endeavor beyond the scope of traditional policing requires thoughtful and detailed due diligence, including budgeting, resource allocation, scope, and many other factors. They may want to seek outside consultation. Police leaders can seek assistance from any sources, which include International Association of Chiefs of Police, The Police Foundation, Police Executive Research Forum, National Institute of Justice, and Center for Problem-Oriented Policing, as well as the National Network for Safe Communities at John Jay College. There are many steps to consider before implementation can take place. Guidelines for that process, compiled from various sources, include:

- Develop a vision and mission for community policing and specific programs.
- Survey residents, businesses, stakeholders, and officers to identify problems and concerns.
- Collect data: crime statistics, patterns, and locations.
- Identify and select leaders in the department and community.
- Identify and examine best practices, explore what other agencies are doing.
- Seek consultation from outside organizations, university researchers, or other agencies.

- Devise strategies, timeline, and budget.
- Form work groups of interested parties.
- Identify and apply for grants and other resources.
- Develop public relations campaigns to encourage participation of stakeholders, community, and police officers.
- Develop partnerships with social service agencies and elected officials.
- Extend police services through nonsworn personnel and citizen volunteers.
- Provide training to the department employees: sworn or nonsworn administrators, supervisors, dispatchers, and volunteers.
- Identify desired goals, outcomes, actions, and assessment.[14, 15]

As noted previously, not all programs result from the systematic planning outlined above. Occasionally, a program may spring organically from an idea or impromptu meeting. Some of the programs illustrated below resulted from that kind of start. Regardless of how a program comes about, there are challenges, especially where communities have great animosity toward the police, or where resources are limited.

COMMUNITY OUTREACH

Because so many factors must be taken into consideration, it is difficult to go from concept to fruition. As was discussed in earlier chapters, community policing works best when the department is fully on board; however, some departments may assign one or more officers to an outreach effort. They may even use civilian or nonsworn members of the department to facilitate the program. The entire program may be assigned to an auxiliary police unit. Whether an outreach endeavor involves detailed planning or comes about spontaneously, the most important aspects are, first, engaging key stakeholders, and second, attaining a level of commitment and accountability toward its success. While citizens cannot be forced to participate, there should be every effort to engage them. Departments may have to use creative ways to pique the interest of citizens and to keep them interested and engaged.

During the early years of policing, citizens did have incidental and tangential roles with law enforcement. Vigilante justice, posse comitatus, and citizens' arrest were three actions where citizens became the enforcers. In the traditional era of policing, police marketed themselves as the experts in crime fighting and did not need the help of the public to do their job. Police officers viewed citizens as the enemy, commonly referred to as the "us versus them" mentality. According to one author, police officers' perspective of citizens fell into three broad categories: suspicious persons, assholes, or no-nothings.[16] The animosity between citizens and police was the single most influential reason for police reform. As part of the move toward community policing, police leaders looked to innovative programing to promote positive relationships. Some of the early efforts, such as Neighborhood Watch Programs, Citizen Police Academies, and youth and senior programs are very much active today.

Historical Citizen Participation

Citizen participation in crime prevention and enforcement is not new. During colonial times, the night watchman system consisted of citizens protecting the community. The expansion Westward and slave patrols in the South led to three areas where citizens had law enforcement powers. **Vigilante justice**, where citizens took the law into their own hands, came about due to an absence of legal avenues. Second, linked in the same historical context, was the practice of **posse comitatus**, whereby citizens were deputized by the sheriff to assist with arrests during times of emergency. Posse comitatus required able-bodied citizens to serve in that role or face fines. Vigilante justice and posse comitatus have negative connotations associated with these practices, specifically linked to slavery and lynching. Third, the common law practice of allowing ordinary citizens to make arrests became codified into American law as a **citizens' arrest**. A citizen may make an arrest of a person for crimes that they, themselves, witnessed. As part of that law, police officers must accept into their custody any person who has been arrested by a citizen.

As formalized policing grew, citizens had decreasing responsibility for their own safety and that of their communities. Due to the rise in crime in the 1960s and problems associated with hostile community relations, police leaders were forced to look for ways to work with citizens in an effort to solve rising crime, to prevent crime, and to promote positive community relations. In part, the hope was for citizens to be empowered to report crime to police and to have some level of accountability for their own safety and quality of life. After all, it was their community. Neighborhood and Block Watch programs provided an avenue for citizens to take a lead in crime prevention efforts. As the concepts of citizen involvement and accountability expanded, the practice of training citizens as to what to look for and what to do led to the idea of Citizen Police Academies. Other endeavors, with a particular focus on youth, led to the creation of Explorer and Cadet Programs and Teen Police Academies. Finally, it was discovered that older citizens had a lot to give and were an untapped resource for police departments. Not only are they retired, but many have prior law enforcement or military experience and have nothing but time on their hands. Senior Police Partners, Citizen Police Academies for Seniors, Seniors and Law Enforcement Together (SALT), and Senior Volunteer Patrol (SVP) are just a few of the programs that benefit the departments in unexpected ways.

Neighborhood Watch Programs. Neighborhood Watch Programs, popular in the 1970s and 1980s, remain in existence today. Some are an extension of a neighborhood association, while other watch programs have less cohesion and are essentially a loosely organized group of residents. The original implementation was the brainchild of the National Sheriffs' Association (NSA) in 1972, with the idea of having citizens take the lead in their neighborhood's crime prevention.[17] In part, the creation of watch programs was a response to civil unrest and increased crime in the 1960s and as a way to mend police–community relations through partnerships. Currently a National Neighborhood Watch Program website serves as a resource for anyone interested in setting up a watch program. Either police departments or interested citizens have the power to set up a watch program.

In general, goals of a watch program included encouraging relationship building between police and citizens; increasing citizen accountability for their role in crime prevention; and police learning more about the needs and priorities of the communities

CASE STUDY 14.1: NEIGHBORHOOD WATCH PROGRAMS ON TRIAL—THE CASE OF GEORGE ZIMMERMAN

In a very publicized case, there was a backlash against Neighborhood Watch Programs when George Zimmerman, the coordinator in his Neighborhood Watch Program for the gated community of Twin Lakes, shot Trayvon Martin to death, whom he believed was responsible for recent burglaries, and was up to no good that night. Zimmerman had a history of reporting problems in the neighborhood to authorities, including four calls about Black men in the neighborhood. Zimmerman, charged with murder, was acquitted for killing Martin. Days of protests followed with demands that federal civil charges be filed.[18] Neighborhood Watch Programs were scrutinized in the aftermath, Questions of authority of civilians to be armed, what rules were taught, and who taught them became the subject of the trial.[19] There was some indication that police departments had a liability for citizen actions. Despite the negative publicity of the Zimmerman case, Neighborhood Watch Programs are still popular, especially in suburban communities.

Discussion Questions:

1. How should George Zimmerman have handled his suspicions of Trayvon Martin's activities as a Neighborhood Watch participant?

2. Should citizens who participate in Neighborhood Watch Programs be allowed to carry a gun? Why or why not?

they serve. Neighborhood Watch Programs were early outreach efforts in the community policing and crime prevention movement as an avenue to build partnerships in the community. Unfortunately, limited resources hampered what could have been beneficial to both police and the community. It was a missed opportunity for police departments in many ways. For example, instead of patrol officers attending the meetings, more often than not, departments substituted low-level administrators or civilian personnel to attend or facilitate rather than the patrol officer who worked that area. In some departments, one officer was assigned to be the liaison officer to all watch programs. Due to limited resource allocations from crime prevention programs in favor of crime control, Neighborhood Watch Programs, in general, are not being used effectively.[20] Evaluations of these programs have shown that they are mostly successful in middle-class areas but not so much in high-crime areas or areas where animosity between police and residents existed. In some communities, there was a suspicion that Neighborhood Watch Programs were actually the police trying to infiltrate their social spheres by deputizing homeowners to be their eyes and ears, to spy on them, and report activity to police.[21] See Case Study 14.1 for a contrary outcome in a Neighborhood Watch Program.

Citizen Police Academies

Another program that grew out of the community policing movement was the Citizen Police Academy (CPA). The purpose of CPA is to educate citizens about what the police do. According to Assistant Chief Bill Weems of the Avon, Indianapolis, Police Department, "Citizen Police Academies help people see first-hand how we operate; learn about our policies, procedures and programs; meet local officers; and understand challenges we face on the job."[22] With greater understanding, citizens view the police more favorably and are more likely to cooperate and work together with police.[23] Modeled after a British program called the Police Night School, the police department in Orlando, Florida, developed the first American version of the program in 1985.[24] Currently, police, sheriffs, and tribal departments are involved in U.S. Charter Citizen Police Academies.

Most of the academies share similar curricular content but programs vary by design program and how they are administered. Academies run from six to ten weeks and cover various topics, including:

- Crime scene investigation
- Active shooter and acts of violence
- Firearms familiarization
- Gangs and graffiti
- Investigations
- Property and evidence
- Communication and dispatch
- Patrol procedures
- Specialized Weapons and Tactics (SWAT) and other special units

Citizen participants have the opportunity to ride along with officers, go to a pistol range, or learn about processing evidence. Like Neighborhood Watch Programs, the budget is meager; however, trainers (usually police officers) either volunteer or earn compensatory time or overtime.[25] Agencies advertise and recruit interested people for upcoming academy sessions by attending Neighborhood Watch meetings and other civic events. An application and background check are part of the process. Academies fill up fast, and therefore, most departments have waiting lists. While these academies are not meant to develop an unpaid reserve program, nor are they used as a conduit for people to become police officers, research has shown people have a greater appreciation for police officers after attending the program.[26]

The increasing number of departments establishing academies attests to the perceived success of such programs. Many police leaders believe these programs are evidence of community policing in action. New types of citizen academies are now being implemented using simulators, other kinds of new technologies, and social media. They are reaching out to younger generations with "next-gen citizen's police academies." These modern CPAs offer microcourses through digital media format, although it does not fully replace the traditional in-person aspects of the program. Benefits include increased outreach to larger geographic areas; greater diversity of participants, including people with disabilities, older people, younger people, and people of color; and affording departments a minimal budgetary impact.[27] One of the best aspects of the new virtual method is its ability to access hard-to-reach populations, allowing participation by people who would likely have been excluded. Critical to objectives of improving relations and opening the lines of communications between residents and police, the CPAs hold much promise.

Youth Programs

Assumptions are often made that forging partnerships refers only to police and adults. However, police and juvenile relationships are as critically important to community policing endeavors as those with adults, not only for building relationships, but also for deterring problematic, risky, and criminal behaviors. Historically, racial-ethnic minority youth have experienced negative interactions with police and have much distrust. Many early programs focused on minority youth and those living in urban areas. However, today, almost every police department has some type of youth outreach program, especially focused on minority youth.

Age as a predictor of criminal behavior is widely accepted throughout research literature. The rate of offending peaks in the teen years from 15 to 19 and then decreases with age.[28] However, this finding fails to take into consideration gender, race/ethnicity, and socioeconomic and environmental variables. Juveniles living in poverty are overly represented in homicide rates, for example.[29] Young men, especially young men of color, living in poor urban communities and absent fathers, contribute to higher rates of delinquency and violent crime.[30] Such findings support outreach efforts on youth in ways that may change the trajectory of their young lives.

Explorer/Cadet Programs. Unlike the CPAs discussed above, Explorer and Cadet Programs are designed to be preparation programs for police careers, or at minimum, to give young people vocational experience. Explorer Programs were originally an offshoot of the Boy Scouts of America but with a stronger focus on law enforcement. Some agencies developed Cadet Programs, and some agencies have both Explorer and Cadet Programs. The Los Angeles Police Department created its first Explorer Program in 1962. Three goals of the program were to prepare young men and women for future careers in law enforcement, provide a forum where young people can perform community service, and provide training for young people to be better citizens and improve physical fitness.[31]

Explorer Programs usually target younger kids between 14 and 20, whereas Cadet Programs are for young people 18–21. The Cadet Program is an apprenticeship program that includes work assignments. Both programs have an application process, background check, and a boot camp. There are physical and mental components similar to what police officers endure in their academy and field training. Participants are exposed to stress and discipline and must adhere to standards and protocol, including strict requirements of uniform, appearance, and behavior. The cadets are unarmed but do work in the field with officers to assist with various tasks. Most agencies place great value on these programs and on the young people who emerge with valuable skills. While some may not specifically connect such programs with community policing objectives, evidence exists that such programs do deter young people from crime, give them a fresh path, create leadership opportunities, and build a solid pool of perspective police officer candidates for the future.

Teen Academy. Similar to Explorer and Cadet Programs, teen police academies recruit young people to participate. While these programs are semistructured, they are not boot camps. They are geared toward ages 13–18 and focus on at-risk youth. Teen and Police Service Academy (TAPS) states that it is designed for youth to:

- Help change their behavior
- Learn responsible decision-making
- Participate in crime prevention projects
- Reduce the social distance between themselves and law enforcement.[32]

These academies are not similar to the CPAs because of their intervention component. For example, one of the objectives of the teen academy is to divert kids from law breaking and provide them with positive pathways and opportunities that they would not have had previously. A most important aspect is the intention to reduce the social distance between these teens and police. One of the events listed on the schedule of the academy was called "Mothers, Sons, and Police," lending weight to the reality that many of these teens have absent fathers.[33] Such programs serve the community by providing strong role models for kids who lack them.

Senior Programs

Police departments are engaging senior volunteers to perform many duties where police officers are better utilized elsewhere. Similar to the Explorer and Cadet Programs, departments benefit from the thousands of volunteer hours they receive from the participants. Started as a program to check on older residents, it soon blossomed into something much more. Police agencies may have their own unique attributes, tasks, and names for their senior programs, but the concept is similar. These senior programs are utilizing senior volunteers' skills, former work experience, availability, and willing spirit to perform welfare checks, vacation checks, traffic control, victim comfort, and many other duties. A number of senior volunteers are retired military or police. In 1997, the San Diego (CA) Police Department began its program by seeking volunteers to check on older residents in the community and soon it found that the volunteers were an extraordinary resource. The program, Retired Senior Volunteer Patrol (RSVP), started as a small group and grew to 48 volunteers, both men and women.[34] The Long Beach (CA) Police Department has a similar program, originally focused on seniors helping seniors, but now includes crime victim assistance, DUI checkpoint assistance, graffiti reporting, safety patrols, issuing parking citations, and special events assistance. The Long Beach Police Department's Senior Police Partners (SPP) was recently featured on the news to highlight its numerous community efforts. One of their oldest members is 87 years old.[35] Like the San Diego program, they are uniformed, unarmed, and patrol in vehicles equipped with lights and sirens. Despite the police-like appearance, they bring comfort to citizens in various circumstances. Both police officers and citizens have praised the program.

Photo 14.1 Members of the Long Beach Police Department's Senior Police Partners (SPPs) gather around one of their patrol cars at the police department in Long Beach, California.

PROGRAMS BASED ON POPULATION TARGETS

In addition to the programs illustrated above, police departments are engaged in various outreach efforts targeting specific populations. The four groups and outreach programs discussed here are Black American, immigrant, homeless, and mentally ill communities. However, significant overlap exists in groups and programming. For example, youth programs are available in most communities regardless of race/ethnicity but are focused on communities situated in urban, lower socioeconomic areas. Additionally, people suffering from mental illness are often part of the homeless community, or any community for that matter. Most communities in the United States are very diverse, and while there may be a concentration of a particular demographic in geographic locals, most communities are not homogeneous adding to the challenges of community outreach.

Outreach to Black Americans

Police relations with Black American communities have been historically adverse. From anecdotal evidence to glaring statistical data, animus is well documented. Black people are more likely to be stopped and searched than other racial groups. In Oakland, California, for example, Black people are 28% of the population but account for 60% of the stops by police, three times the rate of Hispanic people.[36] Young Black men are also overrepresented in jails and prisons. As police are the gatekeepers of the criminal justice system, it matters whether police are inherently biased.[37] Some would argue that Black people commit violent crime at higher rates than other races/ethnicities, and this is why, proportionally, their incarceration rates are higher.[38] Some research has found that crime rates are positively correlated with the percentage of Black people in a neighborhood and that Black people are the most segregated of all races.[39] Just as a reminder, correlation is not the same as causation and research does not support the racist idea that Black people are inherently more violent. Crime rates can be positively correlated with the percentage of Black people who live in a neighborhood without race being the cause of crime. This also does not men that people self-segregate; racially segregated neighborhoods have more to do with exclusion, housing discrimination, red lining, and other issues that people of color have had to endure. Intraracial crime (e.g., Black-on-Black crime) is more common in segregated neighborhoods. In general, there is a higher police presence, more aggressive enforcement tactics, and an increased likelihood of negative encounters.[40]

Highly publicized police shootings of Black men and boys spark racial tensions and deepen the divide between Black communities and police. Not surprisingly, Black Americans have the lowest confidence rates in police.[41] Black Lives Matter, the War on Police, and the Police Abolition Movements are a direct result of continued animus. Some argue that the relationship between the police and Black Americans cannot be healed. In a large part, police brutality and abuse of this population was the impetus for the reform movement in policing, leading to present-day community policing efforts. Both Presidents Johnson in 1968 and Obama in 2015 in their task force reports focused on the need to improve police–community relations specifically in Black communities.

What can be done to mend the relationship, build trust, and form true partnerships in this community? One thing police agencies have tried to do when hiring is to focus on diversity. Hiring of Black officers has been suggested as one avenue for increasing police legitimacy in the Black community. However, there is some evidence that Black officers have the same if not greater bias than White officers. Therefore, while diversity is still a popular concept, it is unlikely to improve relations significantly, or at all.[42] As discussed in Chapter 12, subculture has a powerful influence on officer's perceptions and their behavior, and this does not mean that recruiting a diverse police force should not be a priority. What police leaders are promoting is to be more transparent, to be accountable, to have a positive visibility in the community, and to provide sensitivity training to officers.[43] A few programs that focus on relationship building, especially those aimed at racial-ethnic minority youth, have shown some promise but more must be done. Currently, programming in this community has to do with opening lines of communication. One way is to meet citizens where they are through Walk and Talk patrols, where officers get out of their police cars and walk the neighborhoods. One program that seems to have promise involves officers going into the center of the community and opening up dialog with residents. Another one features police officers playing basketball with neighborhood youth.

Police and Barbers. Barber Shaun "Lucky" Corbett and Detective Garry McFadden put together an opportunity for police and the community to come together in the corner barbershop where Lucky worked. While getting his haircut, Chief Rodney Monroe engaged the barbers in a dialog and learned a lot about what was important to them. Monroe called McFadden to answer some of the barbers' questions. This was at the height of the Michael Brown shooting and Ferguson riots. Barbershops are an important aspect of many Black American communities. What these two men discovered is that it was a perfect place for law enforcement and people in the community to build relationships by talking. Since then, the program has had national interest and has become a foundation to fund young people who want to become barbers.[44]

CASE STUDY 14.2: BASKETBALL COP, SHOOTING HOOPS WITH NEIGHBORHOOD KIDS

In the late 1980s, in Glenarden, Maryland, G. Van Standifer, a retired systems analyst, was convinced that the reason crime was so high in cities was because there was a lack of activities for young men. After a finding that most crime in a low-income, high-crime, racial-ethnic minority community occurred between 10:00 p.m. and 2:00 a.m., an idea that was simple and inexpensive was to open a local basketball court at night and allow young people to play and set up leagues. As part of the program, two uniformed officers would be visible at every game. Since the start of the program, there has been a 30% reduction in crime. The positive results got much publicity and programs sprung up in many urban communities. Unfortunately, the midnight basketball concept was a political point of contention and much of the support for it when away.[45] Getting the kids off the street and knowing where they were was the notion behind midnight basketball; however, it no longer has a significant following.

A resurgence of interest grew when a police officer in Gainesville, Florida, Bobby White, responding to a noise complaint, began playing basketball with some neighborhood boys. The action was caught on video, posted to social media, and immediately went viral. Interest in bringing police and underserved populations of kids, younger and older, together to play basketball was renewed. Since then, a national foundation has been formed, called Basketball Cop Foundation, with a mission to improve police–youth–community relations nationwide.[46]

Outreach to Immigrant Communities

Probably no other issue threatens to undermine the success of community policing endeavors than the problems associated with enforcement of immigration laws. Significant gains were made with regard to police and immigrant relations in the last couple of decades; however, there have been recent setbacks. Those setbacks involve the federal and local law enforcement agreements, or lack thereof. Policies include closing U.S. borders, deporting those who entered illegally, and prosecuting and deporting criminal undocumented immigrants. There are layers of complexity with immigration enforcement. U.S. Immigration and Customs Enforcement (ICE) is responsible for enforcing immigration and customs laws. Section 287(g) of the Immigration and Nationality Act authorizes cooperation between ICE and state and local law enforcement agencies. The agreement proscribes enforcement and detainment decisions. Some states have their own agreement laws authorizing working with ICE, while others have laws restricting local authorities from participating in ICE enforcement.[47] With the Trump administration's tough stance on immigration, much has changed for immigrant communities. Regardless of legal status, uncertainty and fear is pervasive. Many cities, counties, and states are officially declaring sanctuary status to protect the immigrant community. Sanctuary jurisdictions have declined to work with ICE under Section 287(g). As a crackdown against sanctuary cities, Trump passed an executive order refusing federal funds to sanctuary jurisdictions. A federal judge, saying that the order was illegal, issued a nationwide injunction. However, a federal appeals court upheld a portion of the order and found that while funding could not be withheld for failure to cooperate with immigration enforcement, community policing grants could be awarded to cities that cooperated with immigration authorities.[48] Cities that listed immigration enforcement as a priority were granted the community policing funds. Los Angeles, one of the nation's largest sanctuary cities, applied for a community policing grant but was declined because it listed building trust as a priority but not immigration enforcement. Many argue that enforcement is counterproductive to community policing goals.[49] Sanctuary city officials believe they are protecting their communities from aggressive enforcement. However, most people do not recognize the difference between ICE agents and local law enforcement, and therefore, essentially sanctuary or not, immigrants are fearful.

The advent of community policing has brought more focus on police relations with immigrant communities. There are approximately 44.4 million immigrants as of 2017; Mexicans are the largest group, with Asians a close second.[50] Although there has been a drop in recent years of unauthorized persons, the number topped 10.5 million in 2017.[51] It is not unusual for immigrants who share heritage, race/ethnicity, and culture to live in the same community. There is comfort in homogeneity, common culture, and language. When referring to immigrants, however, there are many differences based on legal status, time in America, place of birth, and reasons for coming here. For example, some immigrants are Americanized (e.g., they have lived in this country for most of their lives but are not citizens); others recently arrived; some came into this country illegally or were born elsewhere and brought here illegally as children; and some are refugees who fled persecution in their home countries. Certainly, we can see that immigrant groups are unique and therefore present many challenges to police. Challenges include poor communication due to language barriers, lack of positive interactions with police, bad experiences with police in their home country, cultural barriers, fear of deportation even if they are here legally, and lack of understanding of the laws. The language and cultural barriers interfere with the quality of interactions and can contribute to misunderstanding, confusion, and fear. Second, immigrants, regardless of legal status, typically underreport crimes especially in the area of domestic violence; however, in the in the past two years, crime reporting and service calls have declined significantly in immigrant communities.[52] With the many challenges facing immigrant communities, police agencies are doing everything possible to build relationships and partnerships with immigrant communities.

Aurora, Colorado, Immigration Outreach. The Aurora (CO) Police Department is an example of a police department walking a fine line between building trust with its immigrant and refugee population and working collaboratively with federal authorities on enforcement of immigration laws. Despite the nonsanctuary status of the city, police, city officials, and other stakeholders have come together to address issues particular to immigrant individuals and families who live in the city. In Aurora, one in five residents is foreign born, with most of those coming from Mexico.[53] Community engagement with immigrants and refugees has become one of the police department's top priorities. The first outreach was a public relations campaign in the community to assure residents that police would not enforce civil immigration laws and would not ask the immigration status during police contacts. Police officers were not to ask for any documents of legal residency only for identification as required of all people police encounter during an accident or investigation.

Establishing a sense of safety by allaying the fears of the community was the first step to implementing a program that was not only inclusive but also collaborative.

Even before any outreach strategy was designed, the chief formed a Community Policing Advisory Team (CPAT) with members from schools, youth community leaders, business leaders, immigrant and refugee members, representatives from Asian, Hispanic, lesbian, gay, bisexual, transgender, and queer (LGBTQ), and Black communities, and faith leaders. This group attended a citizen's academy, learning about the policies and practices of the police department, and met monthly to discuss problems, concerns, and ideas for the communities. Additionally, a small group of officers was trained to be Police Area Representatives (PAR), assuming responsibility to bring the community together to resolve problems in their assigned area. Newly hired officers received five hours of sensitivity training to expose them to the lives of refugees and immigrants, their culture, their personal stories, and challenges with understanding English. The goal was not only to communicate to officers the importance the Aurora Police Department places on the people it serves; it also imbues confidence in police by members of the community. The city provides many services to this population as well. Police officers are trained to refer people to the various services. Two other programs are available for the youth in these communities: the Aurora Police Explorer Program and Global Teen Academy. Both programs encourage young people to forge positive relationships with the police as well as providing an opportunity for them to connect with their community. Finally, the police conduct outreach in languages other than English through social media, print material, and bilingual officers.[54]

Outreach to People Experiencing Homelessness

Every major city in the United States is experiencing an increase in the homeless population, with a few exceptions. Approximately half a million people are homeless on a single night in the United States, a majority being male, with an overrepresentation of Black Americans and American Indians. The homeless population represents individuals, families, chronically homeless, veterans, and youth. There is also a regional component, where areas with temperate climates have larger populations; New York, however, is an exception, having the second largest population of next to California.[55]

Over the past decades, police have implemented different approaches to interacting with homeless populations, often with disastrous consequences. Hands-on approaches call for intense enforcement of crimes and ordinance violations while hands-off approaches see the police do little or nothing, for example, letting homeless people camp away from residential and business areas. In 1988, the Santa Ana Police Department launched an enforcement sweep and roundup of homeless people and brought them to an athletic field where they were chained and marked with numbers to process them through the system. The goal was to send a message that homeless people should go elsewhere. As a result of these actions, almost 30 individuals won a lawsuit and were awarded money for civil rights violations.[56] A homeless camp in Ontario, California, started out as a place to move 20 chronically homeless people to an area adjacent to the Ontario Airport, away from residential areas, and it burgeoned to 400 within nine months. Aware of the expanding population and requisite problems, officials attempted to evict some of the tent city's residents. Their plan was to allow people who could prove they were from Ontario to remain in the camp.

Next, they tried to enforce a no-pet order.[57] One resident said, "I gave up my six-year-old son because I was homeless and I'll be damned if I give up my dog, too."[58] The camp had offered them a community, albeit a bad one, with drugs, crime, filth, and disease. As with most homeless people, they have nowhere to go except back on the streets, looking for their next place to live.

Today, there are many tent cities across the United States and officials do not have a solution. The question is: should officials provide assistance, criminalize them, or leave them alone? Businesses and residents in communities where homeless populations are out of control are demanding that something be done. Law enforcement, cities, counties, states, and federal officials have looked for ways to reduce the population, stop crime, stop the spread of diseases, eliminate pollution and filth, and eradicate drug abuse. In the years following the lawsuit involving homeless people living in Santa Ana, Santa Ana police used the hands-off approach. The anticamping ordinance was not enforced, and the Civic Center and courthouse became the home of the city's homeless population, swelling to over 400 people. Now police are afraid to enforce any ordinance for fear of lawsuits. For many years, people reporting for jury duty and the 15,000 government workers had to walk through a gauntlet of tents, trash, urine-soaked sidewalks and walls, feces, and syringes.[59] Something had to be done.

H.E.A.R.T. Program, Santa Ana Police Department. The Homeless Evaluation Assessment Response Team (H.E.A.R.T.) was launched in November 2012, involving a team of trained officers who interface everyday with individuals who are homeless or who are at risk for homelessness. Using a multifaceted approach, the Santa Ana Police Department partners with nonprofit and social service agencies to address many of the problems faced by homeless individuals and families, including a Family Reunification Program, a shelter, children's home, and Salvation Army. The Family Reunification Program helps with families to prevent separation as well as helping individuals reconnect with their families across the United States. The officers serve as a first contact and intervention team member to address mental health and other concerns. They also collect and store personal belongings for 90 days and post in the encampments where the belongings can be claimed. The special detail of officers in and around the Civic Center provide safety and security and enforce the laws. They look for narcotic sales and use, public disorder, and illegal activity. Members of the Psychological Emergency Response Team (P.E.R.T.) are assigned to work with a police officer for a minimum of two days a week and up to 40 hours a week. Additionally, the Orange County Health Care Agency and Santa Ana Police Department have a **Memorandum of Understanding (MOU)** to have a rapid response to mental health crisis by both officers and a P.E.R.T. mental health clinician.[60] An MOU is an agreement between two or more parties, which outlines the responsibilities of each party to achieve a particular goal or set of goals.

Mental Health Outreach

In the examples at the beginning of the chapter, we see that police frequently encounter persons experiencing a mental health crisis. Every one of these encounters has a potential for a negative outcome. Persons suffering from a mental health crisis may include people who are depressed, people with other mental health conditions, people with drug

▶ **Photo 14.2** Some police departments have dedicated homeless liaison officers who can build trust and relationships with the homeless population in the communities they serve. They receive training on how to interact with people with mental issues and the homeless.

addictions, people with disabilities, and older people. Today, with fewer mental health facilities and services available and more people with addictions and mental health issues, including a population of veterans who are returning from active duty with post-traumatic stress disorder (PTSD), the police are experiencing a spike in calls involving suicide, attempted suicide, strange behaviors, violence, and other manifestations of mental health disorders. In the South Bay area of California, for example, a civil grand jury study established that nearly 40% of South Bay police shootings involve someone who is mentally ill.[61] Another study by the Office of Research and Public Affairs found that untreated mentally ill individuals account for one in four of all fatal police encounters' one in five of all jail and prison inmates, and one in ten of all law enforcement responses.[62] Shockingly, another study discovered that mental health disorders are indicated in as many as one in two fatal encounters with police.[63] As was mentioned earlier, police have few strategies for handling these individuals. Some end up being arrested and put in jail only to return to the streets time and time again.

Many of the mentally ill individuals police encounter are also homeless; therefore, programs that deal with both mentally ill and homeless individuals should be considered. One of the most promising community engagement models developed resulted from a mentally ill man being shot to death by police as he was attempting suicide. In Memphis, Tennessee, in 1987, Joseph Dewayne Robinson, a 27-year-old man with mental illness and high on cocaine began stabbing himself in the neck with a foot-long knife. Four officers attempted to subdue Robinson and ordered him to drop the knife. During the encounter, Robinson became agitated and ran at the officers. The officers, feeling threatened, shot and killed him in front of his mother who had called to get help from the police for her mentally ill son.[64] Unlike recent incidents where riots and protests followed, in this case, the community, the police, and mental health professionals came together and formed a task force of experts to find better methods of handling mentally disturbed people in the field.

Crisis Intervention Teams. Through the work of a collaborative task force, the Crisis Intervention Team (CIT) was created. Developed in Memphis, Tennessee, the CIT is now internationally known as the Memphis CIT Model. Many departments across the United States are developing their own models based on the Memphis model. The CIT objectives include educating officers in being able to identify people with mental illness, promoting safety for all involved during the encounters, and diverting individuals whenever possible from the criminal justice process to mental health treatment. The original model called for 40 hours of specialized training for those officers who volunteered to be on the CIT. Educators include mental health clinicians, family advocates, and police trainers.[65] Since 1988, many departments have been requiring all officers to have such training, since all patrol officers are likely to encounter mentally ill persons on their shift.

In Chicago, the police have been using the CIT to deal with people suffering from a mental health crisis. A large part of the program is changing officers' attitudes and behaviors when encountering persons with mental health issues. Second, it recognizes that officers making contact with these individuals need help in the field while from trained mental health clinicians, who may respond to help. Third, the officer is able to take the person into custody and drop them off at an approved psychiatric emergency intake facility. The officer no longer has to take them to jail where there is no treatment available. In addition to the officers, dispatchers are also trained to inform officers about the call they are going into, especially when it involves mental health issues. Dispatchers receive training so that they can give guidance to parties who call. Officers who have a support system in the field and receive the training are better able to handle these calls; however, there is still the potential for things to go wrong as with any situation in the field. Some problems with CIT programs need to be addressed. First, most departments have only a small portion of officers CIT trained, and second, few hospitals are equipped to participate because they are not a psychiatric receiving facility, and facilities that are may not be geographically ideal for transport. CIT programs show promise; however, as mentioned, all officers need the training, because any call could potentially involve a mental health crisis.[66]

Community Outreach: Honorable Mentions

While some programs are large and multifaceted, involving many stakeholders and activities, other outreach efforts may be quite simple but impactful. In a low-income neighborhood in Minneapolis, police officers give vouchers instead of tickets for vehicle equipment repairs, such as broken headlights or turn signals.[67] In San Diego, in an effort to get homeless people connected to services, police officers offer a 30-day stay at a shelter, in lieu of a citation or an arrest.[68] Several departments have adopted the Shop-with-a-Cop program, popular over the holiday season. Walk and Talk programs get officers out of their vehicles and walk and chat with citizens. In fact, some departments are now requiring officers to park their police units for at least an hour, one day a week, and meet people in their geographic areas.[69] Bike patrol became a popular strategy many years ago and is still a popular method to get closer to people in their community, especially juveniles who

THINK ABOUT IT: COMMUNITY POLICING IS A SHAM

A few years ago, Jessica Disu, a humanitarian, rap artist, and peace activist, appeared on Fox News with 29 other people to discuss the police-involved shootings of Black men and the shooting of officers. When Disu was given an opportunity to speak, she stated, "Our police is not working—we need to replace it with something new. It's more than a repair. We need something new . . we need to abolish the police."[70] At the time, she did not know that she was sparking a progressive movement to abolish the police. She asserted that communities could police themselves. Since that time, several grassroots efforts without the police have been underway in Chicago. Communities are providing places where voices can be heard and their needs met—without police.[71] Ironically, these people are doing some of the same things police have been doing under the community policing umbrella (e.g., providing resources to poor, racial-ethnic minority, and at-risk juveniles).

The police abolition movement has emerged and in recent times the Black Lives Matter movement has called for defunding the police as opposed to abolishing the police.[72] Social justice advocates view policing in their communities as a means to continue brutality against marginalized groups such as Black people and LGBTQ people. The activists argue that law enforcement was born out of slavery and it perpetuates inequalities by maintaining capitalistic society ideologies. They believe that police reform is inadequate to address the problems, that policing must be abolished, and communities need to be empowered to govern themselves.[73] As part of that larger movement, the Abolition Research Group, consisting of university students and activists, argue that community policing is a problem, is not effective, and is not a solution for what ails America, especially Black America. They believe that community policing

- is used to bolster the legitimacy of the police when they are undermined by protests or crisis;
- cannot solve the problems that cause crime but can only displace them temporarily;
- is used as an excuse to expand police funding and hiring; and
- extends police presence and surveillance into everyday life, and turns social problems into police problems.[74]

Discussion Questions:

1. Do you think police should be abolished? Why or why not?
2. Taking each of the points of why community policing is a problem, discuss whether you agree or disagree with each.

---

love the bikes. While bike patrol was seen as a good public relations tool, it also became a good crime-fighting tactic, as offenders often did not hear the officers approaching. Coffee-with-a-Cop programs are becoming commonplace, with the goal of getting the community together with police in a positive setting on a bimonthly or monthly schedule. The overall purpose of these programs is to facilitate strong relationships and increase avenues of communication, with the goal of making the community safer and the police more approachable.

SUMMARY

A large part of community policing, as we have learned, is that police officers play a central role in outreach efforts. Patrol officers are the face of the department through their contact with the public. Individual police officers often work quietly behind the scenes doing amazing acts for the common good. In fact, on a day-to-day basis, officers are providing a broad range of services to the community that go unnoticed. While some officers are recognized for heroic acts, like saving a baby from a burning car, all officers are heroes; however, get little recognition from their department or the public. Leadership often fails to officially recognize officers for individual efforts within the community. That failure can damage larger community policing efforts. Leaders need to reward officers for their positive impact on the community.

Every officer will tell you that they want to make a difference, clean up the streets, put bad guys in jail, and save people from harm. Just the same, every officer desires that adrenaline dump during critical calls, when their life or their partners' lives are on the line, and excitement and danger are intermixed. However, at the end of their career, what they will remember most, what means the most to them, are the lives they changed, the lives they saved, and ultimately, the differences they made in the communities they served. Lastly, it should be remembered that police officers are not responsible for the community's well-being.

All stakeholders—including citizens, elected officials, community and business leaders, nonprofit organizations, social services, and the media—all play an important role and must do their part to make communities safe. Community engagement means that everyone participates and has a voice in how they are policed.

KEY TERMS

Citizens' arrest 266
Community engagement 263
Community outreach 264
Memorandum of Understanding (MOU) 276
Posse comitatus 266
Stakeholders 263
Vigilante justice 266

DISCUSSION QUESTIONS

1. What are some of the ways police officers can work with the public?

2. What are some of the avenues citizens have for participating in policing efforts?

3. What populations should police focus on for community engagement and why?

4. How do police fight crime and build relationships in communities they have historically been at odds with?

CHAPTER 15

THE FUTURE OF COMMUNITY POLICING

THE STORY OF FLINT TOWN

In 2018, Netflix aired the documentary *Flint Town*, which focuses on the work of the Flint police agency and the community. Flint, Michigan, is located about 60 miles northwest of Detroit. For two years, the filmmakers followed police officers and community members to show how an understaffed and underfunded police force attempts to deal with violence, drugs, and disorder in a town that went from a flourishing city to a city with the sixth highest violent crime rate in the nation in 2018. In 1908 the car manufacturer General Motors opened a factory in Flint, which produced Pontiac and Buick sedans. For many years, the town thrived as the 18 factories employed 35,000 people. However, in the 1970s, the workforce began to shrink and in 1999 GM closed the Buick City factory, citing a declining demand for big sedans. At the time, the factory employed 2,900 people. The closures of car factories had a great negative effect on the city, its residents, and the police department.[1,2]

Since then Flint has experienced a surge in violent crime, drug use, and disorder, which have made it one of the most dangerous cities in the United States. Table 15.1 shows the violent crime rates from 2003 until 2018.[3]

Since 1999 many people have left Flint. Between 2000 and 2014, the population declined from 124,496 to 99,002. With the loss of the factory jobs, unemployment and poverty have greatly increased. In Genesee County, where Flint is located, employment in manufacturing jobs decreased by 76% between 1990 and 2015. The loss of manufacturing not only resulted in much unemployment but also in a considerable loss of income for the city, mainly due to the loss of taxes. Manufacturing jobs paid an average of $15.66 per hour or about $31,000 per year plus overtime. Jobs have increased in other sectors, but many of these jobs do not pay well. In 2014, the average per capita income in Flint was $16,912 for White people, $13,923 for African

Learning Objectives

1. Discuss the expectations of the public of law enforcement.
2. Discuss the connection between community policing and restorative justice.
3. Explain the concept of restorative policing.
4. Explain restorative practices used by police departments.
5. Discuss potential political influences on policing.
6. Describe how technology will change policing in the future.

Table 15.1 Flint Violent Crime Rate 2003–2018

Year	Flint	United States
2018	1,817.57	380.56
2017	1,945.03	394.86
2016	1,583.84	397.52
2015	1,477.28	373.74
2014	1,708.25	361.55
2013	1,908.13	369.13
2012	2,729.46	387.75
2011	2,336.92	387.06
2010	2,207.88	404.50
2009	2,009.73	431.88
2008	2,024.47	458.61
2007	2,362.44	471.77
2006	2,596.06	479.34
2005	2,260.17	469.04
2004	1,925.74	463.16
2003	1,215.19	475.84

Source: Federal Bureau of Investigation—Crime in the United States.

Americans, $6,833 for Asians, and $13,964 for Hispanics. Accordingly, poverty rates are above 40%, which is the highest among all cities in Michigan. The highest poverty rates in Flint were among those 18 years and younger. It is not hard to imagine what the effects of high unemployment, low salaries, and high poverty rates are on the community overall.[4]

The loss of car manufacturing also had detrimental effects on the city and police force. The city closed its police academy and cut its police force in half. In 2015, Flint elected a new mayor, Karen Weaver, who hired a new police chief, Timothy Johnson. The police department had been downsized to 98 police officers for its total population, or 58 officers per 50,000 residents. This is the lowest police staffing of cities with more than 50,000 residents and very high violent crime rates. In comparison, St. Louis, which has the highest crime rate of cities with more than 50,000 residents, has 277 officers per 50,000 residents.[5]

The lack of police officers creates very difficult working conditions for the police department. They often drive to a crime scene without backup, regularly work double shifts, often get to a crime scene hours after the crime has occurred, and are disliked by the residents. Johnson implemented a zero tolerance policy and officers began cracking down on minor offenses. Johnson also attempted to implement proactive police practices to prevent crimes before they occurred, but the utter lack of trust between the residents and police have made that a fruitless endeavor. Response rates to 911 calls are very slow and citizens feel abandoned by the police who are supposed to protect them. This chapter will discuss public expectations of police officers, the challenges of fulfilling these expectations, and how restorative policing may aid in fulfilling the expectations of the public and create better relationships between the community and police.

PUBLIC EXPECTATIONS AND SATISFACTION WITH LAW ENFORCEMENT

Police have come under much criticism for their practices, including stop-and-frisk procedures, racial profiling, officer-involved shootings, and excessive force. Police have been facing a crisis of legitimacy, which impacts their ability to be effective and efficient crime-fighters, protectors, and service providers. This problem is greatest in impoverished racial-ethnic minority neighborhoods with high crime rates. Responding to crime and protecting the public are very difficult when the relationship between the police and the community is strained. Some believe that the future of policing must be based less on punitive enforcement and more on community engagement to be effective.[6]

Citizens' willingness to obey police decisions and accept police as an authority depends on the perceived legitimacy of law enforcement in general and local police departments specifically. People who believe in the legitimacy of the police will generally comply with police orders voluntarily. Trust in the police also promotes voluntary compliance with police decisions and authority. However, people who don't perceive the police as a legitimate authority are less likely to comply with police decisions voluntarily and more likely to oppose police power. For instance, police may be conducting stop and frisks in the particular neighborhood looking for a suspected robber. A person who perceived the police as a legitimate authority will likely comply with the orders of the police during the stop and frisk without getting upset or angry. In contrast, a person who does not perceive the police power to stop and frisk a person as legitimate may resist, get angry, or even try and run.[7]

One of the critical factors related to perceived police legitimacy is procedural fairness. **Procedural fairness** "concerns public perceptions of how citizens are treated reflecting a consistency of quality in police services"[8] (Nickel, 2018, p. iii). Stated differently, do people feel that they are being treated fairly by the police? If they do, then they are more likely to perceive the police as legitimate.

Citizens' perception of police legitimacy and procedural fairness is closely related to citizens' satisfaction with police. How people are treated by police matters a great deal. Citizens who feel treated fairly are much more satisfied with police than citizens who feel that they are discriminated against or targeted by police. People also feel unsatisfied if the police officer did not give them an opportunity to explain their behavior thoroughly or if the officer was disrespectful.[9]

The Pew Research Center conducted two large surveys of public and police views of the key aspects of law enforcement. They conducted two surveys in 2016, one that was sent to police officers and the other was sent to a representative sample of adults. A total of 7,191 police officers and 4,538 citizens answered the survey, which asked about the role of police

▸ **Photo 15.1** When the public's perception of procedural fairness breaks down, they are less likely to follow police orders.

within the community, the deaths of Black people during officer-involved shootings and other encounters, police officer presence, body-worn cameras, and racial discrimination.

The data from the two surveys indicate that the public expectations of what the key aspects of law enforcement should be diverged widely from the views of police.[10]

Here are a few examples. Whereas 31% of police officers state that they serve as protectors, only 16% of the public agree that police officers are their protectors. Only 8% of police officers see themselves as enforcers, but 29% of the public view the police as enforcers. Most police officers, that is 86%, believe that they need more officers, but the majority of the public, about 57%, does not believe that we need more police officers. The vast majority of the public, that is 83%, state that they understand the risks and challenges of police work. A similarly large percentage of police officers, namely 86%, believe that the public does not understand the trials faced by police in their daily work. These differences are significant because they demonstrate the disconnect between the public and citizens.[11]

Public satisfaction with police depends in large part on the perceived effectiveness of police in fighting crime. In general, people who live in low-crime neighborhoods are more satisfied with the police than people who live in high-crime areas.[12] There is, of course, a racial-ethnic component to this. Non-Hispanic White Americans are substantially more satisfied with police than African Americans and Hispanic Americans. A study by Weitzer and Tuch (2005) found that 48% of Non-Hispanic White Americans, but only 22% of African Americans and 36% of Hispanic Americans, reported being satisfied with police in their city.[13] These racial-ethnic disparities relate closely to the differences in socioeconomic status, community living conditions, and prior experiences with police. Non-Hispanic White Americans are significantly more likely to live in low-crime neighborhoods and thus they experience the police as effective crime-fighters. They also are much less likely to be stopped and questioned by police.

One of the issues that is highly contentious is the number of deaths of African Americans during encounters with police. Several cities have experienced protests surrounding the death of a Black person due to deadly force by police. For instance, in 2014, a White police officer, Darren Wilson, shot 18-year-old Black teenager Michael Brown. The circumstances of the incident were highly contested. During the next few days, citizens from Ferguson and people from across the United States gathered in Ferguson, Missouri, for peaceful protests but also for civil unrest. Some protesters damaged stores, cars, and other property. The protests subsided after a few days, but police reports and videos released from the incident continued to spur tensions between the public and the police. On November 24, a grand jury decided not to indict Officer Wilson. This decision led to renewed public protests and violent incidents in Ferguson and other cities in the United States. The protests spurred movements toward more racial justice across the entire nation.

In 2015, Officer Wilson was also cleared by the Department of Justice of any civil rights violation. However, civil rights leaders and many community members disagreed. Even though the protests have subsided, the tensions between the police and the public, especially people of color, continue to exist. The public does not believe that officer-involved shootings with Black victims are isolated incidents. Rather, 60% of the public believe that there are problems specific to the police and Black communities. Among the police officers, the view of Black officers differs substantially from the view of White officers. Whereas 72% of White officers believe that these are isolated incidents, only 57% of Black officers believe that to be true. This reflects the racial tensions that broiled over during the riots in Ferguson in 2014. The differences in how police officers perceive the racial tensions also show in their opinion about racial progress within the United States. About 92% of White officers, but only 29% of Black officers, believe that racial progress has been made and Black people have equal rights to White people. Among the public, 57% of White people, but only 12% of Black people, believe that racial equality has been achieved. According to the majority of police officers (86%), the protests and public scrutiny have made their work harder.[14] Case Study 15.1 looks at an innovative strategy of the police.

Officer-Involved Fatal Incidents (2015–2019) and Racial Disparities

The Washington Post compiled a database with all people killed by police between 2015 and 2019. The numbers are shown in Table 15.2.

The number of people killed averages between 2.5 and 3 persons per day for each year. That in itself is a high number and cause for concern among the public. But, what has sparked the protest is the staggering number of Black people killed by police. According to the U.S. Census Bureau, the population has 76.5% White people and 13.4% Black people, but about 25% of people killed are Black.[15]

COMMUNITY POLICING AS THE FOUNDATION OF RESTORATIVE JUSTICE

Community policing may offer solutions to current challenges we face as a society. Some may even argue that reducing violence, intercultural conflict, social and economic injustice, resource shortage, and substance abuse can only be addressed effectively by building better relationships between police and citizens, which is a cornerstone of community

CASE STUDY 15.1: EMPLOYING FORMERLY INCARCERATED PEOPLE TO BRIDGE THE POLICE–COMMUNITY DIVIDE

The Washtenaw County Sheriff's Office (WCSO), located in Michigan, is employing formerly incarcerated people to build better working relationships between the community and the police. The program, implemented in 2009, is spearheaded by Sheriff Jerry L. Clayton and Derrick Jackson, the director of community engagement. The WCSO had three goals for the program: (1) reduce re-offending, crime, and victimizations; (2) provide resources to communities; and (3) enhance problem-oriented policing by improving the relationship between the police and the community.

The formerly incarcerated people serve as outreach workers whose task it is to assist with proactive policing strategies. These strategies include referring community members to community resources, such as employment services, alcohol and drug abuse self-help groups, and housing services. They also help residents navigate the different support services, which can be difficult for people who are homeless and/or have mental health and drug abuse issues. Finally, the outreach workers serve as a liaison between the community and the police in an effort to build bridges. The outreach workers are able to choose a specific area of interest, such as homeless people or people with drug abuse issues.

The police profit from the outreach workers as they can communicate with community members who would not talk to the police. The outreach workers also profit by having stable employment and training opportunities, which help them with their reentry into society. Between 2009 and 2018, the police department hired 16 outreach workers. Most of them stay for one year, but there are a few who stay with the department longer and work on special projects.

The outreach program focuses on four neighborhoods with especially high crime rates and strained relationships between the police and community members. The program has been successful in changing how police and the community members interact. For instance, the outreach workers facilitated a meeting between mothers of gun-violence victims. The goal was to prevent further retaliatory violence within the community. Together with the mothers, they developed a plan that would stop the ongoing violence and deescalate further tensions. The outreach workers have become a viable alternative to punitive enforcement by police. They can relate to the offenders, victims, and their families within the community and they receive respect from the residents.[16]

Discussion Questions:

1. Why are formerly incarcerated people effective outreach workers?

2. Do some research on this program and discuss how this program helps formerly incarcerated individuals reintegrate into society.

policing. Policing is very closely tied to the justice system and how victims, offenders, and communities experience the justice system.

Police chiefs and officers are judged by the community on their effectiveness to reduce crime, arrest offenders, and help the victims. This expectation has often resulted in focusing on short-term strategies and strategies approved by citizens and politicians. Community policing, however, is a long-term strategy because it takes time for officers to build the necessary skills and build relationships with community leaders. Building trust

Table 15.2 People Killed by Police Between 2015 and 2019[17]

Year	People Killed	White	Black	Hispanic	Others	Male
2019	896	275	188	145	288	853
2018	992	451	229	165	147	939
2017	986	459	223	179	125	939
2016	962	465	234	160	103	922
2015	994	497	258	172	67	952

Source: Tate, J., Jenkins, J., & Rich, S. (December 19, 2019). Fatal force. *Washington Post*, Retrieved from https://www.washingtonpost.com/graphics/2019/national/police-shootings-2019/

and a working relationship does not happen overnight. Community policing efforts do not have the same visibility as patrol cars and responses to calls for service. But community policing is imperative to reducing the number of calls for service and the number of patrol cars needed to keep citizens safe. The better officers know and understand the residents in their neighborhood, the better they are equipped to prevent crimes from happening. For instance, as discussed in the book, one of the most common and most dangerous calls for service relates to domestic violence. Community policing can significantly reduce domestic violence calls for service by using proactive policing strategies, such as regularly checking in with residents who have a history of domestic violence incidents.

Using reductions in crime as a measure of police effectiveness is convenient because incidents of crime can be counted and reported. Everyone understands that crime rates do not include crimes that are not reported (e.g., dark number), but nonetheless reported crimes serve the purpose of keeping track of crime trends, crime hot spots, and changing crime patterns. This information can then be used by police chiefs to set priorities, determine strategies, and inform the public. It is inherently more difficult to measure fairness, trust, and the effectiveness of community–police relations.[18]

Restorative justice is based on the idea that offenders can repair the harm caused to the victim. Restorative justice allows offenders to return to the community and contribute to the community while repairing the harm to those affected by the criminal conduct. One of the main strengths of restorative justice is that it elevates the role of the victim, which is very important but rarely accomplished by other punishments, such as incarceration or probation.[19]

For example, offenders may be ordered to pay restitution to the victims to compensate the victim for monetary losses suffered due to the crime. Restitution may be ordered by a judge in cases of simple theft and other minor offenses.

Restorative justice is based on three key principles. First, the justice system should focus on repairing the harm caused by the criminal behavior rather than simply punishing the offender. Second, the justice system should enable the victims and offenders of crime to participate in the process in a nonadversarial way. Third, the government has the responsibility to maintain order and build peace.[20] Some academics are proposing that there is another goal closely related to the three key principles. The goal is to share power with the community and enable the community to be more active participants in promoting public safety and justice. Even though it seems that restorative justice programs could be best integrated with community policing, this has not been the case on a consistent basis. If the community, however, becomes a more active participant and cooperation partner with police, the community could be working to promote peace, whereas the police would be responsible for preserving the order. Police must invest in this relationship and help the community develop the skills necessary to respond informally to crime, harm, and conflict. Police also have to work to mobilize the residents of the community by engaging community leaders. This is a difficult task because crime-fighting is traditionally the role of the police. This has been a persistent problem community police officers and communities are facing.[21]

One of the strategies used by police departments is to establish a citizen academy (Case Study 15.2) where citizens can learn about the training, attitude, and demands on police officers and support staff.[22]

Restorative Policing

The involvement of the community and community police officers in restorative justice is different from the typical way in which community policing has been implemented. Applying community policing to restorative justice has also been referred to as **restorative policing**.[23] Community policing asks officers to seek nontraditional solutions to problems and crime. Thus, officers are proactive and focus on solving problems in an effort to prevent crime. Problem solving requires officers to identify problems, analyze the problems, and develop strategies that address the problems and assess whether their strategy was successful. During that process, officers must consider a wide range of alternative responses to criminal behavior. These alternative responses, especially to juvenile offenders, often involve community groups equipped to support offenders and provide supervision in combination with rehabilitation. Restorative policing strategies are different from the typical community policing strategies. The typical activities of community policing include organizing group events, such as neighborhood watch, cleanups, and setting priorities for police activities.[24] In contrast, working together in the context of restorative justice requires police officers to give up their proprietary role as crime-fighters and involve the community in the decision-making process. The community participates in finding effective solutions to crime through restorative conferencing techniques, where police officers and community members make decisions about the obligations of offenders toward the victim and the community. These obligations can include restitution, community service, reparation of damage (e.g., painting a wall to remove graffiti), apologies, and peacemaking pacts.[25]

The main idea of restorative policing is to develop a collective community ownership for all stakeholders, including victims, offenders, their families, and the community residents. This is also referred to as **community building**. Community building is aimed at "mobilizing and enhancing citizen and community groups' skills and confidence in informal responses to crime, harm, and conflict."[26] One of the main goals of community building is to promote informal social control and social support. **Informal social control** aims to promote conformity to norms and laws via informal sanction, such as shaming, peer group pressure, family intervention, or other entities, such as churches. The community helps in promoting peace and the criminal justice system helps to preserve order. This is not an easy task because for

CASE STUDY 15.2: CITIZENS ACADEMIES—BUILDING TRUST, INCREASING LEGITIMACY, CORRECTING MISCONCEPTIONS

Newspaper headlines, TV news, and posts in social media outlets have been very critical of police officers. This is especially true for officer-involved shootings and other incidents where citizens suffer harm due to the acts of the police. But there is rarely a discussion in the media of the daily work of police officers. Citizen academies can help police departments in developing a relationship with the community. Citizen academies have several goals. First, they are meant to create positive interactions and help the community understand what officers do on a daily basis. Second, they bring citizens and police together in a positive environment where citizens interact with officers and support staff for several weeks on a regular basis. The police department hopes that these interactions and the learning experience stop misperceptions and reduce frivolous calls for service. The police department also hopes that citizens will start to correct misperceptions on social media. Finally, it also helps police build partnerships with the faith-based community.[27]

Curriculum of Citizen Academies

The curriculum varies depending on the police department. In general, participants are exposed to the different department units, roles of police officers, support staff, and procedures. For instance, the Tustin Police Department in California developed a 14-week course, which is offered twice a year. The academy is held each week for three hours in the training room of the police department on a weekday evening. The Tustin police serve a population of 78,000 residents and have developed a model program that can be implemented by other police departments across the United States. The program is very comprehensive, spanning all different parts of police officer jobs as well as the support staff. The following is a sample program:

"Week 1: Introduction of the Department

Week 2: Overview of the Criminal Justice System

Week 3: Patrol Procedures

Week 4: Chaplain Services, Patrol Division

Week 5: Communications/Dispatch Operations, Crime Analysis

Week 6: Professional Standards, Ethics, Citizen Complaints, Officer Discipline

Week 7: Use of Force, Levels of Escalation

Week 8: Firearms Familiarization

Week 9: Live Fire at Firing Range

Week 10: Gang Unit, Street Gangs, Gang Graffiti

Week 11: Detective Bureau, Crime Investigation, SWAT

Week 12: School Recourse Officer Program, K-9 Demonstration

Week 13: Property Room, Evidence and Safekeeping, Crime Scene Investigations

Week 14: Narcotics Enforcement and Drug Operations, Ride Along with PTO, Visit to County Morgue, Graduation Dinner"[28]

Participants

The selection of participants is important and departments must decide who they want to admit. Citizens who want to participate have to complete an application form, which includes basic demographic information, employment information, driver's license information, education, and criminal

Continued

CASE STUDY 15.2: CITIZENS ACADEMIES—BUILDING TRUST, INCREASING LEGITIMACY, CORRECTING MISCONCEPTIONS (CONTINUED)

background. It may also ask for references whom the police department can call. Applicants have to certify that all information provided is true. The police department conducts a background check, which includes the criminal history, outstanding warrants, and other relevant information. A criminal record may not automatically disqualify applicants, however. Departments may decide on a case-by-case basis whether a specific applicant is admitted.[29] Some programs also ask applicants to write a short essay about why they want to attend.[30]

Officer Concerns and Discussion Points

Some officers may have concerns about the citizen academy. They may have concerns that citizens could learn sensitive information, but citizen academies are not teaching tactics. The officers must be actively involved in developing the curriculum. There may also be concerns that citizens could become vigilantes or police impersonators. Citizen academies typically give the participants a shirt that has a police department logo. These shirts should not be blue or black, to avoid any impression that they look like police. The onboarding of officers and union is key. It is important to listen to the concerns of officers and actively involve them in the process.

Other discussion points revolve around the selection criteria for participants and who should be teaching in the program. In addition, departments have to decide how many participants to admit. For instance, the Tustin academy admits 24 applicants.[31] Departments may also have budget restraints that limit how many participants they can enroll and how often they can offer the program. For smaller departments, it may be very difficult to extend personnel for this purpose. However, the understanding and trust of citizens garnered by the program can be a great asset in times of officer-involved shootings or other incidents that raise concerns in the community. Departments should identify opportunities to collaborate with other agencies to create a citizen academy.[32]

---

so many years, communities have been told to leave crime control and punishment to the police, prosecutors, and lawyers. Under the concept of restorative policing, crime control and sanctioning unwanted behavior are under the purview of the community. Repairing the harm is more important than punishment of the offender per se. The main question in restorative policing is what is in the best interest of the community.[33]

Some have argued that there is a need for systemic reform that would change the criminal justice system as a whole, including changes in professional roles. Systemic change means more than just a few adjustments; it means that the department undergoes a transformation that begins at the department level and expands outward to the community. The department must have restorative principles available for all incidents of crime. These restorative principles must also be available to the community members who bring cases to the police. It is not sufficient to implement a few restorative programs but rather communities must be saturated with restorative practices.[34]

Systemic change also means that officers apply restorative practices not only to conflict resolution, punishments, and community engagement but also that officers apply restorative practices to all of their functions: crime prevention, peacekeeping, service to the community, and other law enforcement functions. It's an all-encompassing concept. The restorative policing training would therefore also include all of the department staff.[35]

Implementation of Restorative Policing

There are four sectors critical to systemic change: (1) legislative/policy; (2) organizational; (3) individual officer; and (4) community. First, the legislature could tie funding for police departments to the implementation of restorative policing practices. This would put greater emphasis on the role of police as peacekeepers as they would be working as front-end referral agents to restorative programs and work with the community members to use restorative practices. Second, the organizational domain includes police officer job description and training that emphasize restorative practices. It also includes incentives for the use of restorative practices and support by the management/leadership of the police department. As part of systemic change, the entire department structure and culture must revolve around restorative practices. The main goal is to change the priorities of policing toward problem solving, conflict resolution, and building trust with the community. These goals have priority over aggressive policing tactics and coercion. Third, the individual officer domain encompasses hiring officers who have a holistic vision of policing that focuses on conflict resolution and building community. Finally, the community domain includes community involvement and community building. This may well be the essence of the implementation challenge. The community is essential to advancing reforms and helping sustain the commitment to restorative practices. Community members participate not only as victims, offenders, and community supporters but more importantly as volunteers and community builders. They can use informal control mechanisms to respond to crimes. Community members, and especially community leaders, are also imperative as a liaison between the police department and the community.[36]

Restorative Policing: From Theory to Practice

One of the main problems of reforms relates to the actual implementation of restorative police practices. What sounds so enticing in theory has been very difficult to implement. This section will discuss several case studies of programs and police departments that have implemented practices that engage and empower the community in their efforts to reduce crime and improve the living conditions of the residents.

Police-based Restorative Conferencing Programs

Restorative conferencing encompasses a "range of strategies for bringing together victims, offenders, and community members in nonadversarial community-based processes aimed at responding to crime by holding offenders accountable and repairing the harm caused to victims and communities."[37] The main idea is to make the criminal justice process less formal and involve the community in the process. There are four main types of restorative conferencing programs: (1) Victim–Offender Mediation, (2) Community Reparative Boards, (3) Family Group Conferencing, and (4) Circle Sentencing.

Victim–Offender Mediation programs have been used for more than 40 years and they receive thousands of case referrals from courts every year. The main target group is juvenile offenders who have committed nonviolent crimes, especially property crimes. But increasingly, these mediations are also targeting more serious offenders. The victim has the opportunity to tell the offender—through a mediator—about the physical, emotional, and economic harm they have suffered due to the crime. The victim can ask the offender questions that they may otherwise not get an answer to and the victim can actively participate in the punishment of the offender, including a restitution plan. The main goal is not to reach a settlement, however, but rather to engage in a dialogue and support the healing process of the victim.[38]

Research suggests that victims are often less fearful of revictimization after the mediation, that about 95% of mediations result in a mutually agreed upon restitution agreement, the victims are more satisfied with the justice system, and the recidivism rates of offenders involved in mediation dropped from 27% to 18%. Thus, mediation can be an effective program for offenders and victims.[39]

Community reparative boards are also known as youth panels, neighborhood boards, or community diversion boards. They have been used since the 1920s, but only in a few states, including Vermont. They also target offenders who have committed nonviolent crimes. Typically, reparative boards consist of a small group of citizens who have completed intensive training. They meet with the offender and discuss the crime and develop sanction agreements. Following this, they monitor compliance and submit reports to the court. The board members engage in a dialogue with the offender about the consequences of the crime and reparation terms. The offenders must submit progress reports for the board that detail their compliance with the reparation agreement. The goal of the program is not only to make offenders aware of the negative consequences of their crime and repair the damage they have caused but also to engage community members in the criminal justice process and promote citizen ownership of the process. There is very little research on the effectiveness of reparative boards. However, experiential information suggests that these boards can be effective as a response to nonviolent crime.[40]

Family group conferencing stems originally from the traditions of the Maori in New Zealand and focuses on the community members most effected by the crime, including the victim, the offender, the families of the victim and offender, friends, and key supporters. They all come together with the assistance of a trained facilitator to discuss the harm caused to the victim and their family and to find a resolution to the crime. In Australia, family group conferencing is used for nonviolent and violent crimes, such as theft, arson, minor assaults, drug offenses, vandalism, and, in some states, child maltreatment. In the United States, family group conferencing has been used in several states, including Vermont, Minnesota, and Montana. The process is initiated by a facilitator who contacts the victim and offender and asks them to invite key supporters who will also be invited. During the conferencing, the offender is asked to describe the offense and the victim and other members talk about the harm they have suffered. The victim and their family and supporters can ask the offender questions and engage in a dialogue. Following this, the victim states their desired outcomes. The main goal of family group conferencing is to give the victim an opportunity to confront the offender and participate in the sanctioning decisions. The victim participation also increases the offenders' awareness of the consequences of their crime, which may make it less likely that they will re-offend. In addition, by involving the offenders' family and support system, they may be able to steer the offenders' future behavior in a positive direction. In the United States, family group conferencing is used mainly for juveniles who have committed nonviolent crimes. There are very few studies that have assessed the effectiveness of family group conferencing. However, the studies found that the victims were satisfied with the process and experienced less fear following the conferencing. They also found that the offenders and their families also expressed their satisfaction with the process because they experienced it as a more holistic reintegrative process as compared to the standard procedures.[41]

Circle sentencing originally stems from the Aboriginal people in Canada and the American Indians. They are also known as sentencing circles or peacemaking circles, which were used for juvenile and adult offenders. Minnesota adopted circle sentencing as a pilot project in 1996. Similar to the other three practices, circle sentencing involves a dialogue between the victim, offender, and family and friends of both. It also involves criminal justice and social service personnel. This practice is meant to address the offenders' behavior and the needs of the victim and their family. All parties involved discuss which steps should be taken to help the victim heal and prevent future delinquent acts or crimes by the offender. The group engages in deliberations until they have reached a consensus. The main goal is to promote the healing of the victim, offender, and family members. It provides an opportunity for the offender to make amends and empower the victim. Finally, the process is meant to build community and instill a sense of conflict resolution. The community may implement a community justice committee to adjust the process to the community culture. To date, there is very little research on the effectiveness of circle sentencing.[42] However, a study from Canada by Barry Stuart (1996) showed that offenders who had participated in circle sentencing were less likely to recidivate.[43]

Nashville Police Department

The Nashville Police Department faced a similarly bad situation as the police department in Flint: high levels of violent crime and burglaries, open drug use, and residents who were afraid to walk on the street and felt abandoned by the police and the city. Nashville also had a high number of homeless people and a high poverty rate, including high rates of food anxiety. The police department designated six officers to a community engagement task force with the goal to improve the relationships with the residents, gain their trust, and get the communities' help in fighting crime. These six officers engaged in a variety of activities aimed to show their presence and their willingness to help residents who needed support. For instance, the officers were at bus stops in the morning to make sure children were safe. They participated in afterschool programs and helped kids with their homework. They started their own summer camp at West District to get to the kids early and provide a positive role model. In addition, the afterschool program and summer camps were used to bridge the gap with the parents. Officers also brought food to the residents in neighborhoods with high poverty rates and helped the homeless find a place to stay. The police department introduced stakeholders to engage in that community. Local nonprofits, local government, landlords, and faith-based groups came together to discuss what each could do to help make this community safer.[44]

The community engagement resulted in a substantial decrease in crime, and criminals left the neighborhoods as the community members felt safe talking to the police about crimes and criminals. The community began working together with the police to make their neighborhoods safer. As a result, crime rates decreased and so did the calls for service.

The community became less of a strain on the police department. It also resulted in the empowerment of the community. Residents felt safe again to walk on the street.

Baton Rouge Police Department

The Baton Rouge Police Department in Louisiana has one of the highest crime rates in the country. In 2017, Baton Rouge experienced 1,027 violent crimes per 100,000 residents, which is 68% higher than the average violent crime rate for Louisiana and 161% higher than the national violent crime rate.[45] High crime rates and shootings have contributed to the city's bad reputation, which affects the quality of life of residents and the work of police officers. The rise in the homicide rate from 62 in 2016 to 104 in 2017 was an especially great concern for the residents and the police department.[46]

Police chief Paul J. Murphy and the police department developed a new strategy to reduce crime. The strategy focused on building relationships with the community and creating mutual respect between the residents and the police. Chief Murphy stated that police used to refer to the residents as "them" as though they were on the opposing side. That is not true, however, because residents and police have very similar goals, including feeling safe, walking on the street without fear of getting shot, and ensuring the safety of children. The police department determined that the majority of crimes were committed by a few bad actors and that the community was the best resource to get the bad actors to leave the city. In order to get the attention and help of the community, police officers went on community listening sessions where residents could voice their concerns and talk to officers in an environment that was not the scene of a crime or some other adverse event. The officers identified community partners and participated in more than 300 community events in 2018. Officers also met with the families of the victims of crime and told them that they needed their help to catch the offenders because they were hiding in the communities; without help from the residents, it was very difficult for police to arrest them. Over the course of the next year, the families became the voice for the police. They would call the police and provide information about crimes and where to find the offenders. This sent a message to the criminals and many of them left the community because they knew that residents would report their whereabouts. The strategy successfully reduced the crime rates, including a reduction of the number of homicides from 104 in 2017 to 84 in 2018. This is still much higher than the average number of homicides in Louisiana and nationwide, but Chief Murphy is optimistic that the continuation of the community engagement program will further reduce crime and increase the residents' quality of life.[47]

The Baton Rouge Police Department also worked together with the faith-based community, including Reverend Markel Hutchins. The faith-based community is one of the greatest assets for police. The largest body of volunteers in the United States are people who are part of faith-based groups. This relationship began in 2016 with the goal to improve public safety via police–community collaboration. The collaboration also aimed to avert divisive public response to police-involved incidents and turn police agitators into partners for police. Participating congregations "adopted" a partner with their officers to develop strategies for crime prevention and community engagement. Reverend Hutchins emphasized that the community must own the relationship with the police. Typically, the police lead community policing efforts and decide which strategies should be used. However, the faith-based community has an enormous amount of resources, their own infrastructure, and effective problem-solving strategies within their congregation. If these resources could be used for the community as a whole, in collaboration with the police, offenders would find it much more difficult to commit crimes and victimize residents.[48]

National League of Cities: Four Steps to Building Better Working Relationships Between Police and Faith-based Community

Jack Calhoun, a consultant for the National League of Cities, lays out four steps police departments can take to build stronger relationships with faith-based groups. First, police departments must build a trusted link between law enforcement and the faith-based community. Research has shown that cities where a partnership exists typically do not experience riots or other citizen uprisings after officer-involved shootings. Religious leaders have much influence and can calm the community and communicate the concerns of the community with law enforcement. Similarly, law enforcement can communicate with the community through the religious leaders and provide pertinent information. It is very important that the police build this partnership and retain it by having regular meetings and by being fully transparent when incidents occur. For instance, after a shooting in Boston, law enforcement met with Reverend Jeffrey Brown, the community, the clergy, and the National Association for the Advancement of Colored People (NAACP) representative and shared the footage of the video and explained what happened.[49]

Second, start the partnership at the personal level and work toward a more formalized relationship after trust has been built. Many police officers participate in religious services and events. These officers can form relationships with the leaders of their church and provide insight into their daily work. For instance, officers can offer "ride-alongs" to give religious leaders and community members a look inside the work of officers, their struggles with stress and fatigue, and their efforts to keep the community safe. Many officers face dangerous situations on a daily basis and make split-second decisions deciding whether to shoot or not. The chronic stress and threat of harm take a psychological toll. It is important for the faith-based community to understand the circumstances under which police officers work and make decisions. Other police departments, such as in Stockton, California, have moved town hall meetings to a smaller setting, such as churches, living rooms, and community centers, to provide a more comfortable atmosphere. Small meetings are better suited to discuss issues and questions from community members and law enforcement. Large town hall meetings can easily turn into a trial-type atmosphere where the community accuses officers of wrongdoing and the officers are trying to defend themselves. Such meetings do not build trust but rather increase the barriers between community members and police officers because both feel under attack and there is no fruitful conversation about how these issues can be addressed.

Finally, law enforcement must incorporate the faith communities, calling. Faith-based communities are known to speak truth to power. They stand up for social justice and the needs of communities and their residents. Police share many of the same goals, including reducing crime, bettering the life of residents, keeping everyone safe, and building strong communities where residents support one another and employ informal social control mechanisms, especially for youth at risk. They may use different strategies to accomplish these goals, but a partnership allows them to pursue these goals together.[50] (See Case Study 15.3.)

CASE STUDY 15.3: ANTI-POLICE PROTESTER UNDERGOES USE OF FORCE SCENARIO TRAINING

In 2015, Reverend Jarrett Maupin, a vocal critic of law enforcement, participated in three different shoot–don't-shoot scenarios. The training consisted of three separate scenarios in which Maupin had to make a decision on whether to shoot or not to shoot. Each scenario was followed by an evaluation and an interview about when he perceived the encounter as a threat and why he drew his weapon. In the first scenario, Reverend Maupin approached a suspect in the parking lot and was shot almost instantaneously. In the second scenario, Reverend Maupin encountered two men who were fighting. When one of the men charged toward him, he drew his weapon and shot at the man. When he was asked why he shot at the man, he answered that he felt that the man was an imminent threat. He didn't see a weapon, but the behavior of the man suggested that he intended to harm the Reverend. In the third scenario, Reverend Maupin was called to a burglary. He managed to get the suspect to the ground without using his weapon.

Following the training, the Reverend stated that the training had left a last impression. He said, "I didn't understand how important compliance was ... people need to comply with the orders of law enforcement officers, for their own safety."[51]

Discussion Questions:

1. Discuss whether police departments should invite citizens to a use of force training.

2. How can the use of force training help improve the relationship between citizens and the police?

Third, law enforcement and religious leaders should establish larger programs aimed to address problems in the community, such as programs for youths (especially in high crime areas), mentoring programs for offenders who are being released to the community, and programs that help victims of human trafficking and sexual abuse. For instance, the city of Shreveport, Louisiana, built an afterschool program, Community Renewal International, in the city's highest crime neighborhood. People of faith are mentoring youth and helping them with their homework. Since the inception of the program, crime rates in that neighborhood have dropped by 50%. In Portland, Oregon, members of the faith-based community stopped sex trafficking in Holladay Park, which has reduced crime in the park by 50%. The police department supports the volunteers by providing rain coats, umbrellas, vests, and jackets. Everyone profits from these partnerships, including victims of crime and those who could become victims.

POLICE INFLUENCES ON LAW ENFORCEMENT: THE POTENTIAL ROLE OF CONGRESS

Some are proposing that Congress should take a more active role in changing the relationship between police and citizens. The federal government has an interest in improving the relationship between citizens and the police. There have been many incidents of citizen protests, violence, and riots in recent years that are of concern to members of Congress. Critics contend that policing strategies and practices fall under the sole power of state and local authorities. The federal government does not have the power to compel local law enforcement agencies to implement specific programs.

However, there are three main strategies the federal government could use to influence the relationship between police and citizens:

1. Collecting and disseminating data about the use of force by police
2. Developing statutes that will enable the federal government to investigate police misconduct
3. The Department of Justice (DOJ) can support police departments by providing funding and acting as a policy leader.

There are also several specific recommendations how the DOJ can support the implementation of practices that will improve police–community relations. First, the DOJ can place conditions on federal funding to encourage law enforcement agencies to implement policies that will improve law enforcement and community relations. Second, it can support the use of body-worn cameras by providing grants to departments. Third, it can assist law enforcement agencies in investigating instances of excessive force by police. Fourth, it can provide grants through the Community-Oriented Policing Services to enable police departments to hire community policing officers.[52]

POLITICAL INFLUENCES ON POLICING—THE POWER OF THE PEOPLE

George Floyd died on Memorial Day 2020 during an encounter with police in Minneapolis, sparking worldwide protests. On May 25, 2020, police were called to a local Cup Foods, where staff claimed that a customer had paid with a fake $20 note. Shortly after 8 pm, officer Derek Chauvin and three other officers arrived at the store and arrested Floyd. During the arrest, Chauvin pinned Floyd to the ground with his knee on Floyd's neck, basically kneeling on him. He stayed in this position for more than 8 minutes during which Floyd repeatedly said, "I can't breathe"—until he died. The three other officers were standing nearby watching the arrest. Neither attempted to intervene while Floyd was struggling to breathe. The entire encounter was videotaped and released on Facebook a few hours later. The official autopsy by the state examiners officer stated that Floyd's death was caused by cardiopulmonary arrest. A second autopsy conducted after Floyd's family asked for an independent examiner came to a very different conclusion: homicide by asphyxia due to the compression of the neck and back, which inhibited blood flow to the brain. All four officers were fired from the police department, but not immediately charged with a crime.

After Floyd's death, protests against police brutality and racism erupted across the country and around the globe. The words "Say his name" and "I can't breathe" became the chorus of the protests that lasted for several weeks. Another slogan that is now heard often is "defund the police." Numerous videos of police brutality now draw the attention of U.S. citizens and citizens across the globe as police and protesters clash.[53]

CASE STUDY 15.4: COLORADO'S POLICE REFORM BILL

Since the video of George Floyd's death surfaced and protesters rallied against current police practices of use of force, and especially chokeholds and similar practices, some state and city legislatures as well as police departments started thinking about how to address the concerns of the population and civil rights advocates. How can police departments be reformed? The first police reform bill called "Enhance Law Enforcement Integrity" was passed in Colorado and signed by Governor Jared Polis on June 19, 2020. The goal of the bill is to enhance law enforcement integrity by being more transparent to the public in cases where police are accused of misconduct, in cases where a person is seriously injured or killed by police, and by limiting the use of force during public demonstrations. The bill includes three main provisions.

1. Starting July 1, 2023, all officers must wear body cameras and record all incidents. In case of a complaint of misconduct, the tape from the camera must be released to the public. There are sanctions for officers who fail to activate the camera.

2. "Each agency and the Colorado state patrol that employ peace officers shall report to the division:
   a. All use of force by its *peace* officers that results in death or serious bodily injury;
   b. All instances when *a peace* officer resigned while under investigation for violating department policy;
   c. All data relating to *contacts* conducted by its peace officers; and
   d. All data related to the use of an unannounced entry by a peace officer."

3. "In response to a protest or demonstration, a law enforcement agency and any person acting on behalf of the law enforcement agency shall not
   a. Discharge kinetic impact projectiles and all other non- or less-lethal projectiles in a manner that targets the head, pelvis, or back;
   b. Discharge kinetic impact projectiles indiscriminately into a crowd; or
   c. Use chemical agents or irritants, including pepper spray and tear gas, prior to issuing an order to disperse in a sufficient manner to ensure the order is heard and repeated if necessary, followed by sufficient time and space to allow compliance with the order."

The bill also includes a provision related to civil lawsuits against police officers. Persons may bring civil lawsuits against police officers for civil rights violations. Police officers are indemnified against such lawsuits, except "if the peace officer's employer determines the officer did not act upon a good faith and reasonable belief that the action was lawful, then the peace officer is personally liable for 5 percent of the judgment or $25,000, whichever is less, unless the judgment is uncollectible from the officer, then the officer's employer satisfies the whole judgment."[54]

Discussion Questions:

1. How does this bill and the changes to the reporting and use of force address the concerns of the protesters after George Floyd's death?

2. If you were the police chief of a large city, what would you do to address the concerns of the protesters after George Floyd's death?

HOW TECHNOLOGY MAY CHANGE POLICING IN THE FUTURE

Technology is an important part of policing and it may become even more important in the future. What could policing look like in 2043? That is a question that is somewhat hypothetical because we really cannot look that far into the future. Seven years ago, body cams were just starting to become popular. Now, they have become an important part of many police departments.[55]

▶ Photo 15.2
What impact could advancements in technology have on community policing?

Now, many agencies and companies are working on artificial intelligence, self-driving cars, drones, and robots. The discussion about these technologies does not only center on its feasibility, costs, and benefits but also on whether these technologies can be used ethically and whether community members find the use acceptable. Chief Jim Bueerman, the president of the Police Foundation, states:

> Policing can be automated, but the issues that cause police to show up on someone's doorstep won't change. I think there is going to be a point in the not-too-distant future where cops won't be driving cars—you will be sitting there preparing for the call as the car can drive better than you can. There are certain things AI can do better than human beings. The cost of policing is off the charts, so communities are going to have to think differently about how many cops they have and how we leverage what we can afford with the technology at the end of the day.[56]

Augmented Reality

One of the new technologies is augmented reality (AR). Most people know AR from sports events; in football, where it has been used to show a yellow "first down" line. This is a simple version of AR. The advanced version of AR, which is being tested by law enforcement and by the

Naval Research Laboratory, can "overlay virtual (computer-generated) images onto a person's real world field of vision or into a real-world experience in a way that improves and enhances the ability to accomplish a wide range of tasks and assignments."[57] This technology substantially improves situational awareness and a more efficient workplace. The FBI believes that an officer who has access to AR can do the same amount of work as three individuals who do not have AR.

Patrol officers may be able to use AR for real-time intelligence about crimes and criminals in their community. It can also provide recognition data based on voices, faces, and other biometric characteristics. In addition, officers have access to three-dimensional maps of buildings and surrounding areas, which would greatly increase their ability to respond to problems, find hideouts, and avoid traps that could endanger their lives. It could also help officers find victims of crimes faster and save lives. Finally, AR could make driving safer for the officers by providing them with car-operating data and traffic-management information.[58]

AR also enhances the work of SWAT teams, which often face dangerous situations. Situational awareness is detrimental to the success of their missions. AR improves situational awareness, which allows the SWAT team members to coordinate better and increase cohesiveness. Tasks can be carried out more efficiently and SWAT team operations become safer and more effective as a result.[59] SWAT teams also use robots, drones, and other technical support to improve their situational awareness during their operations.[60]

AR also modulates the sounds of gunshots during SWAT operations, which allows the team members to distinguish between friendly and hostile fire. This reduces friendly fire casualties and creates more certainty for officers who they can shoot at. In addition, zoom, infrared, and thermal imaging help SWAT team members find victims and offenders, buried disaster survivors, and so on.[61]

Criminal investigations also benefit from AR as it enables police to collect information, quickly analyze large amounts of information, and visualize the data in real time. This data can then be dispatched to police officers and assist them in locating and arresting criminals. This is especially important in cases of mass shootings and terrorist acts where many lives are at stake and response time is crucial to saving lives. AR also allows police to match voices against a database of voices from criminals, use optical devices to lip-read, visualize blood patterns and stains, and calculate distance and height.[62]

For police training purposes, AR can be used to create realistic scenarios with real-world equipment. Police departments can also monitor patrol activities of the trainees and other officers.[63]

Drones

Police have been using drones for real-time information during in-progress calls for service and in other situations where police need assistance. By 2020, 347 law enforcement agencies in 43 states were using drones for different purposes. Drones are used for search-and-rescue missions, traffic-collision and crime-scene reconstruction, active-shooter incidents, and surveillance of suspects or places. For instance, drones can be used to patrol dangerous areas and provide real-time information that can help officers assess the situation, arrest criminals, and save lives.[64] For instance, the Chula Vista Police Department has been using drones as first responders to certain 911 calls. During a call for service, a police officer can deploy a drone and feed live video to the officers and supervisors. The officers who are responding to the call for service have the live feed for several minutes before they arrive at the scene of the incident. They can identify suspect(s), see if they have a weapon, and what kind of weapon. This information improves the safety of the officers, the public, and the suspect.[65]

SUMMARY

This last chapter discussed the concept of restorative policing and community building with the goal to reduce crime and disorder, strengthen communities, and promote a better working relationship between community members and law enforcement. Citizen academies are one tool to bridge the gap between the residents and police officers. Experiencing policing from the viewpoint of the officer can change a person's attitude toward police and increase cooperation with police. Community policing offers great potential, but it also faces many hurdles, including resource shortages, lack of time, lack of skills, and resistance by recruits. Police departments must work to overcome these hurdles to provide a solid foundation on which community policing can progress and tackle the challenges of the 21st century. Effective leadership and a clear vision are imperative for the progress of community policing and buy-in from police officers and the community. Law and order activities cannot be separated from the communities in which police operate. The effectiveness of police will always be measured by their ability to fight crime and arrest criminals, but the real value of law enforcement for communities lies in their ability to create peace so that people can go about their life without an imminent fear of crime and personal harm.

KEY TERMS

Community building 288
Informal social control 288
Procedural fairness 283
Restorative conferencing 291
Restorative justice 287
Restorative policing 288

DISCUSSION QUESTIONS

1. Discuss how politics impact law enforcement.
2. Discuss public expectations of law enforcement and the difficulties police have in meeting these expectations.
3. Discuss how citizen academies can help change the relationship between citizens and law enforcement.
4. Discuss the differences between community policing and restorative policing.
5. Discuss how restorative policing can promote peace and justice.
6. Discuss how law enforcement can build better working relationships with the faith-based community. Why are these relationships important in combating crime and disorder?

GLOSSARY

Abortion extremists. Oppose abortions and use violence against doctors, nurses, and staff of clinics where abortions are performed.

Aggression. Defined as any form of behavior directed toward the goal of harming or injuring another living being who is motivated to avoid such treatment.

Animal rights/environmental extremism. Classified as special interest extremist groups.

Anti-government/anti-authority extremism. Loosely organized groups with no central leadership who oppose any type of authority, such as a central government, laws, and police.

Barrio 18. A notorious criminal street gang with an estimated 30,000 to 50,000 members across the United States and Canada, founded in Los Angeles.

BDUs. Battle Dress Uniform is a camouflaged combat uniform originally used by United States Armed Forces as a standard uniform, most recently adopted by police departments as an alternative to their normal uniform.

Beats. Geographic areas with set boundaries such as streets or buildings and often by crime statistics, so that beats are not overly busy compared to others.

Binge drinking. Defined as drinking alcohol until the drug alcohol level is above 0.08g/dL.

Bonds. Contribute to community resilience by providing a sense of belonging and connection with others who are similar.

Broken windows theory. A theory that states that if a broken window goes unrepaired, it seems as if no one cares and more windows would be broken, creating an environment ripe for crime.

Bridges. Contribute to community resilience by promoting a sense of belonging and connection with people who are dissimilar in important ways.

Call subjects. Health or social service providers may call the police when they have an individual who is aggressive or refuses to leave.

Chain of command. A hierarchy or line of authority whereby each person at a specific level answers to persons in the level directly above them.

Change management. The ability of leadership to prepare and support employees through organizational change.

Citizens' arrest. Codified into law, a citizen may make an arrest of a person for crimes that they, themselves, witnessed and as part of that law, police officers must accept into their custody any person who has been arrested by a citizen.

Civil asset forfeiture. Allows police to seize property of a person if they believe that the property was used in the commission of a crime and was purchased with money obtained through criminal behavior.

Clearance rates. Calculated by dividing the number of cases resolved by the total number of crimes recorded, often used as a method of determining police effectiveness.

Closed system. A characteristic of classic police organization is that there is little input from the environment and little or nothing is shared with those outside the organization.

Code of Silence. A characteristic of the police subculture which holds that police officers should be loyal to each other by keeping quiet about possible wrong-doing of their brothers in blue. Often referred to as the blue code of silence or wall of silence.

Collective efficacy. Means if a community shares the belief that it can overcome obstacles, then it is more effective in accomplishing its goals.

Consent decrees. Mutually binding agreements between two or more parties, which allow federal courts to require oversight and enforcement of the agreement.

Controlled items. Items police departments can purchase that require a high degree of accountability.

Community building. Defined as developing a collective community ownership for all stakeholders, including victims, offenders, their families, and the community residents.

Community engagement. Working collaboratively with and through groups of people affiliated by geographic proximity, special interest, or similar situations to address issues affecting the wellbeing of those people.

Community outreach. The effort to engage members of the public through proactive strategies with a goal of building partnerships and forging positive relationships.

Community partnerships. Include collaboration among law enforcement agencies, individuals, and organizations they serve to develop solutions to problems and increase trust.

Community policing. A philosophy that promotes organizational strategies that support the systematic use of partnerships and problem-solving techniques to proactively address the immediate conditions that give rise to public safety issues such as crime, social disorder, and fear of crime.

Compstat. Stands for computer statistics and is an analytical managerial accountability system used to better understand crime patterns and to apply this knowledge to solutions.

Consent decree. Court-ordered reform plan for a police department due to misconduct, overseen by an appointed monitor to ensure compliance with a specific timeline.

Crisis intervention teams. Specialized police-based programs based on a systematic response intervention model.

"The Cut." Refers to "in-between" spaces, "indoor-outdoor" spaces, and controlled public spaces.

Decentralized policing. Police organizations are under control by regional divisions, i.e., city, county, and state government, not necessarily under federal authority.

Defund policing. Refers to taking away funds allocated to policing and redistributing those funds to other social services, such as mental health care, drug rehabilitation, domestic violence, and homelessness.

Diversion programs. Seek to divert offenders away from jail and prison by diverting them into community programs.

DLA. Defense Logistics Agency, which is responsible for the transfer of controlled and noncontrolled (or uncontrolled) military equipment.

Domestic terrorism. Defined as "violent, criminal acts committed by individuals and/or groups to further ideological goals stemming from domestic influences, such as those of a political, religious, social, racial, or environmental nature."

Domestic violent extremists. Are inspired by domestic ideologies that are often spread via social media and the Internet.

Drug tolerance. Means that users need a higher amount of the drug or need to take the drug more often to get the same effect from the drug.

Due care. The degree of care that a reasonable person would take in a situation, and in this case the degree of care a police officer would take to accomplish a task.

Emotional intelligence. An individual's ability to recognize and regulate the emotions of self and others toward successful environmental adaption. Often referred to as emotional quotient (EQ).

Fear of crime. Fear of being a victim of crime whether there is a real chance or not, resulting in feelings of anxiety and stress that change behavior and impact quality of life.

Ferguson effect. The impact and loss of police legitimacy on racial-ethnic minority communities that followed the shooting death of Michael Brown, an African American man, by a White police officer Darren Wilson in Ferguson, Missouri, in 2014.

Force option models. Model developed to guide officers when making decisions on use of force.

Gang. An association of three or more individuals who collectively identify using various forms of dress, language, tattoos, or other physical markings, and whose purpose generally involves criminal activity.

G.R.E.A.T. Stands for Gang Resistance Education and Training.

Hate crime. Criminal offense committed against a person, property, or society that is motivated, in whole or in part, by the offender's bias against a race, religion, disability, sexual orientation, or ethnicity/national origin.

Hate crime laws. Basically laws that provide prosecutors with enhanced penalties for crimes committed out of prejudice.

Hate incidents. Acts committed out of prejudice that do not involve a criminal act.

Hate speech. Defined as communication that carries no meaning other than the expression of hatred for some group, especially in circumstances in which the communication is likely to provoke violence.

Heavy drinking. Defined as binge drinking on five or more days during the past month.

Homegrown violent extremists. Are inspired by international terrorist groups.

Human capital. Consists of skills such as reading, writing, critical thinking, engineering a software, and knowing how to build and repair a car, which allow people to find a job that allows them to rent an apartment or buy a house, take care of their children and other dependents, and contribute to the economy.

Hypoxia. A condition where the brain does not receive enough oxygen.

Inconvenience policing. Aims at creating obstacles for buyers and sellers and discouraging novice and casual buyers from buying drugs at the open-air drug markets.

Informal social control. Aims to promote conformity to norms and laws via informal sanction, such as shaming, peer group pressure, family intervention, or other entities, such as churches.

Intelligence. Involves the collection of critical information related to the targeted criminality that provides substantive insight into crime threats and identifies individuals for whom there is a reasonable suspicion of relationship to a crime. Information collection is a constant process, along with ongoing information verification and analysis.

Intelligence-led policing. To prevent crime and terrorist attacks, more efficiently allocate resources, and develop counterterrorism strategies.

Intent. The purpose to use a particular means to achieve some definite result.

International terrorism. Defined as "violent, criminal acts committed by individuals and/or groups who are inspired by, or associated with, designated foreign terrorist organizations or nations (state-sponsored)."

Investigative reporter. In the news media, this type of reporter investigates a story, attempting to unearth evidence particularly in cases of wrongdoing to reveal to the public.

LESO. Law Enforcement Support Office.

Less-than-lethal weapons. Weapons that officers can use to gain compliance without going hands on physically with a subject, and not resorting to lethal force with their firearms.

Linking. Contributes to community resilience by promoting a "connections and equal partnership across vertical power differentials, e.g., government and communities."

Lone actors. Terrorist acts committed by one offender.

"Lost in transit." Defined as prescription drugs being misplaced while they are being transported from one place to another.

Memorandum of understanding (MOU). An agreement between two or more parties that outlines the responsibilities of each party to achieve a particular goal or set of goals.

Motive. Defined as the cause or moving power that impels action to achieve that result.

Motorcycle gangs. ongoing organizations, associations or groups of three or more persons with a common interest or activity characterized by the commission of, or involvement in, a pattern of criminal conduct. Outlaw motorcycle gangs (OMG) refer to themselves as the "one-percenters".

MS-13. Stands for *Mara Salvatrucha*, a violent and notorious gang known for using machetes to kill its victims.

Net widening. Refers to the increase in the number of people having contact with the criminal justice system as the unintentional result of a new practice.

New nets. Created by reforms that transfer intervention authority or jurisdiction from one agency or control system to another.

Officer discretion. Officers are given certain freedom when making decisions about how to handle various calls, with fewer restrictions on minor offenses and greater restrictions on felonious crimes.

Officer-involved domestic violence (OIDV). Domestic violence where the police officer is the offender.

Open system. A system that allows information to flow both ways, exchanging feedback from the environment or in this case the community policing, between police and citizens.

Open-air drug markets. Visible public settings where few barriers to access exist, as individuals unknown to dealers are able to purchase drugs.

Operational intelligence. Imperative operational decisions, such as the surveillance of an individual or a group.

Organizational transformation. Alignment of organizational management, structure, personnel, and information systems to support community partnerships and proactive problem solving.

Outlaw Motorcycle Gangs. Defined as ongoing organizations, associations, or groups of three or more persons with a common interest or activity characterized by the commission of, or involvement in, a pattern of criminal conduct.

Paradigm shifts. The shifts that occur when there is one set of thoughts, ideas, beliefs, values, and practices associated with an organization (in this case policing) that are challenged or discredited, and a new set of thoughts, ideas, beliefs, values, and practices replace the old ways.

Paramilitary. A semimilitary force that is similar to, but not part of a country's military force, characterized by having similar military structure, tactics, training, and subculture.

Perishable skills. Any skill that an officer can lose after a period of nonuse is called a perishable skill, such as pursuit driving, tactical firearms, force options, arrest and control, and verbal communications.

Police legitimacy. The extent to which members of the public believe and trust that the police have authority to enforce the law.

Posse comitatus. The practice of deputizing citizens by the sheriff to assist with arrests or during times of emergency, and the citizens are mandated by law to do so.

Pretextual stop. Police stop a subject using a minor infraction or violation as a means to dig deeper and find unrelated crimes.

Preventive patrol. A police strategy by walking or driving through an area, to prevent or discover potential problems in a neighborhood.

Prison gangs. Defined as criminal organizations that originated within the penal system and have continued to operate within correctional facilities throughout the United States.

Problem solving. Process of engaging in the proactive and systematic examination of identified problems to develop and evaluate effective responses.

Procedural fairness. Concerns public perceptions of how citizens are treated reflecting a consistency of quality in police services.

Procedural justice. The application of fairness and transparency in all police/citizen encounters.

Quality-of-life-policing. Describes a type of policing that targets highly visible crimes, even if they are minor.

Rapid response. A police strategy to respond to crime occurrences quickly in order to increase the likelihood of intervening in the crime and apprehending the offender.

Restorative conferencing. Encompasses a range of strategies for bringing together victims, offenders, and community members in nonadversarial community-based processes aimed at responding to crime by holding offenders accountable and repairing the harm caused to victims and communities.

Restorative policing. Refers to the application of community policing to restorative justice.

Revolving door of incarceration. Refers to individuals returning to jail over and over throughout their lives.

SARA model. A widely used problem-solving tool employed by police to identify problems and develop strategies on how to approach these problems.

Shooting galleries. Hidden indoor locations used by drug dealers and users.

Social capital. The value and trust in relationships between people in a particular community and the police

Social constructionism. The theory that people develop knowledge, perceptions, and beliefs of reality through social interaction, and the media plays a large role in that construction.

Soft skills. Refers to skills of communication, compassion, creativity, and problem solving which are especially effective in police work and other kinds of employment.

Spoils System. A term that refers to the notion "to the victors, go the spoils."

Stakeholders. Individual or groups, such as elected officials, business owners, nonprofit groups, other agencies, and the media, who have an interest or concern about decisions or actions of the police.

Stop and frisk. Officers may stop and pat down a subject when they have reasonable suspicion.

Strategic intelligence. Allows police department to plan and allocate resources and understand specific intelligence targets, including the targets, philosophy, structure, motivation, and characteristics.

Street gangs. Typically deal with large quantities of illegal drugs and are very violent in nature.

Stronger nets. Result of "reforms that increase the state's capacity to control individuals through intensifying state intervention."

SWAT teams. Special units of police that use military-style equipment and tactics.

Swatting. False reporting of emergencies to police in order to get a SWAT team sent to the address.

Tactical intelligence. Typically used for "raid planning" or in crisis situations.

Terrorism. Violent or dangerous crimes that "appear to be intended" to either (1) intimidate or coerce a civilian population, (2) influence government policy by intimidation or coercion, or (3) affect government conduct by mass destruction, assassination, or kidnapping.

Terrorism committed by large groups. Terrorist acts committed by organizations or groups with a large number of followers.

Trusting relationships. Relationships "where citizens voluntarily approach police officers with information or problems because they trust that law enforcement represents their best interests."

Unconscious biases. Typically defined as prejudice or unsupported judgments in favor of or against one thing, person, or group as compared to another, in a way that is usually considered unfair.

Uncontrolled items. Items police departments can purchase that are not tightly controlled, are of small monetary value, and have a short life span.

Unity of command. A classical approach of management, whereby each officer is accountable to one superior officer, except in special assignments or line-of-duty emergencies.

Use of force. The amount of effort required by police to gain compliance and overcome resistance of individuals engaged in unlawful behavior.

Us versus them mentality. Refers to officers viewing themselves as the good guys and the public as the enemy.

Vicarious liabilit. Vicarious liability is a legal concept whereby the city, the department, and supervisors are held accountable and legally liable for their officers' decisions and actions.

Vigilante justice. Citizens take the law into their own hands when there is an absence of legal avenues.

Violence. Generally used as a forensic term for investigations of interpersonal aggression or aggression between intimate partners or family members, also referred to as domestic violence.

Vote of no confidence. A vote by a majority in the organization that reflects nonsupport for the actions or decisions of a leader.

Wide nets. Created by reforms that increase the proportion of subgroups in society differentiated by such factors as age, sex, class, and ethnicity.

NOTES

CHAPTER 1

1. Bennett, D., Lee, J., Cahlan, S. (2020, May 30). The death of George Floyd: What video and other records show about his final minutes. *The Washington Post*. Retrieved from https://www.washingtonpost.com/nation/2020/05/30/video-timeline-george-floyd-death/?arc404=true
2. Fitz-Gibbon, J. (2020, May 28). Here's everything we know about the death of George Floyd. *New York Post*. Retrieved from https://nypost.com/2020/05/28/everything-we-know-about-the-death-of-george-floyd/
3. Campbell, J., Sidner, S., & Levenson, E. (2020, June 4). All Four Former officers involved in George Floyd's killing now face charges. Retrieved from *CNN*.https://www.cnn.com/2020/06/03/us/george-floyd-officers-charges/index.html
4. Andrew, S. (2020, June 11). There's a growing call to defund the police. Here's what it means. *CNN*. Retrieved from https://www.cnn.com/2020/06/06/us/what-is-defund-police-trnd/index.html
5. Wallace, D. (2020, June 8). These cities have begun defunding police in the wake of George Floyd protests. Fox News. Retrieved from https://www.foxnews.com/us/defund-police-george-floyd-protest-reforms-new-york-los-angeles-minneapolis
6. Andone, D. (2020, June 11). Seattle police want to return to vacated precinct in what protesters call an 'autonomous zone'. *CNN*. Retrieved from https://www.cnn.com/2020/06/11/us/seattle-police-autonomous-zone/index.html
7. Sprunt B, (2020, June 8). Read: Democrats release legislation to overhaul policing, https://www.npr.org/2020/06/08/872180672/read-democrats-release-legislation-to-overhaul-policing.
8. Davey, M. (September 7, 2018). A young man working to stop Chicago's gun violence loses his life to it. *The New York Times*. Retrieved from https://www.nytimes.com/2018/09/07/us/chicago-shooting-anti-gun-volunteer.html
9. FBI (2017). *Uniform crime reports*. Retrieved from https://ucr.fbi.gov/crime-in-the-u.s/2017/crime-in-the-u.s.-2017/tables/table-8/table-8-state-cuts/illinois.xlshg
10. Petition for Martial Law in Chicago. Retrieved from https://www.change.org/p/people-of-illinois-declare-martial-law-in-chicago
11. Rohloff, K. (August 7, 2018). *The Daily Signal*. Retrieved from https://www.dailysignal.com/2018/08/07/chicago-democrat-bucks-party-with-plea-for-trumps-help-after-weekend-bloodbath/
12. Riechmann, D., & Tarm, M. (October 9, 2018). Trump calls on Chicago to embrace stop-and frisk policing. *Huffington Post*. Retrieved from https://www.huffingtonpost.com/entry/trump-chicago-stop-and-frisk_us_5bbc79afe4b028e1fe412545
13. Schutz, P. (July 27, 2018). Deal struck for federal oversight of Chicago Police Department. WTTW News. Retrieved from https://news.wttw.com/2018/07/27/deal-struck-federal-oversight-chicago-police-department
14. Kappeler, V. E., & Gaines, L. K. (2015). *Community policing: A contemporary perspective* (7th ed.). New York, NY: Routledge.
15. Kelling, G. L., Pate, T., Dieckman, D., & Brown, C. E. (1974). The Kansas City Preventive Patrol Experiment: A summary report. First published in 1974, Recreated in 2003 for the Police Foundation Web site, www.policefoundation.org. Library of Congress Catalog Number 74-24739. Retrieved from https://www.policefoundation.org/publication/the-kansas-city-preventive-patrol-experiment/
16. Larson, R. (1975). What happened to patrol operations in Kansas City? A review of the Kansas City preventive patrol experiment. *Journal of Criminal Justice, 3*, 267–297. Retrieved from 10.1016/0047-2352(75)90034-3, file:///C:/Users/000002508/Downloads/Patrol-Operations-KS-a-review-1975.pdf
17. Kelling, G. L., Pate, T., Dieckman, D., & Brown, C. E. (1974). *The Kansas City preventive patrol experiment: A summary report*. First published in 1974, Recreated in 2003 for the Police Foundation Web site, www.policefoundation.org. Library of Congress Catalog Number 74-24739. Retrieved from https://www.policefoundation.org/publication/the-kansas-city-preventive-patrol-experiment/
18. Spelman, W., & Brown, D. (1984). Calling the police: Citizen reporting of serious crime (NIJ research report). Washington, D.C.: U.S. Dept. of Justice, National Institute of Justice. Retrieved from https://www.ncjrs.gov/App/publications/Abstract.aspx?id=82276

19. Proposal solicitation: Reducing non-emergency calls to 911: An assessment of four approaches to handling citizen calls for service (July 1997). The National Institute of Justice (NIJ) with support from the Office of Community Oriented Policing Services (COPS). Find request for proposal. Retrieved from https://www.ncjrs.gov/pdffiles/911.pdf
20. Ye Hee Lee, M. (July 12, 2016). Are most job-related deaths of police caused by traffic incidents? *The Washington Post*. Retrieved from https://www.washingtonpost.com/news/fact-checker/wp/2016/07/12/are-most-job-related-deaths-of-police-caused-by-traffic-incidents/?utm_term=.dcf33a57e297
21. NHTSA (National Highway Traffic Safety Administration). U.S. Department of Transportation. January 2011. Retrieved from https://crashstats.nhtsa.dot.gov/Api/Public/ViewPublication/811411 DOT HS 811 411
22. *Law enforcement officer motor vehicle safety*. Retrieved from https://www.cdc.gov/niosh/topics/leo/default.html
23. Woman, 20, dies after officer responding to call collides with vehicle (August 11, 2018). WCVB 5, ABC News Center 5. Retrieved from https://www.wcvb.com/article/police-cruiser-vehicle-involved-in-serious-crash-in-somerset/22695102
24. Police chief: Officer in fatal crash should be fired (September 20, 2018). *USA News*. Retrieved from https://www.usnews.com/news/best-states/indiana/articles/2018-09-20/police-chief-officer-in-fatal-crash-violated-policies
25. Greenwood, P. W., & Petersilia, J. (October 1975). *The criminal investigation process. Volume 1: Summary and policy implications*. Prepared under a grant from the National Institute of Law Enforcement and Criminal Justice, LEAA, Department of Justice. R-1776-DOJ. Retrieved from https://www.ncjrs.gov/pdffiles1/Digitization/148116NCJRS.pdf
26. The Free Dictionary. Legal Dictionary. Definition retrieved from https://legal-dictionary.thefreedictionary.com/due+care
27. Gaines, L. K., & Kappeler, V. E. (2014). *Policing in America* (8th ed.). Waltham. MA: Elsevier (Anderson Publishing).
28. FBI (2017). *Uniform crime reports*. Retrieved from https://ucr.fbi.gov/crime-in-the-u.s/2017/crime-in-the-u.s.-2017/tables/table-29
29. Cooper, C. (October 2, 2017). Community policing: What it's not. *Columbia Daily Tribune*. Retrieved from http://www.columbiatribune.com/news/20171002/community-policing-what-its-not
30. Kappeler, V. E., & Gaines, L. K. (2015). *Community policing: A contemporary perspective* (7th ed.). New York, NY: Routledge.
31. *Community policing defined*, COPS, U.S. Department of Justice. Retrieved from https://ric-zai-inc.com/Publications/cops-p157-pub.pdf
32. *Community policing defined*, COPS, U.S. Department of Justice. Retrieved from https://ric-zai-inc.com/Publications/cops-p157-pub.pdf
33. *Community policing defined*, COPS, U.S. Department of Justice. Retrieved from https://ric-zai-inc.com/Publications/cops-p157-pub.pdf
34. *Community policing defined*, COPS, U.S. Department of Justice. Retrieved from https://ric-zai-inc.com/Publications/cops-p157-pub.pdf
35. Fritsvold, E. (n.d.). 10 innovative police technologies. Law enforcement and public safety leadership. University of San Diego. Retrieved from https://onlinedegrees.sandiego.edu/10-innovative-police-technologies/
36. Trojanowicz, R., & Bucqueroux, B. (1990). *Community policing: A contemporary perspective* (pp. xiii–xv). Cincinnati, OH: Anderson Publishing Co.
37. Reisig, M. (2010). Community and problem-oriented policing. *Crime and Justice*, 39(1), 1–53. doi:10.1086/652384
38. City of Chula Vista News (November 8, 2018, 3:41 p.m.). Chula Vista police receives international award. CVPD awarded for domestic violence reduction. Retrieved from https://www.chulavistaca.gov/Home/Components/News/News/1840/3175
39. Hardyns, W., & Rummens, A. (2018). Predictive policing as a new tool for law enforcement? Recent developments and challenges. *European Journal on Criminal Policy and Research*, 24(3), 201–218. doi: 10.1007/s10610-017-9361-2
40. Dukowitz, Z. (May 10, 2018). 6 ways police departments use drones in their work. *UAV Coach*. Retrieved from https://uavcoach.com/police-drones/
41. Noble, B. (July 2, 2015). 7 ways police departments use social media. *Newsmax Independent American*. Retrieved from https://www.newsmax.com/FastFeatures/police-social-media/2015/07/02/id/653310/
42. Dees, T. (November 15, 2019). Research review: Identifying the benefits of ALPR systems. PoliceOne.com. Retrieved from https://www.policeone.com/police-products/traffic-enforcement/license-plate-readers/articles/research-review-identifying-the-benefits-of-alpr-systems-wYft4lyw4ONt5Wqv
43. News stories about law enforcement ALPR successes: September 2017 to September 2018. (n.d.). Retrieved from https://www.theiacp.org/sites/default/files/ALPR%20Success%20News%20Stories%202018.pdf
44. Roufa, T. (August 3, 2019). Technologies that are changing the way police do business. *the balancecareers*. Retrieved from https://www.thebalancecareers.com/technologies-that-are-changing-the-way-police-do-business-974549

45. Trojanowicz, R., & Bucqueroux, B. (1990). *Community policing: A contemporary perspective* (pp. 209–212). Cincinnati, OH: Anderson Publishing Co.

CHAPTER 2

1. U.S. Marshals Service. U.S. Department of Justice. Retrieved from https://www.usmarshals.gov/history/broad_range.htm
2. CNN (March 22, 2019). DC area sniper fast facts. Retrieved from https://edition.cnn.com/2013/11/04/us/dc-area-sniper-fast-facts/index.html
3. Kelling, G. L., & Moore, M. H. (1988). *The evolving strategy of policing*. U.S. Department of Justice. Retrieved from https://pdfs.semanticscholar.org/a614/21a27a6c4fa0e25962ef30e95a22371c1b9c.pdf
4. Walker, S. (1977). *A critical history of police reform*. Lexington, MA: Lexington Books
5. Ornaghi, A. (May 19, 2018). Civil service reforms: Evidence from U.S. police departments. Retrieved from file:///C:/Users/000002508/Pictures/ariannaornaghi59767pdf.pdf
6. Walker, S. (1996). *The police in America: An introduction*. New York: McGraw Hill.
7. Walker, S. (1977). *A critical history of police reform*. Lexington, MA:[8] Lexington Books.
8. Walker, S. (1977). *A critical history of police reform*. Lexington, MA: Lexington Books.
9. Kelling, G. L., & Moore, M. H. (1988). *The evolving strategy of policing*. U.S. Department of Justice. Retrieved from https://pdfs.semanticscholar.org/a614/21a27a6c4fa0e25962ef30e95a22371c1b9c.pdf
10. Federal Bureau of Investigation (2012). *The Hoover legacy: 40 years after*. Retrieved from https://www.fbi.gov/news/stories/the-hoover-legacy-40-years-after-part-4
11. Kelling, G. L., & Moore, M. H. (1988). *The evolving strategy of policing*. U.S. Department of Justice. Retrieved from https://pdfs.semanticscholar.org/a614/21a27a6c4fa0e25962ef30e95a22371c1b9c.pdf
12. Kelling, G. L. & Moore, M. H. (1988). *The Evolving Strategy of Policing*. U.S. Department of Justice. Available online: https://pdfs.semanticscholar.org/a614/21a27a6c4fa0e25962ef30e95a22371c1b9c.pdf
13. President Johnson's Crime Commission Report on Law Enforcement and Administration (1967). *The challenge of crime in a free society*. United States Government Printing Office, Washington, DC. Retrieved from https://assets.documentcloud.org/documents/3932081/Crimecommishreport.pdf
14. Kelling, G. L., & Moore, M. H. (1988). *The evolving strategy of policing*. U.S. Department of Justice. Retrieved from https://pdfs.semanticscholar.org/a614/21a27a6c4fa0e25962ef30e95a22371c1b9c.pdf
15. Bureau of Justice Assistance. (August 1994). *Understanding community policing: A framework for action*. NCJ 148457 (Monograph). Retrieved from https://www.ncjrs.gov/pdffiles/commp.pdf
16. The Peelian principles. The crime prevention website. Retrieved from https://thecrimepreventionwebsite.com/police-crime-prevention-service-a-short-history/744/the-peelian-principles/
17. Reith, C. (1956). *A new study of police history*. London: Oliver & Boyd. p. 140.
18. Trojanowicz, R., & Bucqueroux, B. (1990). *Community policing: A contemporary perspective*. Cincinnati, OH: Anderson Publishing Co.
19. Trojanowicz, R., & Bucqueroux, B. (1990). *Community policing: A contemporary perspective*. Cincinnati, OH: Anderson Publishing Co.
20. *The challenge of crime in a free society*. A report by the President's Commission on Law Enforcement and Administration of Justice (February 1967). United States Government Printing Office, Washington, D.C. p. 118. Retrieved from https://www.ncjrs.gov/pdffiles1/11430.pdf file:///C:/Users/000002508/Pictures/42.pdf
21. Sherman, L.W., Milton, C.H., & Kelly, T. V. (August 1973). *Team policing seven case studies*. Police Foundation, Washington, D.C. Retrieved from https://www.ncjrs.gov/pdffiles1/11430.pdf
22. *The Newark foot patrol experiment* (1981). The Police Foundation. Washington, D. C. Retrieved from https://www.policefoundation.org/publication/the-newark-foot-patrol-experiment/
23. *The challenge of crime in a free society*. A report by the President's Commission on Law Enforcement and Administration of Justice (February 1967). United States Government Printing Office, Washington, D.C. Retrieved from https://www.ncjrs.gov/pdffiles1/nij/42.pdf
24. *Perspectives on foot patrols: Lessons learned from foot patrol programs and an overview of foot patrol in San Francisco* (November 19, 2007). Interim Report. Prepared for The City and County of San Francisco. Retrieved from https://www.sfcontroller.org/ftp/uploadedfiles/controller/reports/SFPD_Foot_Beat_Report-Final%20_111907.pdf

25. *The Newark foot patrol experiment* (1981). The Police Foundation. Washington, D.C. Retrieved from https://www.policefoundation.org/publication/the-newark-foot-patrol-experiment/
26. Trojanowicz, R., & Bucqueroux, B. (1990). *Community policing: A contemporary perspective*. Cincinnati, OH: Anderson Publishing Co.
27. Holder, S. (January 10, 2018). What happened to crime in Camden? City Lab. Retrieved from https://www.citylab.com/equity/2018/01/what-happened-to-crime-in-camden/549542/
28. Garofalo, J. (Summer 1981). The fear of crime: Causes and consequences. *Journal of Criminal Law and Criminology*. V.72(2) Article 20. Retrieved from 0091-4169/81/7202-0839 and https://scholarlycommons.law.northwestern.edu/cgi/viewcontent.cgi?referer=https://search.yahoo.com/&httpsredir=1&article=6243&context=jclc
29. Skogan, W. G. (2012) Disorder and crime. Chapter 9 in Brandon C. Welsh & David P. Farrington (Eds.), *The Oxford Handbook of Crime Prevention*. Oxford: Oxford University Press. Retrieved from http://www.skogan.org/files/Disorder_and_Crime.in_Welsh_and_Farrington_2012.pdf
30. Kelling, G., & Wilson, J. Q. (March 1982). Broken windows: The police and neighborhood safety. *The Atlantic*. Retrieved from https://www.theatlantic.com/magazine/archive/1982/03/broken-windows/304465/ or https://www.manhattan-institute.org/pdf/_atlantic_monthly-broken_windows.pdf
31. Kelling, G., & Wilson, J. Q. (March 1982). Broken windows The police and neighborhood safety. *The Atlantic*. Retrieved from https://www.theatlantic.com/magazine/archive/1982/03/broken-windows/304465/ or https://www.manhattan-institute.org/pdf/_atlantic_monthly-broken_windows.pdf
32. Sampson, R. J. (2009). Disparity and diversity in the contemporary city: Social (dis)order revisited. *British Journal of Sociology*, *60*, 1–31.
33. Duncan, J., & Banerji, S. (April 5, 2015, 8:42 p.m.). Community policing brings hope to Camden. CBS News. Retrieved from https://www.cbsnews.com/news/community-policing-brings-hope-to-camden/
34. Holder, S. (January 10, 2018). What happened to crime in Camden? City Lab. Retrieved from https://www.citylab.com/equity/2018/01/what-happened-to-crime-in-camden/549542/
35. Duncan, J., & Banerji, S. (April 5, 2015). Community policing brings hope to Camden. CBS News. Retrieved from https://www.cbsnews.com/news/community-policing-brings-hope-to-camden/
36. Zernike, K. (August 31, 2014) Camden turns around with new police force. *The New York Times*. Retrieved from https://www.nytimes.com/2014/09/01/nyregion/camden-turns-around-with-new-police-force.html
37. Holder, S. (January 10, 2018). What happened to crime in Camden? City Lab. Retrieved from https://www.citylab.com/equity/2018/01/what-happened-to-crime-in-camden/549542/
38. What is broken windows theory. Retrieved from https://www.wisegeek.com/what-is-the-broken-windows-theory.htm
39. Pate, A. M., Wycoff, M. A., Skogan, W. G., & Sherman, L. W. (1986). *Reducing fear of crime in Houston and Newark. A summary report*. Police Foundation. National Institute of Justice. Retrieved from https://www.policefoundation.org/wp-content/uploads/2015/07/Pate-et-al.-1986-Reducing-Fear-of-Crime-in-Houston-and-Newark-Summary-Report-.pdf

CHAPTER 3

1. Deak, M. (2018). Penn state frat members gave pledge 18 drinks in less than 90 minutes, video shows. *USA Today*. Retrieved from https://eu.usatoday.com/story/news/nation-now/2017/11/13/penn-state-frat-charges/859524001/
2. National Institute of Justice. Practice profile: Problem oriented policing. Retrieved January 5, 2020 from https://www.crimesolutions.gov/PracticeDetails.aspx?ID=32
3. Eck, J., & Spelman, W. (1987). *Problem solving: Problem-oriented policing in Newport news*. Washington, DC: Police Executive Research Forum.
4. National Institute of Justice. Practice profile: Problem oriented policing. Retrieved January 5, 2020 from https://www.crimesolutions.gov/PracticeDetails.aspx?ID=32
5. Substance Abuse and Mental Health Services Administration (SAMHSA) 2015 National Survey on Drug Use and Health (NSDUH). Table 2.41B—Alcohol use in lifetime, past year, and past month among persons aged 12 or older, by demographic characteristics: Percentages, 2014 and 2015. Retrieved March 6, 2019 from https://www.samhsa.gov/data/sites/default/files/NSDUH-DetTabs-2015/NSDUH-DetTabs-2015/NSDUH-DetTabs-2015.htm#tab2-41b
6. Substance Abuse Service Center Addiction treatment services. Gambling treatment services. Retrieved March 6, 2019 from http://www.sasc-dbq.org/alcohol-and-crime
7. Substance Abuse and Mental Health Services Administration (SAMHSA). 2015 National Survey on Drug Use and Health (NSDUH). Table 2.41B—Alcohol use in lifetime, past year, and past month among persons aged 12 or older, by

demographic characteristics: Percentages, 2014 and 2015. Retrieved March 6, 2019 from https://www.samhsa.gov/data/sites/default/files/NSDUH-DetTabs-2015/NSDUH-DetTabs-2015/NSDUH-DetTabs-2015.htm#tab2-41b

8. Substance Abuse and Mental Health Services Administration (SAMHSA) 2015 National Survey on Drug Use and Health (NSDUH). Table 2.41B—Alcohol use in lifetime, past year, and past month among persons aged 12 or older, by demographic characteristics: Percentages, 2014 and 2015. Retrieved March 6, 2019 from https://www.samhsa.gov/data/sites/default/files/NSDUH-DetTabs-2015/NSDUH-DetTabs-2015/NSDUH-DetTabs-2015.htm#tab2-41b

9. Gorman, D. M., Gruenewald, P. J., & Speer, P. (2001). Spatial dynamics of alcohol availability, neighborhood structure, and violent crime. *Journal of Studies on Alcohol*, *62*(5), 628–646.

10. Murdoch, D., Pihl, R. O., & Ross, D. (1990). Alcohol and crimes of violence: Present issues. *International Journal of the Addictions*, *25*, 1065–1081.

11. Hoaken, P. N. S., & Stewart, S. H. (2003). Drugs of abuse and the elicitation of human aggressive behavior. *Addictive Behaviors*, *28*, 1533–1544.

12. Hoaken, P. N. S., & Stewart, S. H. (2003). Drugs of abuse and the elicitation of human aggressive behavior. *Addictive Behaviors*, *28*, 1533–1544.

13. Community Livability Unit (2018). *Targeting alcohol use and abuse to reduce crime in a university town*. Presentation at the Center for Problem Oriented Policing. November 5, 2018.

14. Gorman, D. M., Gruenewald, P. J., & Speer, P. (2001). Spatial dynamics of alcohol availability, neighborhood structure, and violent crime. *Journal of Studies on Alcohol*, *62*(5), 628–646.

15. Wilson, J. Q., and Kelling, G. L. (1982). Broken windows: The police and neighborhood safety. *Atlantic Monthly*, *March*, 29–38.

16. Livingston M., Chikritzhs, T., & Room, R. (2007). Changing the density of alcohol outlets to reduce alcohol-related problems. *Drug and Alcohol Review*, *26*, 557–566.

17. Burau of Justice Assistance (2006). *Alcohol and violent crime: What is the connection? What can be done?* Office of Justice Programs. Retrieved from http://www.nllea.org/documents/alcohol_and_crime.pdf.

18. Livingston M., Chikritzhs, T., & Room, R. (2007). Changing the density of alcohol outlets to reduce alcohol-related problems. *Drug and Alcohol Review*, *26*, 557–566.

19. 2006 Ohio revised code—2923.15. *Using weapons while intoxicated*. Retrieved from https://law.justia.com/codes/ohio/2006/orc/jd_292315-68af.html.

20. Branas, C. *Alcohol and firearms: Research, gaps in knowledge, and possible interventions*. The National Academies Press. Retrieved December 30, 2018 from https://www.nap.edu/resource/21814/Alcohol-Firearms.pdf.

21. Weinberg, W. (2012). Irvine man convicted of gross vehicular manslaughter while intoxicated. Orange County DUI Lawyer Blog. Retrieved from https://www.orangecountyduilawyerblog.com/irvine_man_convicted_of_gross_1/.

22. Burau of Justice Assistance (2006). Alcohol and violent crime: What is the connection? What can be done? Office of Justice Programs. Retrieved from http://www.nllea.org/documents/alcohol_and_crime.pdf.

23. Treatment Advocacy Center (2019). *Road runners: the role and impact of law enforcement in transporting individuals with severe mental illness, a national survey*. Retrieved from https://www.treatmentadvocacycenter.org/road-runners

24. Wood, J. D., Watson, A. C., & Fulambarker, A. J. (2017). The "gray zone" of police work during mental health encounters: Findings from an observational study in Chicago. *Police Quarterly*, *20*(1), 81–105.

25. Wood, J. D., Watson, A. C., & Fulambarker, A. J. (2017). the "gray zone" of police work during mental health encounters: Findings from an observational study in Chicago. *Police Quarterly*, 20(1), 81–105.

26. Yohanna, D. (2013). Deinstitutionalizing people with mental illness: Causes and consequences. *American Medical Association Journal of Ethics*, *15*(10), 886–891.

27. Treatment Advocacy Center (2016). *Serious mental illness (SMI) in jails and prisons*. A background paper from the Office of Research and Public Affairs. Retrieved from https://www.treatmentadvocacycenter.org/storage/documents/backgrounders/smi-in-jails-and-prisons.pdf

28. National Alliance on Mental Illness (2019). *Jailing people with mental illness*. Retrieved from https://www.nami.org/Learn-More/Public-Policy/Jailing-People-with-Mental-Illness

29. Wood, J. D., Watson, A. C., & Fulambarker, A. J. (2017). The "gray zone" of police work during mental health encounters: Findings from an observational study in Chicago. *Police Quarterly*, *20*(1), 81–105.

30. Wood, J. D., Watson, A. C., & Fulambarker, A. J. (2017). The "gray zone" of police work during mental health encounters: Findings from an observational study in Chicago. *Police Quarterly*, *20*(1), 81–105.

31. Wood, J. D., Watson, A. C., & Fulambarker, A. J. (2017). The "gray zone" of police work during mental health encounters: Findings from an observational study in Chicago. *Police Quarterly*, *20*(1), 81–105.

32. Ellis, H. A. (2014). Effects of a crisis intervention team (CIT) training program upon police officers before and after crisis intervention team training. *Archives of Psychiatric Nursing, 28*, 10–18. Retrieved from https://www.psychiatricnursing.org/article/S0883-9417(13)00121-0/pdf
33. Ornstein, N., & Leifman, S. (August 11, 2017). How mental health training for police can save lives—and taxpayer dollars. *The Atlantic*. Retrieved from https://www.theatlantic.com/politics/archive/2017/08/how-mental-health-training-for-police-can-save-livesand-taxpayer-dollars/536520/
34. Martin, K., Kay, K., & Gilbert, A. (2018). *The Croft unit: Who's behaving badly?* Durham Constabulary, Presentation at the 25th Problem Oriented Policing Conference in Rhode Island.
35. Hwang, K., & Mack, J. L. (2018). Chicago Mercy Hospital shooting: What we know about Tamara O'Neal, other victims, incident. *IndyStar*. Retrieved from https://eu.indystar.com/story/news/2018/11/20/mercy-shooting-chicago-hospital-tamara-oneal-dayna-less-samuel-jimenez/2068177002/.
36. Reaves, B. A. (2017). *Police response to domestic violence, 2006 to 2015*. Bureau of Justice Statistics. U.S. Department of Justice. Retrieved from https://www.bjs.gov/content/pub/pdf/prdv0615.pdf
37. National Center for Victims of Crime (2018). *Intimate Partner Violence*. Retrieved from https://ovc.ncjrs.gov/ncvrw2018/info_flyers/fact_sheets/2018NCVRW_IPV_508_QC.pdf
38. Reaves, B. A. (2017). *Police response to domestic violence, 2006 to 2015*. Bureau of Justice Statistics. U.S. Department of Justice. Retrieved from https://www.bjs.gov/content/pub/pdf/prdv0615.pdf
39. Reaves, B. A. (2017). *Police response to domestic violence, 2006 to 2015*. Bureau of Justice Statistics. U.S. Department of Justice. Retrieved from https://www.bjs.gov/content/pub/pdf/prdv0615.pdf
40. Block, R. (2003). How can practitioners help an abused woman lower her risk of death? In *Intimate Partner Homicide* (pp. 4–8). National Institute of Justice. Retrieved from https://www.ncjrs.gov/pdffiles1/jr000250.pdf
41. Campbell, J. C., Sharps, P. W., & Glass, N. (2000). Risk assessment for intimate partner violence. In G.-F. Pinard & L. Pagani (Eds.). *Clinical Assessment of Dangerousness: Empirical Contributions* (pp. 136–157). New York: Cambridge University Press.
42. Breul, N., & Keith, M. (2015). Deadly calls and fatal encounters. Analysis of U.S. law enforcement line of duty deaths when officers responded to dispatched calls for service and conducted enforcement (2010–2014). Community Oriented Policing Services. U.S. Department of Justice.
43. Breul, N., & Keith, M. (2015). Deadly calls and fatal encounters. Analysis of U.S. law enforcement line of duty deaths when officers responded to dispatched calls for service and conducted enforcement (2010–2014). Community Oriented Policing Services. U.S. Department of Justice.
44. Oehme, K., Prost, S. G., & Saunders, D. G. (2016). Police responses to cases of officer involved domestic violence: The effects of a brief Web-based training. *Policing: A Journal of Police and Practice, 10*(4), 391-407.
45. Oehme, K., Prost, S. G., & Saunders, D. G. (2016). Police responses to cases of officer involved domestic violence: The effects of a brief Web-based training. *Policing: A Journal of Police and Practice, 10*(4), 391-407.
46. National Center for Women and Policing (2019[VK1]). *Police family violence fact sheet*. Retrieved from http://womenandpolicing.com/violencefs.asp [VK1]
47. Oehme, K., Prost, S. G., & Saunders, D. G. (2016). Police responses to cases of officer involved domestic violence: The effects of a brief Web-based training. *Policing: A Journal of Police and Practice, 10*(4), 391-407.
48. Sadusky, J. (2003). *Community policing and domestic violence. Five promising practices*. The Battered Women's Justice Project. Retrieved from https://www.bwjp.org/assets/documents/pdfs/community_policing_and_domestic_violence.pdf
49. Sadusky, J. (2003). *Community policing and domestic violence. Five promising practices*. The Battered Women's Justice Project. Retrieved from https://www.bwjp.org/assets/documents/pdfs/community_policing_and_domestic_violence.pdf
50. Sadusky, J. (2003). *Community policing and domestic violence. Five promising practices*. The Battered Women's Justice Project. Retrieved from https://www.bwjp.org/assets/documents/pdfs/community_policing_and_domestic_violence.pdf
51. Sadusky, J. (2003). *Community policing and domestic violence. Five promising practices*. The Battered Women's Justice Project. Retrieved from https://www.bwjp.org/assets/documents/pdfs/community_policing_and_domestic_violence.pdf

CHAPTER 4

1. Alcoke, M. (2019). *The evolving and persistent terrorism threat to the homeland*. Washington Institute for Near East Policy Counterterrorism Lecture Series. Federal Bureau of Investigation. Retrieved from https://www.fbi.gov/news/speeches/the-evolving-and-persistent-terrorism-threat-to-the-homeland-111919

2. Braziel, R., Straub, F., Watson, G., & Hoops, R. (2016). *Bringing calm to chaos: A critical incident review of the San Bernardino Public Safety response to the December 2, 2015, terrorist shooting incident at the Inland Regional Center.* Critical Response Initiative. Washington, DC: Office of Community Oriented Policing Services.
3. Braziel, R., Straub, F., Watson, G., & Hoops, R. (2016). *Bringing calm to chaos: A critical incident review of the San Bernardino Public Safety response to the December 2, 2015, terrorist shooting incident at the Inland Regional Center.* Critical Response Initiative. Washington, DC: Office of Community Oriented Policing Services.
4. International Association of Chiefs of Police (2014). *Using community policing to counter violent extremism: Five key principles for law enforcement.* Washington, DC: Office of Community Oriented Policing Services.
5. US Code 18 USC 113B: *Terrorism.* Retrieved February 16, 2020 from https://uscode.house.gov/view.xhtml?path=/prelim@title18/part1/chapter113B&edition=prelim
6. Federal Bureau of Investigation. *Unabomber.* Retrieved February 24, 2020 from https://www.fbi.gov/history/famous-cases/unabomber
7. Federal Bureau of Investigation. *Terrorism.* Retrieved February 16, 2020 from https://www.fbi.gov/investigate/terrorism.
8. Alcoke, M. (2019). *The evolving and persistent terrorism threat to the homeland.* Washington Institute for Near East Policy Counterterrorism Lecture Series. Federal Bureau of Investigation. Retrieved from https://www.fbi.gov/news/speeches/the-evolving-and-persistent-terrorism-threat-to-the-homeland-111919
9. Federal Bureau of Investigation. *Terrorism.* Retrieved February 16, 2020 from https://www.fbi.gov/investigate/terrorism.
10. Alcoke, M. (2019). *The evolving and persistent terrorism threat to the homeland.* Washington Institute for Near East Policy Counterterrorism Lecture Series. Federal Bureau of Investigation. Retrieved from https://www.fbi.gov/news/speeches/the-evolving-and-persistent-terrorism-threat-to-the-homeland-111919
11. Ellis, R., Frantz, A., Faith, K., & McLaughlin, E. C. (June 13, 2016). Orlando shooting: 49 killed, shooter pledged ISIS allegiance. CNN. Retrieved from https://www.cnn.com/2016/06/12/us/orlando-nightclub-shooting/index.html
12. Alcoke, M. (2019). *The evolving and persistent terrorism threat to the homeland.* Washington Institute for Near East Policy Counterterrorism Lecture Series. Federal Bureau of Investigation. Retrieved from https://www.fbi.gov/news/speeches/the-evolving-and-persistent-terrorism-threat-to-the-homeland-111919
13. Donague, E. (2020). Racially-motivated violent extremists elevated to "national threat priority," FBI director says. CBS News. Retrieved from https://www.cbsnews.com/news/racially-motivated-violent-extremism-isis-national-threat-priority-fbi-director-christopher-wray/
14. Lavendera, E. (August 9, 2019). El Paso suspect told police he was targeting Mexicans, affidavit Says. CNN. Retrieved from https://www.cnn.com/2019/08/09/us/el-paso-shooting-friday/index.html
15. McGarrity, M. C., & Shivers, C. A. (2019). *Confronting white supremacy.* Statement Before the House Oversight and Reform Committee, Subcommittee on Civil Rights and Civil Liberties, Federal Bureau of Investigation, Washington, D.C. Retrieved from https://www.fbi.gov/news/testimony/confronting-white-supremacy
16. Alcoke, M. (2019). *The Evolving and persistent terrorism threat to the homeland.* Washington Institute for Near East Policy Counterterrorism Lecture Series. Federal Bureau of Investigation. Retrieved from https://www.fbi.gov/news/speeches/the-evolving-and-persistent-terrorism-threat-to-the-homeland-111919
17. Federal Bureau of Investigation (2010). *Domestic terrorism. Anarchist extremism: A primer.* Retrieved from https://archives.fbi.gov/archives/news/stories/2010/november/anarchist_111610/anarchist_111610
18. Federal Bureau of Investigation (2004). *Testimony before the Senate Judiciary Committee.* John E. Lewis, deputy assistant director, Federal Bureau of Investigation, Washington, DC. Retrieved from https://archives.fbi.gov/archives/news/testimony/animal-rights-extremism-and-ecoterrorism
19. Federal Bureau of Investigation (2004). *Testimony before the Senate Judiciary Committee.* John E. Lewis, deputy assistant director, Federal Bureau of Investigation, Washington, DC. Retrieved from https://archives.fbi.gov/archives/news/testimony/animal-rights-extremism-and-ecoterrorism
20. National Consortium for the Study of Terrorism and Responses to Terrorism (2013). *An overview of bombing and arson attacks by environmental and animal rights extremists in the United States, 1995–2010.* Final Report to the Resilient Systems Division, Science and Technology Doctorate, U.S. Department of Homeland Security. Retrieved from https://www.dhs.gov/sites/default/files/publications/OPSR_TP_TEVUS_Bombing-Arson-Attacks_Environmental-Animal%20Rights-Extremists_1309-508.pdf
21. Federal Bureau of Investigation. What are known violent extremist groups? Retrieved February 28, 2020 from https://www.fbi.gov/cve508/teen-website/what-are-known-violent-extremist-groups
22. Jacobson, M. and Royer, H. (2010). *Aftershocks: The impact of clinic violence on abortion services.* RAND Corporation.

23. Murphy, K. (April 3, 2013). Kansas abortion clinic reopens four years after doctor's murder. Reuters. Retrieved from https://www.reuters.com/article/us-usa-abortion-kansas/kansas-abortion-clinic-reopens-four-years-after-doctors-murder-idUSBRE93305020130404
24. Federal Bureau of Investigation. What are known violent extremist groups? Retrieved February 28, 2020 from https://www.fbi.gov/cve508/teen-website/what-are-known-violent-extremist-groups
25. Davis, E. (July 10, 2013). *Testimony of the police commissioner Boston Police Department before the Senate Committee on Homeland Security and Government Affairs Committee.* Retrieved from at: https://www.hsgac.senate.gov/imo/media/doc/Testimony-Davis-2013-07-10-REVISED.pdf
26. Bureau of Justice Assistance (2008). *Reducing crime through intelligence-led policing.* U.S. Department of Justice. Retrieved from https://www.bja.gov/publications/reducingcrimethroughilp.pdf
27. Bullock, K. (2010). Generating and using community intelligence: The case of neighbourhood policing. *International Journal of Police Science & Management, 12*(1), 1–11.
28. Department of Homeland Security (2013). *Lessons learned: Boston Marathon bombings: The positive effects of planning and preparation on response.* Federal Emergency Management Agency.
29. Boston Globe Media Partners (2018). *Timeline of Boston Marathon bombing event.* Retrieved from https://www.boston.com/news/local-news/2015/01/05/timeline-of-boston-marathon-bombing-events
30. Loyka, S. A., Faggiani, D. A., & Karchmer, C. (2005). *Protecting your community from terrorism: The strategies for local law enforcement series.* Police Executive Research Forum, Washington, DC 20036, United States of America.
31. Loyka, S. A., Faggiani, D. A, & Karchmer, C. (2005). *Protecting your community from terrorism: The strategies for local law enforcement series.* Police Executive Research Forum, Washington, DC 20036, United States of America.
32. Gonzales, A. R., Schofield, R. B., & Herraiz, D. B. (2005). *Intelligence-led policing: A new intelligence architecture.* Bureau of Justice Assistance. U.S. Department of Justice.
33. Loyka, S. A., Faggiani, D. A., & Karchmer, C. (2005). *Protecting your community from terrorism: The strategies for local law enforcement series.* Police Executive Research Forum, Washington, DC 20036, United States of America.
34. CBS (April 20, 2015). We have a terror recruiting problem in Minnesota. Retrieved from https://www.cbsnews.com/news/six-minnesota-men-arrested-trying-to-join-isis-identified/
35. Lyons, W. (2002). Partnerships, information and public safety: community policing in a time of terror. *Policing: An International Journal of Police Strategies and Management, 25*(3), 530–542.
36. Klausen, J. (2009). British counter-terrorism after 7/7: adapting community policing to the fight against domestic terrorism. *Journal of Ethnic and Migration Studies, 35*(3), 403–420.
37. Office of Community Oriented Policing Services. About. Retrieved 2018 from https://cops.usdoj.gov/about.
38. Office of Community Oriented Policing Services. About. Retrieved 2018 from https://cops.usdoj.gov/about.
39. Wasserman, R. and Ginsburg, Z. (2014). *Building relationships of trust: Moving to implementation.* Tallahassee, FL: Institute for Intergovernmental Research.
40. Davies, H. J., & Murphy, G. R. (2004). *Protecting your community from terrorism: The strategies for local law enforcement series. Vol. 2: Working with diverse communities.* Community Oriented Policing Services. U.S. Department of Justice.
41. Davies, H. J., & Murphy, G. R. (2004). *Protecting your community from terrorism: The strategies for local law enforcement series. Vol. 2: Working with diverse communities.* Community Oriented Policing Services. U.S. Department of Justice.
42. International Association of Chiefs of Police (2014). *Using community policing to counter violent extremism: Five key principles for law enforcement.* Washington, DC: Office of Community Oriented Policing Services.
43. Spalek, B. (2013). *Terror crime prevention with communities.* London: Bloomsbury Academic.
44. Cherney, A., & Hartley, J. (2017). Community engagement to tackle terrorism and violent extremism: Challenges, tensions, and pitfalls. *Policing and Society, 27*(7), 750–763.
45. Rodgers, L., Qurashi, S., & Connor, S. (July 3, 2015). 7 July London bombings: What happened that day? BBC News.
46. Innes, M. (2006). Policing uncertainty: Countering terror through community intelligence and democratic policing. *The Annals of the American Academy of Political and Social Science, 605,* 221–241.
47. Henderson, N. J., Ortiz, C. W., Sugie, N. F., & Miller, J. (2006). *Law enforcement and Arab American community relations after September 11, 2001. Engagement in a time of uncertainty.* Vera Institute of Justice.
48. De Bord, M. (September 11, 2017). 23 daunting photos from the September 11 attack that we will never forget. *Business Insider.* Retrieved from http://www.businessinsider.com/haunting-photos-from-september-11th-attacks-2016-9
49. Henderson, N. J., Ortiz, C. W., Sugie, N. F., & Miller, J. (2006). *Law enforcement and Arab American community relations after September 11, 2001. Engagement in a time of uncertainty.* Vera Institute of Justice.
50. Wasserman, R. and Ginsburg, Z. (2014). *Building relationships of trust: Moving to implementation.* Tallahassee, FL: Institute for Intergovernmental Research.

51. Kennedy, D. M. (2011). *Don't shoot: One man, a street fellowship, and the end of violence in inner-city America*. New York: Bloomsbury, USA.
52. Wasserman, R., & Ginsburg, Z. (2014). *Building relationships of trust: Moving to implementation*. Tallahassee, FL: Institute for Intergovernmental Research.
53. Wasserman, R., & Ginsburg, Z. (2014). *Building relationships of trust: Moving to implementation*. Tallahassee, FL: Institute for Intergovernmental Research.
54. Balko, R. (2006). *Overkill: The rise of paramilitary police raids in America*. CATO Institute. Retrieved from research.policyarchive.org/5806.pdf.
55. Sun, I. Y., & Wu, Y. (2015). Arab American's confidence in police. *Crime & Delinquency*, *61*(4), 483–508.
56. Cherney, A., & Hartley, J. (2015). Community engagement to tackle terrorism and violent extremism: Challenges, tensions, and pitfalls. *Policing and Society*, *27*(7), 750–763.
57. Cherney, A., & Hartley, J. (2015). Community engagement to tackle terrorism and violent extremism: Challenges, tensions, and pitfalls. *Policing and Society*, *27*(7), 750–763.
58. Vermeulen, F. (2014). Suspect communities—Targeting violent extremism at the local level: Policies of engagement in Amsterdam, Berlin, and London. *Terrorism and Political Violence*, *26*(2), 286–306.
59. Vermeulen, F. (2014). Suspect communities—Targeting violent extremism at the local level: Policies of engagement in Amsterdam, Berlin, and London. *Terrorism and Political Violence*, *26*(2), 286–306.
60. Palmer, P. (2012). Dealing with the exceptional: Pre-crime anti-terrorism policy and practice. *Policing and Society*, *22*(4), 519–537.
61. Klausen, J. (2009). British counter-terrorism after 7/7: Adapting community policing to the fight against domestic terrorism. *Journal of Ethnic and Migration Studies*. *35*(3), 403–420.
62. Klausen, J. (2009). British counter-terrorism after 7/7: Adapting community policing to the fight against domestic terrorism. *Journal of Ethnic and Migration Studies*. *35*(3), 403–420.
63. Murphy, K., & Cherney, A. (2011). Fostering cooperation with the police: How do ethnic minorities in Australia respond to procedural justice-based policing. *Australian and New Zealand Journal of Criminology*, *44* (2), 235–257.
64. Pickering, S., McCulloch, J., & Wright-Neville, D. (2008). *Counter terrorism policing: Community, cohesion and security*. New York: Springer.
65. Homeland Security Committee. (2016). *Final Report of the Task Force on Combating Terrorist and Foreign Fighter Travel*. New York, NY: Skyhorse Publishing.
66. Cherney, A., & Hartley, J. (2015). Community engagement to tackle terrorism and violent extremism: Challenges, tensions, and pitfalls. *Policing and Society*, *27*(7), 750–763.
67. Vermeulen, F. (2014). Suspect communities—Targeting violent extremism at the local level: Policies of engagement in Amsterdam, Berlin, and London. *Terrorism and Political Violence*, *26*(2), 286–306.
68. Bergengruen, V., & Hennigan, W. J. (August 8, 2019). We are being eaten from within. Why America is losing the battle against white nationalist terrorism. *Time*. Retrieved from https://time.com/5647304/white-nationalist-terrorism-united-states/
69. Sanborn, J. (February 26, 2020). *Confronting the rise in anti-Semitic domestic terrorism*. Statement Before the House Committee on Homeland Security, Subcommittee on Intelligence and Counterterrorism Washington, D.C. Retrieved from https://www.fbi.gov/news/testimony/confronting-the-rise-in-anti-semitic-domestic-terrorism
70. Mercy, J. A., Rosenberg, M. L., Powell, K. E., Broome, C. V., & Roper, W. L. (1993). Public health policy for preventing violence. *Health Affairs*, *12*, 7–29. Retrieved from http://dx.doi.org/10.1377/hlthaff.12.4.7
71. Ellis, B.H., & Abdi, S. (2017). Building community resilience to violent extremism through genuine partnerships. *American Psychologist*, *72*(3), 289–300.
72. Sampson, R. J., Raudenbush, S. W., & Earls, F. (1997). Neighborhoods and violent crime: A multilevel study of collective efficacy. *Science*, *277*, 918–924. Retrieved from http://dx.doi.org/10.1126/science.277.5328.918
73. Ellis, B. H., & Abdi, S. (2017). Building community resilience to violent extremism through genuine partnerships. *American Psychologist*, *72*(3), 289–300.
74. Ellis, B. H., & Abdi, S. (2017). Building community resilience to violent extremism through genuine partnerships. *American Psychologist*, *72*(3), 289–300.
75. Ellis, B.H., & Abdi, S. (2017). Building community resilience to violent extremism through genuine partnerships. *American Psychologist*, *72*(3), 289–300.
76. Smith, B. L., Damphousse, K. R., & Roberts, P. (2006). Pre-incident indicators of terrorist incidents: The identification of geographic and temporal patterns of preparatory conduct. US Department of Justice. NIJ Award Number 2003-DT-CX-0003. Retrieved December 10, 2007 from http://www.ncjrs.gov/pdffiles1/nij/grants/214217.pdf.

77. Department of Homeland Security (2018). *Budget in brief. Fiscal year 2018*. U.S. Department of Homeland Security. Retrieved from https://www.dhs.gov/sites/default/files/publications/DHS%20FY18%20BIB%20Final.pdf
78. Horn, S., & Kuhn, J. (2018). Legislative alert: White House releases the FY 2018 proposed budget. *The Police Chief Magazine*. Retrieved from http://www.policechiefmagazine.org/legislative-alert-white-house-releases-fy-2018-proposed-budget/
79. Office of Community Oriented Policing Services. (2004). COPS Grants Awarded in FY 2004. Retrieved from http://www.cops.usdoj.gov/mime/open.pdf?Item=800

CHAPTER 5

1. Fedschun, T. (January 11, 2018). Female MS-13 gang members, 18, told teen girl "don't forget my name" in video confession of brutal killing. Fox News. Retrieved from http://www.foxnews.com/us/2018/01/11/ms-13-gang-member-told-teen-girl-dont-forget-my-name-in-video-confession-brutal-killing.html
2. Fedschun, T. (January 11, 2018). Female MS-13 gang members, 18, told teen girl "don't forget my name" in video confession of brutal killing. Fox News. Retrieved from http://www.foxnews.com/us/2018/01/11/ms-13-gang-member-told-teen-girl-dont-forget-my-name-in-video-confession-brutal-killing.html
3. U.S. Department of Justice (2015). *About violent gangs*. Retrieved from https://www.justice.gov/criminal-ocgs/about-violent-gangs
4. National Gang Intelligence Center (2009). *2009 National Gang Threat Assessment*.
5. U.S. Department of Justice (2015). About violent gangs. Retrieved from https://www.justice.gov/criminal-ocgs/about-violent-gangs.
6. Griffith, K. (August 6, 2017). Barrio 18: Meet the terrifying gang with 50,000 foot-soldiers across the US and so unashamedly violent it rivals MS-13. *Daily Mail*, Retrieved from http://www.dailymail.co.uk/news/article-4764744/Barrio-18-Meet-terrifying-gang-rivals-MS-13.html.
7. U.S. Department of Justice (2015).About violent gangs. Retrieved from https://www.justice.gov/criminal-ocgs/gallery/prison-gangs.
8. U.S. Department of Justice (2015). About violent gangs. Retrieved from https://www.justice.gov/criminal-ocgs/gallery/prison-gangs.
9. Southern Poverty Law Center (n.d.). *Aryan brotherhood*. Retrieved from https://www.splcenter.org/fighting-hate/extremist-files/group/aryan-brotherhood.
10. Southern Poverty Law Center (n.d.). *Aryan brotherhood*. Retrieved from https://www.splcenter.org/fighting-hate/extremist-files/group/aryan-brotherhood.
11. National Gang Intelligence Center (2015). *National Gang Report 2015*.
12. Bosmia, A. N., Quinn, J. F., Peterson, T. B., Griessenauer, C. J., & Shane Tubbs, R. (2014). Outlaw motorcycle gangs: Aspects of the one-percenter culture for emergency department personnel to consider. *Western Journal of Emergency Medicine*, *15*(4), 523–528. Retrieved from https://www.ncbi.nlm.nih.gov/pmc/articles/PMC4100862/.
13. National Gang Intelligence Center (2015). *National Gang Report 2015*.
14. Bosmia, A. N., Quinn, J. F., Peterson, T. B., Griessenauer, C. J., & Shane Tubbs, R. (2014). Outlaw motorcycle gangs: Aspects of the one-percenter culture for emergency department personnel to consider. *Western Journal of Emergency Medicine*, *15*(4), 523–528. Retrieved from https://www.ncbi.nlm.nih.gov/pmc/articles/PMC4100862/.
15. National Gang Intelligence Center (2015). *National Gang Report 2015*.
16. Park, M., & Hamasaki, S. (2017). Grand jury indicts 11 members of Hells Angels. CNN. Retrieved from https://edition.cnn.com/2017/11/21/us/hells-angels-indictments/index.html.
17. National Gang Intelligence Center (2015). *National Gang Report 2015*.
18. National Gang Intelligence Center (2015). *National Gang Report 2015*.
19. *New York Times* (January 7, 2014). Seeking clues to gangs and crime, detectives monitor internet rap videos. Retrieved from http://www.nytimes.com/2014/01/08/nyregion/seeking-clues-to-gangs-and-crime-detectives-monitor-internet-rap-videos.html?_r=
20. Gordon, M. (April 23, 2017). Their crime made Charlotte feel less safe. now Doug and Debbie London's killers face a reckoning. *The Charlotte Observer*. Retrieved from https://www.charlotteobserver.com/news/local/crime/article146335124.html.
21. Woody, C. (December 17, 2016). The strange way one of America's largest street gangs got its name. *Business Insider*. Retrieved from https://www.businessinsider.de/origins-of-ms-13-mara-salvatrucha-gang-name?r=US&IR=T.

22. The White House (2018). President Donald J. Trump is dedicated to combating MS-13. Fact Sheet. Retrieved from https://www.whitehouse.gov/briefings-statements/president-donald-j-trump-dedicated-combating-ms-13/.
23. Al Jazeera (2018). The teenagers trapped in Trump's M-13 crackdown. Retrieved from https://www.aljazeera.com/indepth/features/teenagers-trapped-trump-ms13-crackdown-180410070837978.html.
24. Department of Justice (2011). Five alleges east side Los Guada blood gang members and associates indicted in Arizona on federal racketeering and attempted murder charges. Office of Public Affairs. Retrieved from https://www.justice.gov/opa/pr/five-alleged-east-side-los-guada-bloods-gang-members-and-associates-indicted-arizona-federal.
25. Leland, E. (May 31, 2017). Blood gang leader used prison cellphone to order hit on prosecutor's father. *The Charlotte Observer*. Retrieved from https://www.charlotteobserver.com/news/local/crime/article152334207.html.
26. Walker, M. A. (February 24, 2014). Detroit police chief answers gang member's death threat. *USA Today*. Retrieved from https://eu.usatoday.com/story/news/nation/2014/02/24/detroit-police-chief-death-threat/5772937/.
27. NBC (2014). Rogers bathroom graffiti threatens "cop will be killed on New Years." Kark.com. Retrieved from https://www.kark.com/news/rogers-bathroom-graffiti-threatens-cop-will-be-killed-on-new-years/199715176.
28. National Gang Intelligence Center (2015). *National Gang Report 2015*.
29. *New York Times* (January 7, 2014). Seeking clues to gangs and crime, detectives monitor internet rap videos. Retrieved from http://www.nytimes.com/2014/01/08/nyregion/seeking-clues-to-gangs-and-crime-detectives-monitor-internet-rap-videos.html?_r=0.
30. National Gang Intelligence Center (2015). *National Gang Report 2015*.
31. Miller, M. E. (June 11, 2018). "A ticking time bomb": MS-13 threatens middle school, warn teachers, parents, students. *The Washington Post*. Retrieved from https://www.washingtonpost.com/local/a-ticking-time-bomb-ms-13-threatens-a-middle-school-warn-teachers-parents-students/2018/06/11/7cfc7036-5a00-11e8-858f-12becb4d6067_story.html?utm_term=.b8fad129579e.
32. G.R.E.A.T. (2018). Gang Resistance and Education Training. Retrieved from https://www.great-online.org/Home/About/ResearchBasis.
33. Green, A. (June 10, 2013). New details: Brothers shot 13-year old in back, bludgeoned him, over Portland girl. *The Oregonion*. Retrieved from https://www.oregonlive.com/portland/index.ssf/2013/06/new_details_brothers_shot_13-y.html.
34. Green, E. (2016). Youth and the gang life: Their stories in their words. *Street Roots News*. Retrieved from https://news.streetroots.org/2016/05/05/youths-and-gang-life-their-stories-their-words.
35. Green, E. (2016). Youth and the gang life: Their stories in their words. *Street Roots News*. Retrieved from https://news.streetroots.org/2016/05/05/youths-and-gang-life-their-stories-their-words.
36. Green, E. (2016). Youth and the gang life: Their stories in their words. *Street Roots News*. Retrieved from https://news.streetroots.org/2016/05/05/youths-and-gang-life-their-stories-their-words.
37. Botvin, G., Baker, E., Dusenbury, L., Botvin, E., & Diaz. T. (1995). Long-term follow-up results of a randomized drug abuse prevention trial, *JAMA, 273*, 1106–1112.
38. Los Angeles Police Department (2018). Why young people join gangs. Retrieved from http://www.lapdonline.org/get_informed/content_basic_view/23473.
39. Green, E. (2016). Youth and the gang life: Their stories in their words. *Street Roots News*. Retrieved from https://news.streetroots.org/2016/05/05/youths-and-gang-life-their-stories-their-words.
40. Howell, K. B. (2011). Fear itself: The impact of allegations of gang affiliation in pre-trial detention. *St. Thomas Law Review, 23*, 620–659. Retrieved from https://academicworks.cuny.edu/cgi/viewcontent.cgi?article=1090&context=cl_pubs
41. Ladd, D. (2018). Dangerous, growing, yet unnoticed: The rise of America's white gangs. *The Guardian*. Retrieved from https://www.theguardian.com/society/2018/apr/05/white-gangs-rise-simon-city-royals-mississippi-chicago
42. Greene, J., & Pranis, K. (2007). *Gang wars: The failure of enforcement tactics and the need for effective public safety strategies*. Justice Policy Institute. Retrieved from http://www.justicepolicy.org/research/1961
43. Greene, J., & Pranis, K. (2007). *Gang wars: The failure of enforcement tactics and the need for effective public safety strategies*. Justice Policy Institute. Retrieved from http://www.justicepolicy.org/research/1961
44. Howell, K. B. (2011). Fear itself: The impact of allegations of gang affiliation in pre-trial detention. *St. Thomas Law Review, 23*, 620–659. Retrieved from https://academicworks.cuny.edu/cgi/viewcontent.cgi?article=1090&context=cl_pubs
45. Ladd, D. (2018). Dangerous, growing, yet unnoticed: The rise of America's white gangs. *The Guardian*. Retrieved from https://www.theguardian.com/society/2018/apr/05/white-gangs-rise-simon-city-royals-mississippi-chicago
46. Snyder, H. N., & Sickmund, M. (1999). *Juvenile offenders and victims: 1999 national report*. Washington, DC: U.S. Department of Justice, Office of Juvenile Justice and Delinquency Prevention.

47. Woods, J. (2012). Systemic racial bias and RICO's application to criminal street and prison gangs. *Michigan Journal of Race and Law, 17*(2), 304–357.
48. Ladd, D. (2018). Dangerous, growing, yet unnoticed: The rise of America's white gangs. *The Guardian*. Retrieved from https://www.theguardian.com/society/2018/apr/05/white-gangs-rise-simon-city-royals-mississippi-chicago
49. National Gang Center (Accessed March 8, 2020). National Youth Gang Survey Analysis, 2009. Retrieved from https://www.nationalgangcenter.gov/survey-analysis.
50. Kobayashi, F. (1999). *Model minority stereotype reconsidered*. New York, NY: ERIC Clearinghouse on Urban Education. (ERIC Document Reproduction Service No. ED434167).
51. Shelden, R. G., Tracy, S. K., & Brown, W. B. (2013). *Youth gangs in American society*. Belmont, CA: Wadsworth, Cengage Learning.
52. Toy, C. (1992). A short history of Asian gangs in San Francisco. *Justice Quarterly, 9*(4), 647–665.
53. United Gangs. *Asian gangs*. Retrieved March 8, 2020 from https://unitedgangs.com/category/asian-gangs/
54. Miller, M. E. (June 11, 2018). "A ticking time bomb": MS-13 threatens middle school, warn teachers, parents, students. *The Washington Post*. Retrieved from https://www.washingtonpost.com/local/a-ticking-time-bomb-ms-13-threatens-a-middle-school-warn-teachers-parents-students/2018/06/11/7cfc7036-5a00-11e8-858f-12becb4d6067_story.html?utm_term=.b8fad129579e.
55. Rodriguez, R. (2020). Court records show how Fresno street gang planned mass shooting that killed four. *Fresno Bee*. Retrieved from https://www.fresnobee.com/news/local/crime/article238921758.html
56. Schmidt, M. (2012). U.S. gang alignment with Mexican drug trafficking organization. *Combating Terrorism Center at West Point, 5*(3), 18–20.
57. Schmidt, M. (2012). U.S. gang alignment with Mexican drug trafficking organization. *Combating Terrorism Center at West Point, 5*(3), 18–20.
58. Schmidt, M. (2012). U.S. gang alignment with Mexican drug trafficking organization. *Combating Terrorism Center at West Point, 5*(3), 18–20.
59. Schmidt, M. (2012). U.S. gang alignment with Mexican drug trafficking organization. *Combating Terrorism Center at West Point, 5*(3), 18–20.
60. Johnson, D. (January 4, 1990). Teen-agers who won't join when drug dealers recruit. *New York Times*. Retrieved from https://www.nytimes.com/1990/01/04/us/teen-agers-who-won-t-join-when-drug-dealers-recruit.html/.
61. National Gang Intelligence Center (2015). *National Gang Report 2015*.
62. U.S. Attorney's Office (2018). *Thirty-seven gang members charged in crackdown of North County heroin, methamphetamine and firearms traffickers*. U.S. Department of Justice. Southern District of California. Retrieved from https://www.justice.gov/usao-sdca/pr/thirty-seven-gang-members-charged-crackdown-north-county-heroin-methamphetamine-and
63. Decker, S. (2013). What is the role of police in preventing gang membership? In T. Simon, N. Ritter, & R. Mahendra (Eds.). *Changing Course: Preventing Gang Membership* (pp. 51–62). Washington, DC: National Institute of Justice. Retrieved from https://www.ncjrs.gov/pdffiles1/nij/239234.pdf.
64. Decker, S. (2013). What is the role of police in preventing gang membership? In T. Simon, N. Ritter, & R. Mahendra (Eds.). *Changing Course: Preventing Gang Membership* (pp. 51–62). Washington, DC: National Institute of Justice. Retrieved from https://www.ncjrs.gov/pdffiles1/nij/239234.pdf.
65. Office of Juvenile Justice and Delinquency Prevention (2010). *Best practices to address community gang problems. OJJDP's comprehensive gang model*. U.S. Department of Justice. Office of Justice Systems. Retrieved from https://www.ncjrs.gov/pdffiles1/ojjdp/231200.pdf
66. Martin-Mollard, M., & Becker, M. (2009). Key Components of hospital-based violence intervention programs. Youth ALIVE! Oakland, CA. Retrieved from file:///Users/janinekremling/Documents/Books/Book%20-%20Community%20Policing/Chapter%205%20-%20Community%20Policing,%20Gangs,%20and%20Drugs%20%20/Articles/Key%20Components%20of%20emergency%20room%20based%20violence%20intervention%20programs.pdf.
67. Office of Juvenile Justice and Delinquency Prevention (2010). *Best practices to address community gang problems. OJJDP's comprehensive gang model*. U.S. Department of Justice. Office of Justice Systems. Retrieved from https://www.ncjrs.gov/pdffiles1/ojjdp/231200.pdf
68. G.R.E.A.T. (2018). Gang resistance and education training. Retrieved from https://www.great-online.org/Home/About/ResearchBasis.
69. G.R.E.A.T. (2018). Gang resistance and education training. Retrieved from https://www.great-online.org/Home/About/ResearchBasis.

CHAPTER 6

1. Robertson, C. (2019). Texas executes White supremacist for 1998 dragging death of James Byrd Jr. *The New York Times*. Retrieved from https://www.nytimes.com/2019/04/24/us/james-byrd-jr-john-william-king.html
2. Federal Bureau of Investigation (2007). *Hate crime statistics 2009*. Washington, DC: U.S. Department of Justice. Retrieved from https://www2.fbi.gov/ucr/hc2009/index.html
3. U.S. Department of Justice (2019). *Hate crime laws*. Retrieved from https://www.justice.gov/crt/hate-crime-laws.
4. Police Foundation (2018). *Releasing open data on hate crimes. A best practices guide for law enforcement agencies*. Washington, D.C. Retrieved from https://www.policefoundation.org/wp-content/uploads/2018/01/PF_Releasing-Open-Data-on-Hate-Crimes_Final.pdf.
5. US Legal (2018). *Hate speech law and definition*. Retrieved from https://definitions.uslegal.com/h/hate-speech/.
6. Bauer-Wolf, J. (February 25, 2019). Hate incidents on campus still rising. *Inside Higher Education*. Retrieved from https://www.insidehighered.com/news/2019/02/25/hate-incidents-still-rise-college-campuses.
7. Bauer-Wolf, J. (October 23, 2017). Lessons from Spencer's Florida speech. *Inside Higher Education*. Retrieved from https://www.insidehighered.com/news/2017/10/23/nine-lessons-learned-after-richard-spencers-talk-university-florida.
8. Toppo, G. (September 24, 2017). Milo Yannopolous holds brief rally after cancelling free speech week. *USA Today*, Sept. 24. Retrieved from https://eu.usatoday.com/story/news/2017/09/24/milo-yiannopoulos-holds-brief-rally-after-uc-berkeley-cancels-free-speech-week/698418001/.
9. U.S. Department of Justice (2017). *Hate Crime Statistics Act. As Amended, 28 U.S.C. §534*. Retrieved from https://ucr.fbi.gov/hate-crime/2017/resource-pages/hate-crime-statistics-act.pdf
10. Police Foundation (2018). *Releasing open data on hate crimes. A best practices guide for law enforcement agencies*. Washington, D.C. Retrieved from https://www.policefoundation.org/wp-content/uploads/2018/01/PF_Releasing-Open-Data-on-Hate-Crimes_Final.pdf.
11. Shively, M. (2005). *Study of literature and legislation on hate crime in America*. U.S. Department of Justice. Retrieved from https://www.ncjrs.gov/pdffiles1/nij/grants/210300.pdf.
12. *Chaplinsky v. New Hampshire* (1942). 315 U.S. 568.
13. *Virginia v. Black* (2003). 538 U.S. 343.
14. *Watts v. U.S.* (1969). 394 U.S. 705
15. Bell, J. (2002). Deciding when hate is a crime: The First Amendment, police detectives, and the identification of hate crime. Articles by Maurer Faculty. *Rutgers Race and the Law Review*, 32–76. Retrieved from https://pdfs.semanticscholar.org/d7f5/da64c1df15eeff359e42d5c648529172b891.pdf
16. Police Foundation (2018). *Releasing open data on hate crimes. A best practices guide for law enforcement agencies*. Washington, D.C. Retrieved from https://www.policefoundation.org/wp-content/uploads/2018/01/PF_Releasing-Open-Data-on-Hate-Crimes_Final.pdf.
17. Shively, M. (2005). *Study of literature and legislation on hate crime in America*. U.S. Department of Justice. Retrieved from https://www.ncjrs.gov/pdffiles1/nij/grants/210300.pdf.
18. Police Foundation (2018). *Releasing open data on hate crimes. A best practices guide for law enforcement agencies*. Washington, D.C. Retrieved from https://www.policefoundation.org/wp-content/uploads/2018/01/PF_Releasing-Open-Data-on-Hate-Crimes_Final.pdf.
19. Shively, M. (2005). *Study of literature and legislation on hate crime in America*. U.S. Department of Justice. Retrieved from https://www.ncjrs.gov/pdffiles1/nij/grants/210300.pdf.
20. Levin, B., Grisham, K., & Reitzel, J. (May 18, 2018). *Hate crimes rise in U.S. cities and counties in time of division and foreign interference. Compilation of official data (38 jurisdictions)*. Center for the Study of Hate & Crime. California State University at San Bernardino. Retrieved from http://www.ochumanrelations.org/wp-content/uploads/2018/05/2018-Hate-Report-CSU-San-Bernardino.pdf.
21. Masucci, M., & Langton, L. (2017). *Hate crime victimization, 2004–2015*. Special report. Bureau of Justice Statistics. U.S. Department of Justice.
22. Police Foundation (2018). *Releasing open data on hate crimes. A best practices guide for law enforcement agencies*. Washington, D.C. Retrieved from https://www.policefoundation.org/wp-content/uploads/2018/01/PF_Releasing-Open-Data-on-Hate-Crimes_Final.pdf.
23. Masucci, M., & Langton, L. (2017). *Hate crime victimization, 2004–2015*. Special report. Bureau of Justice Statistics. U.S. Department of Justice.

24. Federal Bureau of Investigation (November 13, 2018). 2017 Hate crime statistics released. Report shows more departments reporting hate crime statistics. FBI News. Retrieved from https://www.fbi.gov/news/stories/2017-hate-crime-statistics-released-111318
25. The Mississippi Civil Rights Movement. *The murder of Chaney, Goodman, and Schwerner*. Retrieved July 6, 2019 from https://mscivilrightsproject.org/neshoba/event-neshoba/the-murder-of-chaney-goodman-and-schwerner/.
26. U.S. Department of Justice (2019). Criminal interference with right to fair housing. Retrieved from https://www.justice.gov/crt/criminal-interference-right-fair-housing.
27. Deitle, C. M. Esq. (March 6, 2019). The legacies of James Byrd Jr. and Matthew Shepard: Two decades later. *Police Chief Online*. Retrieved from https://www.policechiefmagazine.org/legacies-byrd-and-shepard/.
28. U.S. Department of Justice (2019). *Hate crime laws*. Retrieved from https://www.justice.gov/crt/hate-crime-laws.
29. U.S. Department of Justice. *Laws and policies*. Retrieved 2019 from https://www.justice.gov/hatecrimes/laws-and-policies
30. U.S. Department of Justice. *Criminal interference with right to fair housing*. Retrieved 2019 from https://www.justice.gov/crt/criminal-interference-right-fair-housing
31. U.S. Department of Justice. *Laws and policies*. Retrieved 2019 from https://www.justice.gov/hatecrimes/laws-and-policies
32. Department of Justice (2018). Texas man sentenced to 24 years for hate crime in burning down mosque in Victoria, Texas. Office of Public Affairs. Retrieved from https://www.justice.gov/opa/pr/texas-man-sentenced-almost-25-years-hate-crime-burning-down-mosque-victoria-texas.
33. U.S. Department of Justice (Retrieved 2019). Laws and policies. Retrieved 2019 from https://www.justice.gov/hatecrimes/laws-and-policies
34. Anti-Defamation League (2012). *Hate crime laws—The ADL approach*. Retrieved from https://www.adl.org/sites/default/files/documents/assets/pdf/combating-hate/Hate-Crimes-Law-The-ADL-Approach.pdf
35. Shively, M. (2005). *Study of literature and legislation on hate crime in America*. U.S. Department of Justice. Retrieved from https://www.ncjrs.gov/pdffiles1/nij/grants/210300.pdf.
36. Morsch, J.(1991). The problem of motive in hate crimes: The argument against presumptions of racial motivation. *Journal of Criminal Law and Criminology*, *82*(3), p. 658–689.
37. U.S. Department of Justice. Laws and policies. Retrieved 2019 from https://www.justice.gov/hatecrimes/laws-and-policies
38. Anti-Defamation League (2012). *Hate crime laws—The ADL approach*. Retrieved from https://www.adl.org/sites/default/files/documents/assets/pdf/combating-hate/Hate-Crimes-Law-The-ADL-Approach.pdf
39. Shively, M. (2005). *Study of literature and legislation on hate crime in America*. U.S. Department of Justice. Retrieved from https://www.ncjrs.gov/pdffiles1/nij/grants/210300.pdf
40. Alongi, B. (2016). The negative ramifications of hate crime legislation: It's time to reevaluate whether hate crime laws are beneficial to society. *Pace Law Review*, *37*(1), 325–351.
41. Bell, J. (2002). Deciding when hate is a crime: The First Amendment, police detectives, and the identification of hate crime. Articles by Maurer Faculty. *Rutgers Race and the Law Review*, 32–76. Retrieved from https://pdfs.semanticscholar.org/d7f5/da64c1df15eeff359e42d5c648529172b891.pdf
42. Morsch, J.(1991). The problem of motive in hate crimes: The argument against presumptions of racial motivation. *Journal of Criminal Law and Criminology*, *82*(3), 658–689.
43. Morsch, J.(1991). The problem of motive in hate crimes: The argument against presumptions of racial motivation. *Journal of Criminal Law and Criminology*, *82*(3), 658–689.
44. Morsch, J.(1991). The problem of motive in hate crimes: The argument against presumptions of racial motivation. *Journal of Criminal Law and Criminology*, *82*(3), 658–689.
45. Office for Equity, Diversity, and Inclusion (2019). *Unconscious bias*. Vanderbilt University. Retrieved from https://www.vanderbilt.edu/diversity/unconscious-bias/.
46. Office for Equity, Diversity, and Inclusion (2019). *Unconscious bias*. Vanderbilt University. Retrieved from https://www.vanderbilt.edu/diversity/unconscious-bias/.
47. Lantz, B., Gladfelter, A. S., & Ruback, R. B. (2017). Stereotypical hate crimes and criminal justice processing: A multi-dataset comparison of bias crime arrest patterns by offender and victim race. *Justice Quarterly*, 1–32.
48. Morsch, J.(1991). The problem of motive in hate crimes: The argument against presumptions of racial motivation. *Journal of Criminal Law and Criminology*, *82*(3), 658–689.
49. Swiffen, A. (2018). New resistance to hate crime legislation and the concept of law. *Law, Culture, and the Humanities*, *14*(1), 121–139.
50. Morsch, J. (1991). The problem of motive in hate crimes: The argument against presumptions of racial motivation. *Journal of Criminal Law and Criminology*, *82*(3), 658–689.

51. Wolff, K. B. & Cokely, C. L. (2007). To serve and protect? An exploration of police conduct in relation to the gay, lesbian, bisexual, and transgender community. *Sexuality & Culture*, *11*(2), 1–23.
52. National Coalition of Anti-Violence Programs (2008). *Hate violence against lesbian, gay, bisexual, and transgender people in the United States in 2008*. New York: National Coalition of Anti-Violence Programs, 2009. Retrieved from https://avp.org/wp-content/uploads/2017/04/2008_NCAVP_HV_Report.pdf.
53. National Coalition of Anti-Violence Programs (2008). *Hate violence against lesbian, gay, bisexual, and transgender people in the United States in 2008*. New York: National Coalition of Anti-Violence Programs, 2009. Retrieved from https://avp.org/wp-content/uploads/2017/04/2008_NCAVP_HV_Report.pdf.
54. Not in Our Town. Interview with Bernard Malekian, director for the U.S. Department of Justice Office of Community Oriented Policing Services (COPS). Retrieved July 4, 2019 from https://www.niot.org/niot-video/community-policing-and-hate-crimes-interview-doj-cops-director-bernard-melekian
55. U.S. Department of Justice. *Laws and policies*. Retrieved 2019 from https://www.justice.gov/hatecrimes/laws-and-policies
56. Edwards, J. (2013). Oak Creek: Leading a community in the aftermath of a tragedy. Not in Our Town. Community Oriented Policing Services. Retrieved from https://www.niot.org/cops/casestudies/oak-creek-leading-community-aftermath-tragedy

CHAPTER 7

1. Bickel, K. (2013). Will the growing militarization of police doom community policing. *Newsletter of the COPS Office*, *6*(12). 1–3.
2. American Civil Liberties Union (2014). War comes home. The excessive militarization of american policing. Retrieved from https://www.aclu.org/sites/default/files/assets/jus14-warcomeshome-report-web-rel1.pdf
3. ABC News (January 14, 2018). California man Tyler Barriss charged in Kansas after fatal call of duty "swatting" hoax. Retrieved from http://www.abc.net.au/news/2018-01-14/california-man-charged-in-kansas-for-fatal-hoax-call/9327198.
4. KWCHZ (2018). Kan. Governor signs Anti-Swatting Bill. April 12, Retrieved from http://www.kwch.com/content/news/Kan-governor-signs-anti--479570733.html.
5. Adachi, J. (2016–17). Police militarization and the war on civil rights. *Human Rights Magazine*, *42*(1). Available online at: https://www.americanbar.org/publications/crsj-human-rights-magazine/vol--42/vol-42-no-1/police-militarization-and-the-war-on-citizens.html.
6. Balko, R. (2013). *Rise of the warrior cop. The militarization of America's police force*. New York: Public Affairs.
7. Nordheimer, J. (July 14, 1975). Tough elite police units useful but controversial, *New York Times*. Retrieved from https://www.nytimes.com/1975/07/14/archives/tough-elite-police-units-useful-but-controversial-elite-police.html.
8. Balko, R. (2013). *Rise of the warrior cop. The militarization of America's police force* (p. 304). New York: Public Affairs.
9. Drug Policy Alliance. A brief history of the drug war. Retrieved June 24 from http://www.drugpolicy.org/issues/brief-history-drug-war.
10. Powers, T. (September 4, 1977). Nixon's drug crusade. *New York Times*. Retrieved from https://www.nytimes.com/1977/09/04/archives/nixons-drug-crusade-drug.html.
11. Balko, R. (2013). *Rise of the warrior cop. The militarization of America's police force*. New York: Public Affairs.
12. Balko, R. (2013). *Rise of the warrior cop. The militarization of America's police force*. New York: Public Affairs.
13. The White House (1986). *National Security Decision Directive Number 221*. Retrieved from https://www.hsdl.org/?abstract&did=463177
14. 99th Congress (1988). H.R. 5484—Anti-Drug Abuse Act of 1986. Retrieved from https://www.congress.gov/bill/99th-congress/house-bill/5484
15. 99th Congress (1988). H.R. 5210—Anti-Drug Abuse Act of 1986. Retrieved from https://www.congress.gov/bill/100th-congress/house-bill/05210
16. Murch, D. (2015). Crack in Los Angeles: Crisis, militarization, and Black response to the late twentieth century war on drugs. *The Journal of American History*, *102*(1), 162–173. Retrieved from https://academic.oup.com/jah/article/102/1/162/686732
17. Dansky, K. (2014). *Another day, another 124 violent SWAT raids*. American Civil Liberties Unit. Retrieved from https://www.aclu.org/blog/smart-justice/another-day-another-124-violent-swat-raids.
18. American Civil Liberties Union (2014). *War comes home. The excessive militarization of American policing*. Retrieved from https://www.aclu.org/sites/default/files/assets/jus14-warcomeshome-report-web-rel1.pdf

19. American Civil Liberties Union (2014). War comes home. The excessive militarization of American policing. Retrieved from https://www.aclu.org/sites/default/files/assets/jus14-warcomeshome-report-web-rel1.pdf
20. Rizer, A., & Hartman, J. (November 7, 2011). How the war on terror has militarized the police. *The Atlantic.*
21. Griffith, D. (August 3, 2017). SWAT in the 21st century. *Police Law Enforcement Bulletin.* Retrieved from https://www.policemag.com/342295/swat-in-the-21st-century
22. Griffith, D. (August 3, 2017). SWAT in the 21st century. *Police Law Enforcement Bulletin.* Retrieved from https://www.policemag.com/342295/swat-in-the-21st-century
23. Griffith, D. (August 3, 2017). SWAT in the 21st century. *Police Law Enforcement Bulletin.* Retrieved from https://www.policemag.com/342295/swat-in-the-21st-century
24. Phillips, S. W. (2017). Police militarization. *FBI Law Enforcement Bulletin.* United States Department of Justice. Retrieved from https://leb.fbi.gov/articles/featured-articles/police-militarization
25. Phillips, S. W. (2017). Police militarization. *FBI Law Enforcement Bulletin.* United States Department of Justice. Retrieved from https://leb.fbi.gov/articles/featured-articles/police-militarization
26. General Accounting Office (2017). *DOD excess property. Enhanced controls needed for access to excess controlled property.* GAO 17_532. Retrieved from https://www.gao.gov/assets/690/685916.pdf.
27. General Accounting Office (2017). *DOD excess property. Enhanced controls needed for access to excess controlled property.* GAO 17_532. (p. 6). Retrieved from https://www.gao.gov/assets/690/685916.pdf.
28. General Accounting Office (2017). *DOD excess property. Enhanced controls needed for access to excess controlled property.* GAO 17_532. Retrieved from https://www.gao.gov/assets/690/685916.pdf.
29. General Accounting Office (2017). *DOD excess property. Enhanced controls needed for access to excess controlled property.* GAO 17_532 (p. 22). Retrieved from https://www.gao.gov/assets/690/685916.pdf.
30. General Accounting Office (2017). *DOD excess property. Enhanced controls needed for access to excess controlled property.* GAO 17_532 (p. 25). Retrieved from https://www.gao.gov/assets/690/685916.pdf
31. Bickel, K. (2013). Will the growing militarization of police doom community policing. *Newsletter of the COPS Office, 6*(12), 1–3.
32. Bickel, K. (2013). Will the growing militarization of police doom community policing. *Newsletter of the COPS Office, 6*(12), 1–3.
33. Konnikova, M. (December 3, 2014). Dressed to suppress. *The New Yorker.* Retrieved from https://www.newyorker.com/science/maria-konnikova/will-decreasing-police-use-military-gear-prevent-another-ferguson
34. American Civil Liberties Union (2014). War comes home. The excessive militarization of American policing. Retrieved from https://www.aclu.org/sites/default/files/assets/jus14-warcomeshome-report-web-rel1.pdf
35. Shropshire, T. (February 3, 2016). Georgia family gets 3.6 million after cops threw grenade at baby. *Atlanta Daily World.* Retrieved from https://atlantadailyworld.com/2016/03/02/georgia-family-gets-3-6-million-after-cops-throw-grenade-at-baby/.
36. Bickel, K. (2012). BDUs and community policing? *Community Policing Dispatch, 5*(11).
37. Sack, K. (March 18, 2017). Murder of self-defense if officer is killed in raid. *New York Times.*
38. Sack, K. (March 18, 2017). Murder of self-defense if officer is killed in raid. *New York Times.*
39. Johnson, R. R. (2005). Police uniform color and citizen impression formation. *Journal of Police and Criminal Psychology, 20*(2), 58–66.
40. Quill, P. (2016). *The police uniform and its impact on public perception.* A leadership white paper submitted in partial fulfillment required for graduation from the Leadership Command College. Combes Police Department Combes, Texas. Retrieved from https://shsu-ir.tdl.org/shsu-ir/bitstream/handle/20.500.11875/2163/1687.pdf?sequence=1&isAllowed=y.
41. Baldassari, S. (August 16, 2012). Tactical vs. traditional uniforms. *Law Officer.* Retrieved from http://lawofficer.com/archive/tactical-vs-traditional-uniforms/.
42. Massachusetts police get black uniforms to instill sense of "fear." *Fox News,* April 24.
43. Baldassari, S. (August 16, 2012). Tactical vs. traditional uniforms. *Law Officer.* Retrieved from http://lawofficer.com/archive/tactical-vs-traditional-uniforms/.
44. Shane, J. (2010). Key administrative and operational differences in the police quasi military model. *Law Enforcement Executive Forum, 10,* 75–106.
45. Bickel, K. (2013). Recruit training: Are we preparing officers for a community oriented department? *Community Policing Dispatch, 6*(6). Retrieved from http://cops.usdoj.gov/html/dispatch/06-2013/preparing_officers_for_a_community_oriented_department.asp.

46. Sgt. Glenn French (August 12, 2013). Police militarization and an argument in favor of Black helicopters. *PoliceOne*. Retrieved from http://www.policeone.com/SWAT/articles/6385683-Police-militarization-An-argument-for-black-helicopters/ (last visited March 19, 2014).
47. Bickel, K. (2013). Recruit Training: Are we preparing officers for a community oriented department? *Community Policing Dispatch*, 6 (6). Retrieved from http://cops.usdoj.gov/html/dispatch/06-2013 /preparing_officers_for_a_community_oriented_department.asp .
48. Earle, H., H. (1973). *Police recruit training stress vs. non-stress: A revolution in law enforcement career programs.* Springfield, IL: Charles C. Thomas.
49. Massachusetts Governor's Board (1989). *Report of the governor's panel to review police training programs (p. 72)*.The Commonwealth of Massachusetts.
50. Massachusetts Governor's Board (1989).*Report of the governor's panel to review police training programs (p. 72)*. The Commonwealth of Massachusetts.
51. Conti, N. (2009). A Visigoth system: shame, honor, and police socialization. *Journal of Contemporary Ethnography*. Retrieved from http://jce.sagepub.com/content/early/2009/03/17/0891241608330092
52. Sgambelluri, R. (2000). Police culture, police training, and police administration: Their impact on violence in police families. *In* D. C. Sheehan (Ed.), *Domestic Violence by Police Officers* (p. 309–322). Washington, D.C.: U.S. Government Printing Office.
53. Padilla, K. (2016). *Stress and maladaptive coping among police officers.* Master's Thesis. Arizona State University. Retrieved from https://repository.asu.edu/attachments/170591/content/Padilla_asu_0010N_15916.pdf.
54. Hustmyre, C. (unknown). *NOPD versus Hurricane Katrina.* Hendon Media Group. Retrieved July 19, 2018 from http://www.hendonpub.com/resources/article_archive/results/details?id=3626.
55. Goldman, A. (August, 28, 2017). Trump reverses restrictions on military hardware for police. *The New York Times*. Retrieved from https://www.nytimes.com/2017/08/28/us/politics/trump-police-military-surplus-equipment.html.
56. McIntyre, D. (2017). A new report on the San Bernardino terrorist attack details the shootout with police. *The Sun*. https://www.sbsun.com/2017/05/25/a-new-report-on-the-san-bernardino-terrorist-attack-details-the-shootout-with-police/
57. General Accounting Office (2017). *DOD excess property. Enhanced controls needed for access to excess controlled property.* GAO 17_532. (pp. 36–37). Retrieved from https://www.gao.gov/assets/690/685916.pdf.
58. General Accounting Office (2017). *DOD excess property. Enhanced controls needed for access to excess controlled property.* GAO 17_532. (p. 38). Retrieved from https://www.gao.gov/assets/690/685916.pdf.
59. Buchanan, L., Fessenden, F., Rebecca Lai, K. K., Park, H., Parlapiano, A., Tse, A., Wallace, T., Watkins, D., & Yourish, K. (August 10, 2015). What happened in Ferguson? *New York Times*. Retrieved from https://www.nytimes.com/interactive/2014/08/13/us/ferguson-missouri-town-under-siege-after-police-shooting.html.
60. Goldman, A. (August 28, 2017). Trump reverses restrictions on military hardware for police. *The New York Times*. Retrieved from https://www.nytimes.com/2017/08/28/us/politics/trump-police-military-surplus-equipment.html
61. Jackman, T. (August 27, 2017). Trump to restore program sending surplus military weapons, equipment to police. *The Washington Post*.
62. Goldman, A. (August, 28, 2017). Trump reverses restrictions on military hardware for police. *The New York Times*. Retrieved from https://www.nytimes.com/2017/08/28/us/politics/trump-police-military-surplus-equipment.html
63. Paul, J., & Birzer, M. (2008). The militarization of the American police force: A critical assessment. *Critical Issues in Justice and Politics*, *1*(1), 15–29.
64. Friedmann, R. R., & Cannon, W. J. (2007). Homeland Security and community policing: Competing or complementing safety policies. *Journal of Homeland Security and Emergency Management*, *4*(4), 1–20.

CHAPTER 8

1. Jamison, P. (December 18, 2018). Falling out. A generation of African American drug users is dying in the opioid epidemic nobody talks about. The nation's capital is ground zero. *Washington Post*. Retrieved from https://www.washingtonpost.com/graphics/2018/local/opioid-epidemic-and-its-effect-on-african-americans/?utm_term=.9b983e98358b.
2. National Institute of Drug Abuse (2019). *What is fentanyl*. Retrieved from https://www.drugabuse.gov/publications/drugfacts/fentanyl.
3. National Institute of Drug Abuse (2018) *What is heroin*. Retrieved from https://www.drugabuse.gov/publications/opioid-facts-teens/opioids-heroin

4. National Institute of Drug Abuse (2019). *What is fentanyl*. Retrieved from https://www.drugabuse.gov/publications/drugfacts/fentanyl.
5. Jamison, P. (December 18, 2018). Falling out. A generation of African American drug users is dying in the opioid epidemic nobody talks about. The nation's capital is ground zero. *Washington Post*. Retrieved from https://www.washingtonpost.com/graphics/2018/local/opioid-epidemic-and-its-effect-on-african-americans/?utm_term=.9b983e98358b.
6. National Institute of Drug Abuse (2018). What naloxone. Retrieved from https://www.drugabuse.gov/related-topics/opioid-overdose-reversal-naloxone-narcan-evzio
7. P&T Community (2018). Online ordering of life-saving opioid overdose drug naloxone now available for texas businesses and organizations. Retrieved from https://www.ptcommunity.com/wire/online-ordering-life-saving-opioid-overdose-drug-naloxone-now-available-texas-businesses-and.
8. U.S. Department of Justice (2018). *2018 National drug threat assessment*. Drug Enforcement Administration. Retrieved from https://www.dea.gov/sites/default/files/2018-11/DIR-032-18%202018%20NDTA%20final%20low%20resolution.pdf.
9. Bipartisan Policy Center (2019). *Tracking federal funding to combat the opioid crisis*. Retrieved from https://bipartisanpolicy.org/wp-content/uploads/2019/03/Tracking-Federal-Funding-to-Combat-the-Opioid-Crisis.pdf.
10. Bernstein, L. (July 17, 2019). Most fentanyl is now trafficked across Mexico-US border, not China. WJLA. Retrieved from https://wjla.com/news/nation-world/most-fentanyl-is-now-trafficked-across-us-mexico-border-not-from-china.
11. Bernstein, L. (July 17, 2019). Most fentanyl is now trafficked across Mexico-US border, not China. WJLA. Retrieved from https://wjla.com/news/nation-world/most-fentanyl-is-now-trafficked-across-us-mexico-border-not-from-china.
12. U.S. Department of Justice (2018). *2018 National drug threat assessment*. Drug Enforcement Administration. Retrieved from https://www.dea.gov/sites/default/files/2018-11/DIR-032-18%202018%20NDTA%20final%20low%20resolution.pdf.
13. Reardon, S. (2019). The US opioid epidemic is driving a spike in infectious diseases. *Nature*. Retrieved from https://www.nature.com/articles/d41586-019-02019-3.
14. U.S. Department of Justice (2018). *2018 National drug threat assessment*. Drug Enforcement Administration. Retrieved from https://www.dea.gov/sites/default/files/2018-11/DIR-032-18%202018%20NDTA%20final%20low%20resolution.pdf.
15. Bipartisan Policy Center (2019). *Tracking federal funding to combat the opioid crisis*. Retrieved from https://bipartisanpolicy.org/wp-content/uploads/2019/03/Tracking-Federal-Funding-to-Combat-the-Opioid-Crisis.pdf.
16. Alcohol and Drug Abuse Institute (2018). *Washington state syringe exchange health survey: 2017 Results*. University of Washington. Retrieved from http://adai.uw.edu/pubs/pdf/2017syringeexchangehealthsurvey.pdf.
17. Nguyen, T. Q., Weir, B. W., Des Jarlais, D. C., Pinkerton, S. D., & Holtgrave, D. R. (2014). Syringe exchange in the United States: A national level economic evaluation of hypothetical increases in investment. *Aids and Behavior*, *18*(11), 2144–2155.
18. Greenstone, S. (July 28, 2019). Afraid of enabling drug use, Washington cities push back against needle exchanges. *The Seattle Times*. Retrieved from https://www.seattletimes.com/seattle-news/homeless/afraid-of-enabling-drug-use-washington-cities-push-back-against-needle-exchanges/.
19. Mangat, M. and Perkins, M. (2019). *Community–police partnership fight the opioid epidemic where it lives*. LISC. Retrieved from http://www.lisc.org/our-stories/story/community-police-partnerships-fight-opioid-epidemic-where-it-lives.
20. De Vaan, M., & Stuart, T. (2019). Does intra-household contagion cause an increase in prescription opioid use? *American Sociological Review*, *84*(4), 577–608. Retrieved from https://www.asanet.org/sites/default/files/attach/journals/aug19asrfeature.pdf.
21. Addictions and Recovery (2019). *Drug and alcohol withdrawal*. Retrieved from https://www.addictionsandrecovery.org/withdrawal.htm.
22. Levenson, M. (July 30, 2019). Death in police custody underscore dangers for people arrested with drug issues. *Boston Globe*. Retrieved from https://www.bostonglobe.com/metro/2019/07/30/deaths-police-custody-underscore-dangers-for-people-arrested-with-drug-issues/QP9yv3wDCHFOkWYzY5hLwJ/story.html.
23. Police Executive Research Forum (2014). *New challenges for police: A heroin epidemic and changing attitudes towards marijuana*. Critical Issues in Policing. Retrieved from https://www.policeforum.org/assets/docs/Critical_Issues_Series_2/a%20heroin%20epidemic%20and%20changing%20attitudes%20toward%20marijuana.pdf
24. Police Executive Research Forum (2014). *New challenges for police: A heroin epidemic and changing attitudes towards marijuana*. Critical Issues in Policing. Retrieved from https://www.policeforum.org/assets/docs/Critical_Issues_Series_2/a%20heroin%20epidemic%20and%20changing%20attitudes%20toward%20marijuana.pdf

25. U.S. Department of Justice (2018). *2018 National drug threat assessment*. Drug Enforcement Administration. Retrieved from https://www.dea.gov/sites/default/files/2018-11/DIR-032-18%202018%20NDTA%20final%20low%20resolution.pdf.
26. National Conference of State Legislators (2019). *Marijuana overview*. Retrieved from http://www.ncsl.org/research/civil-and-criminal-justice/marijuana-overview.aspx.
27. Police Executive Research Forum (2014). *New challenges for police: A heroin epidemic and changing attitudes towards marijuana*. Critical Issues in Policing. Retrieved from https://www.policeforum.org/assets/docs/Critical_Issues_Series_2/a%20heroin%20epidemic%20and%20changing%20attitudes%20toward%20marijuana.pdf
28. Police Executive Research Forum (2014). *New challenges for police: A heroin epidemic and changing attitudes towards marijuana*. Critical Issues in Policing. Retrieved from https://www.policeforum.org/assets/docs/Critical_Issues_Series_2/a%20heroin%20epidemic%20and%20changing%20attitudes%20toward%20marijuana.pdf
29. Monte, A. A., Shelton, S. K., Mills, E., Saben, J., Hopkinson, A., Sonn, B., Devivo, M., Chang, T., Fox, J., Brevik, C., Williamson, K., & Abbott, D. (2019). Acute illness associated with cannabis use, by route of exposure: An observational study. *Annals of Internal Medicine*, *170*(8), 531–537.
30. Gliha, L. J. (2018). Exclusive: Killer blames marijuana edible for wife's murder; new research explores edibles and psychiatric effects. Rocky Mountain PBS. Retrieved from http://www.rmpbs.org/blogs/news/exclusive-killer-speaks-out-for-the-first-time-blames-marijuana-edible-for-wifes-murder/.
31. Colorado Official State Web Portal (2019). *Colorado marijuana MED bulletin: New packaging and labeling rules and effective dates*. Retrieved from https://www.colorado.gov/pacific/marijuana/news/med-bulletin-new-packaging-and-labeling-rules-and-effective-dates
32. Police Executive Research Forum (2014). *New challenges for police: A heroin epidemic and changing attitudes towards marijuana*. Critical Issues in Policing. Retrieved from https://www.policeforum.org/assets/docs/Critical_Issues_Series_2/a%20heroin%20epidemic%20and%20changing%20attitudes%20toward%20marijuana.pdf
33. Twargowski, M. A., Link, M. M., and Twardowski, N. M. (2019). Effects of cannabis use on sedation requirements for endoscopic procedures. *The Journal of the American Osteopathic Association*, *119*, 307–311. Retrieved from https://jaoa.org/article.aspx?articleid=2731067.
34. U.S. Department of Justice (2018). *2018 National drug threat assessment*. Drug Enforcement Administration. Retrieved from https://www.dea.gov/sites/default/files/2018-11/DIR-032-18%202018%20NDTA%20final%20low%20resolution.pdf.
35. U.S. Department of Justice (2018). *2018 National drug threat assessment*. Drug Enforcement Administration. Retrieved from https://www.dea.gov/sites/default/files/2018-11/DIR-032-18%202018%20NDTA%20final%20low%20resolution.pdf.
36. U.S. Department of Justice (2018). *2018 National drug threat assessment*. Drug Enforcement Administration. Retrieved from https://www.dea.gov/sites/default/files/2018-11/DIR-032-18%202018%20NDTA%20final%20low%20resolution.pdf.
37. U.S. Attorney's Office (2019). Twenty-seven defendants charged in takedown of Newark open-air drug market. U.S. Department of Justice. District of New Jersey. Retrieved from https://www.justice.gov/usao-nj/pr/twenty-seven-defendants-charged-takedown-newark-open-air-drug-market
38. Kerr, T., Small, W., and Wood, E. (2005). The public health and social impacts of drug market enforcement: A review of the evidence. *International Journal of Drug Policy*, *16*, 210–220.
39. Meyers, R. (March 30, 2017). How the opioid boom transformed policing: we're not just making arrests—we're caretakers. *Vox*. Retrieved from https://www.vox.com/first-person/2017/3/30/15115066/opioid-epidemic-heroin-crisis-ohio-police.
40. Otterman, S., & Correal, A. (May 29, 2019). The opioid crisis in the Bronx claims tiny victim: 1-year-old. *New York Times*. Retrieved from https://www.nytimes.com/2019/05/29/nyregion/baby-overdose-bronx-darwin-santana-gonzalez.html.
41. Meyersohn, J., & Weiss, J. (May 30, 2006). Kids are the collateral damage in meth epidemic. ABC News. Retrieved from https://abcnews.go.com/Primetime/FosterCare/story?id=1991958&page=1.
42. Meyers, R. (May 30, 2017). *How the opioid boom transformed policing: We're are not just making arrests—we're caretakers. Vox*. Retrieved from https://www.vox.com/first-person/2017/3/30/15115066/opioid-epidemic-heroin-crisis-ohio-police.
43. Harocopos, A., & Hough, M. (2010). *Drug dealing in open-air markets*. Problem Oriented Guides for Police. Problem Specific Guides Series. Guide No. 31. Center for Problem Oriented Policing. Retrieved from https://popcenter.asu.edu/sites/default/files/drug_dealing_in_open-air_markets_2012_update.pdf.
44. Evans, W., Garthwaite, C., & Moore, T. (2018). *Guns and violence: The enduring impact of crack cocaine markets on young Black males*. Northwestern Institute of Policy Research. Retrieved from https://www.ipr.northwestern.edu/publications/docs/workingpapers/2018/wp-18-17.pdf.

45. Harocopos, A. and Hough, M. (2010). *Drug dealing in open-air markets*. Problem Oriented Guides for Police. Problem Specific Guides Series. Guide No. 31. Center for Problem Oriented Policing. Retrieved from https://popcenter.asu.edu/sites/default/files/drug_dealing_in_open-air_markets_2012_update.pdf.
46. Green, L. (1995). Cleaning up drug hot spots in Oakland, California: The displacement and diffusion effects. *Justice Quarterly, 12*, 737–754.
47. Kerr, T., Small, W., & Wood, E. (2005). The public health and social impacts of drug market enforcement: A review of the evidence. *International Journal of Drug Policy, 16*, 210–220.
48. Kerr, T., Small, W., and Wood, E. (2005). The public health and social impacts of drug market enforcement: A review of the evidence. *International Journal of Drug Policy, 16*, 210–220.
49. Kerr, T., Small, W., and Wood, E. (2005). The public health and social impacts of drug market enforcement: A review of the evidence. *International Journal of Drug Policy, 16*, 210–220.
50. Saunders, J., Ober, A., Barnes-Proby, D., & Brunson, R. D. (2016). Police legitimacy and disrupting overt drug markets. *Policing: An International Journal of Police Strategies & Management. 39*(4), 667–679.
51. Donohoo, J., Hattie, J., & Eells, R. (2018). The power of collective efficacy. *Educational Leadership, 75*(6), 40–44. Retrieved from http://www.ascd.org/publications/educational-leadership/mar18/vol75/num06/The-Power-of-Collective-Efficacy.aspx.
52. Saunders, J., Ober, A., Barnes-Proby, D., & Brunson, R. D. (2016). Police legitimacy and disrupting overt drug markets. *Policing: An International Journal of Police Strategies & Management. 39*(4), 667–679.
53. Kennedy, D. M., & Wong, L.-S. (2006). *The high point drug market intervention strategy*. Center for Crime Prevention and Control. John Jay College of Criminal Justice. Retrieved from https://ric-zai-inc.com/Publications/cops-p166-pub.pdf.
54. Kennedy, D. M., & Wong, L.-S. (2006). *The high point drug market intervention strategy*. Center for Crime Prevention and Control. John Jay College of Criminal Justice. Retrieved from https://ric-zai-inc.com/Publications/cops-p166-pub.pdf.

CHAPTER 9

1. National Institute of Justice. *Practice profile. Juvenile awareness program (Scared Straight)*. Retrieved January 29, 2020 from https://www.crimesolutions.gov/PracticeDetails.aspx?ID=4
2. Center on Juvenile and Criminal Justice (2001). *Widening the net in juvenile justice and the dangers of prevention and early intervention*. Retrieved from http://www.cjcj.org/uploads/cjcj/documents/widening.pdf
3. Prichard, J. (2010). Net-widening and the diversion of young people from court: A longitudinal analysis with implications for restorative justice. *The Australian and Tasmanian Journal of Criminology, 43*(1), 112–129.
4. Prichard, J. (2010). Net-widening and the diversion of young people from court: A longitudinal analysis with implications for restorative justice. *The Australian and Tasmanian Journal of Criminology, 43*(1), 112–129.
5. Austin, J., & Krisberg, B. (1981). Wider, stronger and different nets: The dialectics of criminal justice reform. *Journal of Research in Crime and Delinquency, 18*(1), 165–196.
6. Austin, J., & Krisberg, B. (1981). Wider, stronger and different nets: The dialectics of criminal justice reform. *Journal of Research in Crime and Delinquency, 18*(1), 165–196.
7. Austin, J., & Krisberg, B. (1981). Wider, stronger and different nets: The dialectics of criminal justice reform. *Journal of Research in Crime and Delinquency, 18*(1), 165–196.
8. Austin, J., & Krisberg, B. (1981). Wider, stronger and different nets: the dialectics of criminal justice reform. *Journal of Research in Crime and Delinquency, 18*(1), 165–196.
9. Center on Juvenile and Criminal Justice (2001). *Widening the net in juvenile justice and the dangers of prevention and early intervention*. Retrieved from http://www.cjcj.org/uploads/cjcj/documents/widening.pdf
10. Shelden, R. G. (1999). *Detention diversion advocacy: An evaluation. Bulletin*. Washington, D.C.: U.S. Department of Justice, Office of Justice Programs, Office of Juvenile Justice and Delinquency Prevention.
11. Austin, J., & Krisberg, B. (1981). Wider, stronger and different nets: The dialectics of criminal justice reform. *Journal of Research in Crime and Delinquency, 18*(1), 165–196.
12. O'Hear, M. M. (2009). Rethinking drug courts: Restorative justice as a response to racial injustice. *Stanford Law and Policy Review, 20*(2), 463–500.
13. Lassiter, M. D. (2015). Impossible criminals: The suburban imperatives of America's war on drugs. *The Journal of American History, June*, 126–140.
14. Hoffman, M. B. (2001). The rehabilitative ideal and the drug court reality. *Federal Sentencing Reporter, 14*, 172–178.

15. O'Hear, M. M. (2009). Rethinking drug courts: Restorative justice as a response to racial injustice. *Stanford Law and Policy Review, 20*(2), 463–500.
16. O'Hear, M. M. (2009). Rethinking drug courts: Restorative justice as a response to racial injustice. *Stanford Law and Policy Review, 20*(2), 463–500.
17. Spaletti, S. (2014). The economics of education in Adam Smith's *Wealth of Nations. Journal of World Economic Research, 3*(5), 60–64.
18. Goldin, C. (2019). *Human capital. Handbook of cliometrics*. Berlin: Springer Verlag.
19. O'Hear, M. M. (2009). Rethinking drug courts: Restorative justice as a response to racial injustice. *Stanford Law and Policy Review, 20*(2), 463–500.
20. O'Hear, M. M. (2009). Rethinking drug courts: Restorative justice as a response to racial injustice. *Stanford Law and Policy Review, 20*(2), 463–500.
21. O'Hear, M. M. (2009). Rethinking drug courts: Restorative justice as a response to racial injustice. *Stanford Law and Policy Review, 20*(2), 463–500.
22. O'Hear, M. M. (2009). Rethinking drug courts: Restorative justice as a response to racial injustice. *Stanford Law and Policy Review, 20*(2), 463–500.
23. O'Hear, M. M. (2009). Rethinking drug courts: Restorative justice as a response to racial injustice. *Stanford Law and Policy Review, 20*(2), 463–500.
24. Wilson, J. Q., & Kelling, G. L. (1982). Broken windows. *Atlantic Monthly* (March), pp. 29–38.
25. Skogan, W. G. (1990). *Disorder and decline: Crime and the spiral of decay in american neighborhoods*. Berkeley, CA: University of California Press.
26. Harcourt, B. E. (2001). *Illusion of order: The false promise of broken windows policing*. Cambridge, MA: Harvard University Press.
27. Braga, A. A., Weisburd, D., Waring E. J., Mazerolle, L. G., Spelman, W., & Gajewski, F. (1999). Problem-oriented policing in violent crime places: a randomized controlled experiment. *Criminology, 37*(3), 541–580.
28. Katz C. M., Webb, V. J., & Schaefer, D. R. (2001). An assessment of the impact of quality-of-life policing on crime and disorder. *Justice Quarterly, 18*(4), 825–876.
29. Elliott, L., Golub, A., & Dunlap, E. (2012). Off the street and into the "The Cut:" Deterrence and displacement in NYC's quality of life marijuana policing. *International Journal of Drug Policy, 23*(3), 210–219.
30. Elliott, L., Golub, A., & Dunlap, E. (2012). Off the street and into the "The Cut": Deterrence and displacement in NYC's quality of life marijuana policing. *International Journal of Drug Policy, 23*(3), 210–219.
31. Elliott, L., Golub, A., & Dunlap, E. (2012). Off the street and into the "The Cut": Deterrence and displacement in NYC's quality of life marijuana policing. *International Journal of Drug Policy, 23*(3), 210–219.
32. Elliott, L., Golub, A., & Dunlap, E. (2012). Off the street and into the "The Cut": Deterrence and displacement in NYC's quality of life marijuana policing. *International Journal of Drug Policy, 23*(3), 210–219.
33. Elliott, L., Golub, A., and Dunlap, E. (2012). Off the street and into the "The Cut": Deterrence and displacement in NYC's quality of life marijuana policing. *International Journal of Drug Policy, 23*(3), 210–219.
34. Connor, D. L. (2019). *Asset forfeiture policy manual*. Money Laundering and Asset Recovery Section, Criminal Division, Department of Justice. Retrieved from https://www.justice.gov/criminal-afmls/file/839521/download
35. Golub, A., Johnson, B. D., & Taylor, A. (2003). Quality-of-life policing: Do offenders get the message? *Policing, 26*(4), 690–707.
36. Bratton W. (1996). Cutting crime and restoring order: What America can learn from New York's finest. Heritage Foundation. Retrieved from https://www.heritage.org/crime-and-justice/report/cutting-crime-and-restoring-order-what-america-can-learn-new-yorks-finest.
37. Harcourt, B.E. (2001). *Illusion of order: The false promise of broken windows policing*. Cambridge, MA: Harvard University Press.
38. Golub, A., Johnson, B. D., Taylor, A., & Eterno, J. (2004). Does quality-of-life policing widen the net? A partial analysis. *Justice Research and Policy, 6*(1), 19–42.
39. Elliott, L., Golub, A., & Dunlap, E. (2012). Off the street and into the "The Cut": Deterrence and displacement in NYC's quality of life marijuana policing. *International Journal of Drug Policy, 23*(3), 210–219.
40. Casella, S. D. (2015). Asset forfeiture law in the US. *The Palgrave handbook of criminal and terrorism financing law* (pp. 427–446). Retrieved from http://assetforfeiturelaw.us/wp-content/uploads/2016/10/Chapter-for-Colin-King.pdf
41. Sallah, M., O'Harrow Jr., R., Rich, S., & Silverman, G. (September 6, 2014). Stop and seize. Aggressive take hundreds of millions of dollars from motorists not charged with crimes. *The Washington Post,*. Retrieved from https://www.washingtonpost.com/sf/investigative/2014/09/06/stop-and-seize/?utm_term=.c366f5bd2cb7

42. Sallah, M., O'Harrow Jr., R., Rich, S., & Silverman, G. (September 6, 2014). Stop and seize. Aggressive take hundreds of millions of dollars from motorists not charged with crimes. *The Washington Post*. Retrieved from https://www.washingtonpost.com/sf/investigative/2014/09/06/stop-and-seize/?utm_term=.c366f5bd2cb7
43. *Timbs v. Indiana*, 586 US ____ (2019).
44. *Timbs v. Indiana*, 586 US ____ (2019).

CHAPTER 10

1. Kelly Thomas trial: Officers acquitted in homeless man's death (January 14, 2014). *Fox News*. Retrieved from https://www.foxnews.com/
2. Kelly Thomas trial: Officers acquitted in homeless man's death (January 14, 2014). *Fox News*. Retrieved from https://www.foxnews.com/
3. Ponsi, L., & Emery, S. (November 23, 2015). Police-beating case: $4.9 million settlement reached in Kelly Thomas wrongful-death lawsuit. *The Press Enterprise*. Retrieved from https://www.pe.com/
4. McCarthy, M., Porter, L., Townsley, M., & Alpert, G. (2019). The effect of community-oriented policing on police use of force: Does community matter? *Policing: An International Journal*, *42*(4), 556–570. DOI: 10.1108/PIJPSM-10-2018-0148
5. McCarthy, M., Porter, L., Townsley, M., & Alpert, G. (2019). The effect of community-oriented policing on police use of force: Does community matter? *Policing: An International Journal*, *42*(4), 556–570. DOI: 10.1108/PIJPSM-10-2018-0148
6. Lynch, L. E. (October 5, 2016). Loretta E. Lynch: Community policing can make us all safer. *The Virginia-Pilot*. Retrieved from https://www.pilotonline.com
7. Sir Robert Peel's policing principles. Law Enforcement Action Partnership. Website can be found online: https://lawenforcementactionpartnership.org/peel-policing-principles/
8. Gaines, L., & Kappeler, V. (2015). *Policing in America* (8th ed.). Waltham, MA: Elsevier (Anderson Publishing).
9. Tyler, T. R. (2006). *Why people obey the law*. Princeton: Princeton University Press.
10. Gau, J., Corsaro, N., Stewart, E., & Brunson, R. (2012). Examining macro-level impacts on procedural justice and police legitimacy. *Journal of Criminal Justice*, *40*(4), 333–343. DOI: 10.1016/j.jcrimjus.2012.05.00
11. McEwen, J. (1996). National data collection on police use of force (Discussion paper (United States. Bureau of Justice Statistics)). Washington, DC: National Institute of Justice : U.S. Department of Justice, Office of Justice Programs, Bureau of Justice Statistics.
12. Baćak, V., Mausolf, J., & Schwarz, C. (2019). How comprehensive are media-based data on police officer-involved shootings? *Journal of Interpersonal Violence*. DOI: 10.1177/0886260519860897
13. People have been shot and killed by police in 2019. (October 4, 2019). *The Washington Post*. Retrieved from https://www.washingtonpost.com
14. Mapping police violence. (n.d.). Retrieved from https://mappingpoliceviolence.org/
15. Johnson, R. R. (July 2016). Dispelling the myths surrounding police use of lethal force. Retrieved October 6, 2019 from https://www.dolanconsultinggroup.com/wp-content/uploads/2019/02/Dispelling-the-Myths-Surrounding-Police-Use-of-Lethal-Force.pdf
16. National use-of-force data collection. FBI (n.d.). Retrieved from https://www.fbi.gov/services/cjis/ucr/use-of-force
17. Mapping Police Violence. (n.d.). Retrieved from https://mappingpoliceviolence.org
18. Mapping Police Violence. (n.d.). Retrieved from https://mappingpoliceviolence.org
19. Khan, A. (2019). Getting killed by police is a leading cause of death for young Black men in America. *Los Angeles Times*. Retrieved from https://www.latimes.com
20. Gramlich, J. (May 21, 2019). From police to parole, Black and White Americans differ widely in their views of criminal justice system. Pew Research Center. Retrieved from https://www.pewresearch.org
21. Law enforcement and violence: the divide between Black and White Americans. (n.d.). The Associated Press–NORC Center for Public Affairs Research. Retrieved from http://www.apnorc.org
22. Braimah, A. (August 17, 2017). Abner Louima (1966–). *BlackPast*. Retrieved from https://www.blackpast.org
23. Del Barco, M. (August 11, 2005). Decades after the L.A. Riots, Watts still suffers. *NPR*. Retrieved from https://www.npr.org/
24. Istook, E. (August 18, 2014). "No justice; No peace!" chants in Ferguson are actually saying, "We want revenge!" *The Washington Times*. Retrieved from https://www.washingtontimes.com
25. International Association of the Chiefs of Police. Police use of force in America 2001. Alexandria, Virginia. Retrieved from https://www.theiacp.org/sites/default/files/2018-08/2001useofforce.pdf

26. Gorman, A. (June 27, 2004). Bratton vows to review police use of flashlights. *Los Angeles Times*. Retrieved from https://www.latimes.com/archives
27. Gorman, A. (June 27, 2004). Bratton vows to review police use of flashlights. *Los Angeles Times*. Retrieved from https://www.latimes.com/archives
28. Marik, P. (March 18, 2016). 15 use-of-force cases every cop needs to know. PoliceOne.com. Retrieved from https://www.policeone.com
29. Bulzomi, M. J. (January 1, 2011). Off-duty officers and firearms. FBI Law Enforcement Bulletin. Retrieved from https://leb.fbi.gov/articles/legal-digest/legal-digest-off-duty-officers-and-firearms
30. Girl killed in botched holdup at McDonald's. (June 17, 1997). *The Los Angeles Times*. Retrieved from https://www.latimes.com
31. Winton, R., Shalby, C., & Carcamo, C. (September 25, 2019). Costco shooting: LAPD officer won't face criminal charges; security video released. *Los Angeles Times*. Retrieved from https://www.latimes.com
32. Winton, R. (September 26, 2019). Parents wounded by LAPD officer in Costco shooting are outraged by lack of charges. *Los Angeles Times*. Retrieved from https://www.latimes.com
33. Fieldstadt, E. (August 27, 2019). Parents of disabled man killed in Costco said they "begged" off-duty officer not to shoot. *NBC News*. Retrieved from https://www.nbcnews.com
34. Perishable skills program. (n.d.). Commission on Peace Officers Standards and Training. Retrieved from https://post.ca.gov/perishable-skills-program
35. *Popow v. Margate*, 476 F. Supp. 1237 (N.J. 1979).
36. *Popow v. Margate*, 476 F. Supp. 1237 (N.J. 1979).
37. Grossi, D. (June 23, 2011). Police firearms training: How often should you be shooting? PoliceOne.Com. Retrieved from https://www.policeone.com
38. Ford, A. (April 8, 1988). Police Laser village theme: "Go ahead, make my Ray." *Los Angeles Times*. Retrieved from https://www.latimes.co
39. Griffith, D. (March 1, 2018). Training simulators: Going beyond Shoot, Don't Shoot. *Police Law Enforcement Solutions*. Retrieved from https://www.policemag.com/
40. The use-of-force continuum (August 3, 2009). National Institute of Justice. Retrieved from http://nij.ojp.gov/topics/articles/use-force-continuum
41. Smith, M.R., Kaminski, R. J., Alpert, G. P., Fridell, L. A., MacDonald, J., & Kubu, B. (July 2010). A multi-method evaluation of police use of force outcomes: Final report to the National Institute of Justice. NCJ231176. Retrieved from http://www.ncjrs.gov/pdffiles1/nij/grants/231176.pdf Supported by grant number 2005-IJ-CX-0056 from National Institute of Justice.
42. National Institute of Justice. Study of deaths following electro muscular disruption, Special Report, Washington, D.C.: U.S. Department of Justice, National Institute of Justice, 2011: 3, NCJ 233432.
43. Holder, E. H., Robinson, L. O., & Laub, J. H., (May 2011). Research in brief: Police use of force, Tasers, and other less-lethal weapons. U.S. Department of Justice, Office of Justice Programs, Washington, DC. Retrieved from: https://permanent.access.gpo.gov/gpo9141/232215.pdf
44. *Plakas v. Drinski* 811 F. Supp. 1356 (1993)
45. McNeff, J. (April 1, 2017). Please remove force continuum from vocabulary. *Law Enforcement Today*. Retrieved from https://www.lawenforcementtoday.com
46. Miller, L. (2015). Why cops kill: The psychology of police deadly force encounters. *Aggression & Violent Behavior, 22*, 97–112. Retrieved from https://doi.org/10.1016/j.avb.2015.04.007
47. Brandl, S., & Stroshine, M. (2013). The role of officer attributes, job characteristics, and arrest activity in explaining police use of force. *Criminal Justice Policy Review, 24*(5), 551–572. DOI: 10.1177/0887403412452424
48. Terrill, W. (2002). Situational and officer-based determinants of police coercion. *JQ: Justice Quarterly*, 19(2), 215-249. DOI: 10.1080/07418820200095221
49. Terrill, W., & Mastrofski, S. (2002). Situational and officer-based determinants of police coercion. *Justice Quarterly, 19*(2), 215–248. DOI: 10.1080/07418820200095221. See also: Paoline, E., &Terrill, W. (2005). Women police officers and the use of coercion. *Women & Criminal Justice, 15*(3–4),97–119. DOI: 10.1300/J012v15n03_05
50. Garner, Joel H. (2002). Characteristics associated with the prevalence and severity of force used by the police. *JQ: Justice Quarterly, 19*(4), 705–747. DOI: 10.1080/07418820200095401
51. Sun, I., & Payne, B. (2004). Racial differences in resolving conflicts: A comparison between Black and White police officers. *Crime & Delinquency, 50*(4), 516–541. DOI: 10.1177/0011128703259298

52. Paoline, E. A. (2007). Police Education, Experience, and the Use of Force. *Criminal Justice & Behavior, 34*(2), 179–197. DOI: 10.1177/0093854806290239
53. Brandl, S., & Stroshine, M. (2013). The role of officer attributes, job characteristics, and arrest activity in explaining police use of force. *Criminal Justice Policy Review, 24*(5), 551–572. DOI: 10.1177/0887403412452424
54. Adams, K. (1999). What we know about police use of force. In K. Adams (Ed.), *Use of force by police: Overview of national and local data* (pp. 1–14). Washington, D.C.: U.S. Department of Justice, Office of Justice Programs, National Institute of Justice.
55. Toch, H. (1995). The "violence-prone" police officer. In W. Geller & H. Toch (Eds). *And justice for all: Understanding and controlling police abuse of force.* (pp. 99–112). Washington, D.C.: Police Executive Research Forum.
56. Klinger, D. (1997). Negotiating order in patrol work: An ecological theory of police response to deviance. *Criminology, 35*(2), 277. Retrieved from https://doi.org/10.1111/j.1745-9125.1997.tb00877.x
57. Roufa, T. (February 11, 2019) Why don't police aim for the arms or legs? Retrieved from https://www.thebalancecareers.com/why-dont-police-shoot-people-in-the-arms-or-legs-974607
58. A.B. 392, 2019 Weber, 2019 Reg. Sess. (CA 2019). Bill retrieved from http://leginfo.legislature.ca.gov
59. Pryor, C., Boman, J., Mowen, T., & Mccamman, M. (2019). A national study of sustained use of force complaints in law enforcement agencies. *Journal of Criminal Justice, 64*, 101623. doi: 10.1016/j.jcrimjus.2019.101623
60. Mccarthy, M., Porter, L., Townsley, M., & Alpert, G. (2019). The effect of community-oriented policing on police use of force: Does community matter? *Policing: An International Journal, 42*(4), 556–570. doi: 10.1108/PIJPSM-10-2018-0148
61. Martin, M. (August 24, 2019). Law professor on California's new police use-of-force law. (Transcript of interview with law professor Seth Stoughton). *NPR.* Retrieved from https://www.npr.org
62. Ortiz, J. L. (August 20, 2019). California's new police use-of-force law marks a "significant" change in law enforcement. Here's why. *USA Today.* Retrieved from https://www.usatoday.com
63. Kelling, G., Wasserman, R., & Williams, H. (1989). Police accountability and community policing (Perspectives on Policing; no. 7). Washington, D.C.: U.S. Dept. of Justice, Office of Justice Programs, National Institute of Justice. Retrieved from https://www.ncjrs.gov/pdffiles1/nij/114211.pdf
64. Terrill, W., & Mastrofski, S. (2002). Situational and officer-based determinants of police coercion. *Justice Quarterly, 19*(2), 215–248. DOI: 10.1080/07418820200095221

CHAPTER 11

1. McCleary, K., & Vera, A. (April 15, 2018). A video of black men being arrested at Starbucks. Three very different reactions. CNN. Retrieved from https://www.cnn.com/2018/04/14/us/philadelphia-police-starbucks-arrests/index.html
2. Williams, J., & Browne Dianis, J. (May 29, 2018). Starbucks' incident proves "Whites Only" spaces still exist. CNN. Retrieved from https://www.cnn.com/2018/05/28/opinions/starbucks-white-space-opinion-williams-dianis/index.html
3. Surette, R. (2015). *Media, crime, and criminal justice: Images, realities, and policies* (5th ed.). Stamford, CT: Cengage Learning.
4. Surette, R. (2015). *Media, crime, and criminal justice: Images, realities, and policies* (5th ed., p. 114). Stamford, CT: Cengage Learning.
5. Gramlich, J. & Parker, K. (January 25, 2017). Most officers say the media treat police unfairly. Pew Research Center. Retrieved from https://www.pewresearch.org/fact-tank/2017/01/25/most-officers-say-the-media-treat-police-unfairly/
6. Morelli, A. (December 23, 2015). The media's war on police officers. The Odyssey Online. Retrieved from https://www.theodysseyonline.com/medias-war-police-officers
7. McLaughlin, E. (April 21, 2015). We're not seeing more police shootings, just more news coverage. Retrieved from https://www.cnn.com/2015/04/20/us/police-brutality-video-social-media-attitudes/
8. Fagan, J., & Tyler, T. (December 2004). Policing, order maintenance and legitimacy. NCJRS. Document No. 207975. Retrieved from https://www.ncjrs.gov/pdffiles1/nij/Mesko/207975.pdf
9. Surette, R. (2015). *Media, crime, and criminal justice: Images, realities, and policies* (5th ed.). Stamford, CT: Cengage Learning.
10. Gerbner, G., Gross, L., Morgan, M., & Signorielli, N. (1986). Living with television: The dynamics of the cultivation process. In J. Bryant & D. Zillman (Eds.), *Perspectives on media effects* (pp. 17-40). Hillsdale, NJ: Erlbaum.
11. Rafter, N. (2007). Crime, film and criminology: recent sex-crime movies, *Theoretical Criminology, 11*(3), 403–420. doi: 10.1177/1362480607079584
12. Ripped from the headlines. (n.d). *Fandom.* Retrieved from http://lawandorder.wikia.com/wiki/Rippedfromtheheadlines
13. Dunham, R. G., & Alpert, G. P. (1997). Police shootings: Myths and realities. In P. Cromwell & R. Dunham (Eds.), *Crime and justice in America: Present realities and future prospects* (pp. 115–123). Newbury Park, CA: Sage.

14. Soulliere, D. M. (2004). Policing on prime-time: A comparison of television and real-world policing. *American Journal of Criminal Justice, 28*(2), 215–233, p.217. DOI: 10.1007/BF02885873
15. List of Police television dramas. (n.d.). In Wikipedia. Retrieved September 10, 2018 from https://en.wikipedia.org/wiki/List_of_police_television_dramas
16. Tops of 2013: TV and social media. (December 17, 2013). Nielsen.com. Retrieved from https://www.nielsen.com/us/en/insights/news/2013/tops-of-2013-tv-and-social-media.html
17. Donovan K., & Klahm IV, C.F. (2015) The role of entertainment media in perceptions of police use of force. *Criminal Justice and Behavior, 42*(12), 1261–1281. doi: 10.1177/0093854815604180.
18. Ripped from the headlines. (n.d.). *Fandom*. Retrieved from http://lawandorder.wikia.com/wiki/Rippedfromtheheadlines
19. Weitzer, R. (2002). Incidents of police misconduct and public opinion. *Journal of Criminal Justice. 30*(5), 397–408. doi: 10.1016/S0047-2352(02)00150-2.
20. Lupton, D. 1999. "Something really nasty". Audience responses to crime in the mass media. *Australian Journal of Communication, 26*(1), 41–54.
21. Surette, R. (2015). *Media, crime, and criminal justice: Images, realities, and policies* (5th ed.). Stamford, CT: Cengage Learning.
22. Giblin, P. (July 28, 2007). 2 News helicopters crash in Phoenix, killing all 4 on board. *The New York Times*. Retrieved from https://www.nytimes.com/2007/07/28/us/28chopper.html
23. Raw video: Naked man leads police on foot chase after abandoning stolen truck in LA. | ABC7. (August 31, 2018). YouTube. Retrieved from https://www.youtube.com/watch?v=Ce0d0G1J8TI
24. Fry, H. (August 31, 2018). Naked man taken into custody following police pursuit through East Los Angeles. *Los Angeles Times*. Retrieved from http://www.latimes.com/local/lanow/la-me-ln-pursuit-20180831-story.html
25. Frank, T. (July 30, 2015). High-speed police chases have killed thousands of innocent bystanders. *USA Today*. Retrieved from https://www.usatoday.com/story/news/2015/07/30/police-pursuits-fatal-injuries/30187827/
26. The thrill and price of police chases. (January 5, 2006). abcNews. Retrieved from https://abcnews.go.com/2020/story?id=132636&page=1
27. Carson, A., & Farhall, K. (2018). Understanding collaborative investigative journalism in a "post-truth" age. In L. Bennett, S. Allan, & M. Berry (Guest Eds.). *Journalism Studies: The Future of Journalism. 19*(13), 1899–1911. doi: 10.1080/1461670X.2018.1494515
28. Eschholz, S., Chiricos, T., & Gertz, M. (2003). Television and fear of crime: Program types, audience traits, and the mediating effect of perceived neighbourhood racial composition. *Social Problems. 50*(3), 395–415.
29. Heath, L., & Gilbert, K. (1996). Mass media and fear of crime. *American BehavioralScientist, 39*(4), 379–386. https://doi.org/10.1177%2F0002764296039004003
30. Garland, D. (2001). The culture of control. *Crime and social order in contemporary society*. Oxford: Oxford University Press.
31. Schlesinger, P., & Tumber, H. (1994). *Reporting crime: The media politics of criminal justice* (Clarendon studies in criminology). Oxford, New York: Clarendon Press, Oxford University Press.
32. The Newark foot patrol experiment. (1981). The Police Foundation, Washington, D.C. Retrieved from https://www.policefoundation.org/wp-content/uploads/2015/07/144273499-The-Newark-Foot-Patrol-Experiment.pdf
33. Bureau of Justice Assistance (1994a). *Neighborhood-oriented policing in rural communities: A program planning guide*. Washington, DC: Bureau of Justice Assistance. See also: Bureau of Justice Assistance (1994b). *Understanding community policing: A framework for action*. Washington, DC: Bureau of Justice Assistance.
34. Sunghoon, R., & Oliver, W. M. (2005). Effects of community policing upon fear of crime: Understanding the causal linkage. *Policing: An International Journal of Police Strategies & Management, 28*(4), 670–683. doi: https://doi.org/10.1108/13639510510628758
35. Crowl, J. N. (2017). The effect of community policing on fear and crime reduction, police legitimacy and job satisfaction: an empirical review of the evidence. *Police Practice and Research, 18*(5), 449–462, DOI: 10.1080/15614263.2017.1303771
36. Dietz, A.S. (1997). Evaluating community policing: quality police service and fear of crime. *Policing: An International Journal of Police Strategies & Management, 20*(1), 83–100, https://doi.org/10.1108/13639519710162024 Jackson, J., Bradford, B., Hohl, K., & Farrall, S. (2009) Does the fear of crime erode public confidence in policing? *Policing: a Journal of Policy and Practice, 3*(1), 100–111. doi: 10.1093/police/pan079
37. CNN wire, Gould, J., & Wynter, K. (September 3, 2018). 8 shot during dice game at San Bernardino apartment complex; 17-year-old in grave condition: Police. KTLA5. Retrieved from https://ktla.com/2018/09/03/10-shot-at-san-bernardino-apartment-complex-police/
38. Moule, R. K. (2020) Under siege? Assessing public perceptions of the "War on Police." *Journal of Criminal Justice, 66*, 101631

39. Nolte, J. (May 11, 2015). Cop killings nearly double after media launch hate campaigns against police. *Breitbart*. Retrieved from https://www.breitbart.com/big-journalism/2015/05/11/cop-killings-nearly-double-after-media-launch-hate-campaigns-against-police/
40. 47 officers in the US have been fatally shot in the line of duty this year. (April 26, 2018). CNN. Retrieved from https://www.cnn.com/2018/02/12/us/officer-shooting-deaths-2018-trnd/index.html
41. Chermak, S., & Weiss, A. (2005). Maintaining legitimacy using external communication strategies: An analysis of police–media relations. *Journal of Criminal Justice, 33*, 501–512. doi: 10.1016/j.jcrimjus.2005.06.001
42. Tankebe, J. (2012). Viewing things differently: The dimensions of public perceptions of police legitimacy. *Criminology, 51*(1), 103–135. doi:10.1111/j.1745-9125.2012.00291
43. Chakraborty, B. (October 5, 2018). Jason Van Dyke trial: Chicago cop found guilty of second degree murder of Laquan McDonald. Fox News. Retrieved from https://www.foxnews.com/us/jason-van-dyke-trial-chicago-cop-found-guilty-of-second-degree-murder-of-laquan-mcdonald
44. Alba, D. (July 31, 2015). Should body-cam footage always go public? It's complicated. *Wired*. Retrieved from https://www.wired.com/2015/07/body-cam-videos-always-go-public-complicated/
45. Singh, A. (2017). Prolepticon: Anticipatory citizen surveillance of the police. *Surveillance & Society,15*(5): 676–688. doi: 10.24908/ss.v15i5.6418
46. Phillips, B. (2012, June 18). Rodney King and the birth of citizen journalism. Throughline. Retrieved from http://www.mrmediatraining.com/2012/06/18/rodney-king-and-the-birth-of-citizen-journalism/
47. Bukszpan, D. (November 21, 2011). America's most destructive riots of all time. CNBC. Retrieved from https://www.cnbc.com/2011/02/01/Americas-Most-Destructive-Riots-of-All-Time.html?slide=11
48. Bialik, K., & Matsa, K. E. (October 4, 2017). Key trends in social and digital news media. Pew Research Center. Retrieved from http://www.pewresearch.org/fact-tank/2017/10/04/key-trends-in-social-and-digital-news-media/
49. Larry DePrimo, NYPD cop, buys homeless man boots (Photo). (November 29, 2012). *Huffington Post*. Retrieved from https://www.huffingtonpost.com/2012/11/29/larry-deprimo-nypd-cop-gives-homeless-boots_n_2209178.html
50. Photo of NYPD Officer giving boots to barefoot man becomes online sensation. (November 29, 2012). 4nbc New York. Retrieved from https://www.nbcnewyork.com/news/local/NYPD-Homeless-Man-Boots-Police-Officer-Times-Square-Photo-181369011.html
51. Garske, M. (August 7, 2014). Teen, last person to see slain officer alive, pays respect. 7 *San Diego*. Retrieved from https://www.nbcsandiego.com/news/local/Daveon-Scott-Remembers--Slain-SDPD-Officer-Jeremy-Henwood-270415031.html
52. Photo of NYPD officer giving boots to barefoot man becomes online sensation. (November 29, 2012). 4nbc New York. Retrieved from https://www.nbcnewyork.com/news/local/NYPD-Homeless-Man-Boots-Police-Officer-Times-Square-Photo-181369011.html
53. Rokos, B. (December 20, 2013). Riverside: Police, community give presents to needy kids. *The Press-Enterprise*. Retrieved from https://www.pe.com/2013/12/20/riverside-police-community-give-presents-to-needy-kids/
54. Reaves, B. A. (July 2015). Local police departments, 2013: Equipment and technology. U.S. Department of Justice. Office of Justice Programs Bulletin. Retrieved from https://www.bjs.gov/content/pub/pdf/lpd13et.pdf
55. Surette, R. (2015). Performance crime and justice. *Current Issues in Criminal Justice: Crime, Media and New Technologies, 27*(2), 195–216. doi: 10.1080/10345329.2015.12036041
56. Bever, L. (April 25, 2017). A Thai man hanged his infant daughter on Facebook Live—then hanged himself, reports say. *The Washington Post*. Retrieved from https://www.washingtonpost.com/
57. Alldredge, J. (2015). The "CSI effect" and its potential impact on juror decisions, *Themis: Research Journal of Justice Studies and Forensic Science. 3*, Article 6. https://scholarworks.sjsu.edu/themis/vol3/iss1/6/
58. Tim. T., Pan, S., Bahri, S., & Fauzi, A. (2017). Digitally enabled crime-fighting communities: Harnessing the boundary spanning competence of social media for civic engagement. *Information & Management, 54*(2). https://doi.org/10.1016/j.im.2016.05.006
59. "Wheel of Fugitive" becomes new weapon for law enforcement. (March 14, 2017). *CBS News*. Retrieved from https://www.cbsnews.com/news/wheel-of-fugitive-brevard-county-sheriff-tips-arrests/
60. Clancy, P. (November 2016). Spotlight on community policing: The benefits of social media. Officer.com. Retrieved from https://www.officer.com/command-hq/technology/communications/article/12263074/spotlight-on-community-policing-the-benefits-of-social-media

CHAPTER 12

1. Mitchell, J. L., & Hubler, S. (April 20, 1994). Rodney King gets award of $3.8 million. *Los Angeles Times*. Retrieved from: https://www.latimes.com/local/california/la-me-king-award-19940420-story.html
2. Los Angeles Riots Fast Facts. CNN Library. (April 22, 2019). Retrieved from https://www.cnn.com/2013/09/18/us/los-angeles-riots-fast-facts/index.html
3. Morison, K.P. (2017). Hiring for the 21st century law enforcement officer: Challenges, opportunities, and strategies for success. Washington, DC: Office of Community Oriented Policing Services.
4. Katzenbach, N. (1967). The challenge of crime in a free society: A report by the President's Commission on Law Enforcement and Administration of Justice. Washington: U.S. Government Printing Office. Retrieved from https://www.hsdl.org/?view&did=709498
5. Pelfrey Jr., W. V. (2004) The inchoate nature of community policing: Differences between community policing and traditional police officer. *Justice Quarterly*, 21(3), 579–601, doi: 10.1080/07418820400095911
6. Zhao, J., Thurman, Q., & Lovrich, N. (1995). Community-oriented policing across the US: Facilitators and impediments to implementation. *American Journal of Police*, 14(1), 11–28. doi: 10.1108/07358549510799143
7. United States Congress House Committee on the Judiciary. Subcommittee on Crime, Terrorism, Homeland Security. (2012). 21st century law enforcement: How smart policing targets criminal behavior: Hearing before the Subcommittee on Crime, Terrorism, and Homeland Security of the Committee on the Judiciary, House of Representatives, One Hundred Twelfth Congress, first session, November 4, 2011. Washington, DC: U.S. G.P.O. Retrieved from https://www.govinfo.gov/content/pkg/CHRG-112hhrg71056/pdf/CHRG-112hhrg71056.pdf
8. Alpert, G. P., & Piquero. (2000). *Community policing contemporary readings* (2nd ed., pp. 9–10). Prospect Heights, IL: Waveland.
9. Westley, W. A. (1970). *Violence and the police: A sociological study of law, custom, and morality*. Cambridge, MA: MIT Press.
10. Britz, M. (1997). The police subculture and occupational socialization: Exploring individual and demographic characteristics. *American Journal of Criminal Justice*, 21(2), 127–146.
11. Herbert, S. (1998). Police subculture reconsidered. *Criminology*, 36(2), 343–370. First published: March 7, 2006. Retrieved from https://doi.org/10.1111/j.1745-9125.1998.tb01251.x
12. Alpert, G. P. and Dunham, R. G. (1997). *Policing urban America* (3rd ed., p. 40). Prospect Heights, IL: Waveland.
13. United States. Congress. House. Committee on the Judiciary. Subcommittee on Crime, Terrorism, Homeland Security. (2012). 21st century law enforcement: How smart policing targets criminal behavior : Hearing before the Subcommittee on Crime, Terrorism, and Homeland Security of the Committee on the Judiciary, House of Representatives, One Hundred Twelfth Congress, first session, November 4, 2011. Washington, DC: U.S. G.P.O. Retrieved from https://www.govinfo.gov/content/pkg/CHRG-112hhrg71056/pdf/CHRG-112hhrg71056.pdf
14. Law enforcement officers killed & assaulted (2018). *Uniform crime reports*. Federal Bureau of Justice, Department of Justice. Retrieved from https://ucr.fbi.gov/leoka/2018
15. Law enforcement officers killed 01/01/2020-03/31/2020. (2020, March 31). *Uniform crime reports*. Federal Bureau of Justice, Department of Justice. Retrieved from http://ucr.fbi.gov/leoka/2020
16. National Census of Fatal Occupational Injuries in 2018. (December 17, 2019). New Release, Bureau of Labor Statistics. U.S. Department of Labor. Retrieved from https://www.bls.gov/news.release/pdf/cfoi.pdf
17. Heyman, M., Dill, J., & Douglas, R. (April 2018). Ruderman white paper: Mental health and suicide of first responders. Ruderman Foundation. Retrieved from https://rudermanfoundation.org/
18. Alpert, G. P. and Dunham, R. G. (1997). *Policing urban America* (3rd ed., p. 40). Prospect Heights, IL: Waveland.
19. POST Commissions on Peace Officer Standards and Training. See Website: https://post.ca.gov/
20. President's Task Force on 21st Century Policing. (2015). *Final report of the President's Task Force on 21st Century Policing*. Washington, DC: Office of Community Oriented Policing Services.
21. President's Task Force on 21st Century Policing. 2015. *Final report of the President's Task Force on 21st Century Policing*. Washington, DC: Office of Community Oriented Policing Services.
22. Ethnic, and gender differences in perceptions of the police: The salience of officer race within the context of racial profiling. *Journal of Contemporary Criminal Justice*, 28(2), 206–227. https://doi.org/10.1177/1043986211425726
23. Weston J., Morrow, W. J., Vickovic, S.G., Dario, L. M., & Shjarback, J. A. (2019) Examining a Ferguson effect on college students' motivation to become police officers. *Journal of Criminal Justice Education*. doi: 10.1080/10511253.2019.1619793
24. Morin, R., & Stepler, R. (September 29, 2016). The racial confidence gap in police performance. Pew Research Center, Social & Demographic Trends. Retrieved from https://www.pewsocialtrends.org/2016/09/29/the-racial-confidence-gap-in-police-performance/

25. Mozingo, J. (July 17, 2016). Deadly attacks in Dallas and Baton Rouge echo a more dangerous time for police. *Los Angeles Times*. Retrieved from https://www.latimes.com/nation/la-na-baton-rouge-history-20160717-snap-story.html
26. Lester, D. (1983). Why do people become police officers: A study of reasons and their predictions of success. *Journal of Police Science and Administration, 11*(2), 170–174.
27. Ridgeway, G., Lim, N., Gifford, B., Koper, C., Matthies, C.F., Hajiamiri, S., & Huynh, A. K. (2008). Strategies for improving officer recruitment in the San Diego Police Department. Santa Monica, CA: RAND Corporation. https://www.rand.org/pubs/monographs/MG724.html
28. Weitzer, R., & Tuch, S. (2005). Racially biased policing: determinants of citizen perceptions. *Social Forces, 83*(3), 1009–1030. (p. 1010). Retrieved from http://www.jstor.org/stable/3598267
29. Lord, V. B., & Friday, P. C. (2003). Choosing a career in police work: A comparative study between applicants for employment with a large police department and public high school students. *Police Practice & Research, 4*(1), 63. https://doi.org/10.1080/1561426032000059196
30. Schroedel, J., Frisch, S., August, R. Kalogris, C., & Perkins, A. (1994). The invisible minority: Asian-American police officers. *State & Local Government Review, 26*(3), 173–180. Retrieved from http://www.jstor.org/stable/4355102
31. Reaves, B. (May 2015). Local police departments, 2013: Personnel, policies, and practices. Washington, DC: Bureau of Justice Statistics, NCJ 248677. https://www.bjs.gov/content/pub/pdf/lpd13ppp.pdf
32. Sun, I. Y. (2002). Police officers' attitudes toward peers, supervisors, and citizens: a comparison between field training officers and regular officers. *American Journal of Criminal Justice, 27*, 69–83, doi: 10.1007/BF02898971
33. Roberg, R., & Bonn, S. (2004). Higher education and policing: Where are we now? *Policing: An International Journal of Police Strategies & Management, 27*(4), 469–486. doi: 10.1108/13639510410566226
34. Schuck, A. M. (2014). Female representation in law enforcement: the influence of screening, unions, incentives, community policing, CALEA, and size. *Police Quarterly, 17*(1), 54–78. doi: 0.1177/1098611114522467
35. Decker, L. K., & Huckabee, R. G. (2002). Raising the age and education requirements for police officers: Will too many women and minority candidates be excluded? *Policing, 25*, 789–801. doi: 10.1108/13639510210450695
36. Reaves, B. (2015, May). Local police departments, 2013: Personnel, policies, and practices. Washington, DC: Bureau of Justice Statistics, NCJ 248677. https://www.bjs.gov/content/pub/pdf/lpd13ppp.pdf
37. Lowmaster, S. E. (2012). Predicting law enforcement officer job performance with the personality assessment inventory. *Journal of Personality Assessment, 94*(3), 254–262. doi: 10.1080/00223891.2011.648295
38. Tawney, M. (December 2008). Integrity testing: The selection tool of the future. *Law and Order*. Retrieved from http://www.hendonpub.com/resources/article_archive/results/details?id=2439#comments
39. O'Connor Boes, J., Chandler, C. J., & Timm, H. (September 1997). Police integrity: Use of personality measures to identify corruption-prone officers. Project partially funded by Grant No. 96-IJ-CX-A056 awarded by the National Institute of Justice, Office of Justice Programs, U. S. Department of Justice. Retrieved from file:///C:/Users/000002508/Downloads/452548%20(3).pdf
40. Prenzler, T., & Ronken, C. (2001). Police integrity testing in Australia. *Criminology & Criminal Justice, 1*(3), 319–342. doi: 10.1177/1466802501001003004
41. New York. Knapp Commission. (1973). *The Knapp Commission report on police corruption*. New York: G. Braziller.
42. Tawney, M. (December 2008). Integrity testing: The selection tool of the future. *Law and Order*. Retrieved from http://www.hendonpub.com/resources/article_archive/results/details?id=2439#comments
43. Sereni-Massinger, C., & Wood, N. (2016). Improving law enforcement cross cultural competencies through continued education. *Journal of Education and Learning, 5*(2), 258–264.
44. Ali, O., Garner, E., & Magadley, I. (2012). An exploration of the relationship between emotional intelligence and job performance in police organizations. *Journal of Police and Criminal Psychology, 27*(1), 1–8. DOI: 10.1007/s11896-011-9088-9
45. Seal, C., & Andrews-Brown, A. (2010). An integrative model of emotional intelligence: Emotional ability as a moderator of the mediated relationship of emotional quotient and emotional competence. *Organization Management Journal, 7*(2), 143–152. doi: 10.1057/omj.2010.22
46. Humphrey, R. (2002). The many faces of emotional leadership. *The Leadership Quarterly, 13*(5), 493–504. DOI: 10.1016/S1048-9843(02)00140-6
47. CONARC Soft Skills Training Conference. (1972). Retrieved from https://apps.dtic.mil/dtic/tr/fulltext/u2/a099612.pdf
48. Miller, A., Devaney, J., & Butler, M. (August 2019). Emotional intelligence: Challenging the perceptions and efficacy of "soft skills" in policing incidents of domestic abuse involving children. *Journal of Family Violence, 34*(6), 577–588. https://doi.org/10.1007/s10896-018-0018-9
49. Miller, S. (1999). *Gender and community policing: Walking the talk* (Northeastern series on gender, crime, and law). Boston: Northeastern University Press.

50. Griffith, J. (April 22, 2019). Man lands job after officer gives him a ride to interview instead of a ticket. NBC News. Retrieved from https://www.nbcnews.com/news/us-news/man-lands-job-after-officer-gives-him-ride-interview-instead-n997146
51. Colquitt, J. A. (2000). Toward an integrative theory of training motivation: A meta-analytic path analysis of 20 years of research. *Journal of Applied Psychology*, *85*(5), 678–708. doi: 10.1037/0021-9010.85.5.678
52. Godin, S. (January 31, 2017). Let's stop calling them "soft skills". Retrieved from https://itsyourturnblog.com/lets-stop-calling-them-soft-skills-9cc27ec09ecb
53. Spradlin, S. S. (2015). A quantitative analysis of emotional intelligence and special operations candidates' selection to military special operations training schools (Order No. 10006445). Available from ABI/INFORM Collection; ProQuest Dissertations & Theses A&I; ProQuest Dissertations & Theses Global: The Humanities and Social Sciences Collection. (1762748420). Retrieved from http://libproxy.lib.csusb.edu/login?url=https://search.proquest.com/docview/1762748420?accountid=10359
54. Terrill, W., Paoline, E. A., III, & Manning, P. K. (2003). Police culture and coercion. *Criminology*, *41*, 1003–1034. Retrieved from https://doi.org/10.1111/j.1745-9125.2003.tb01012.x
55. The White House, Office of the Press Secretary. (May 18, 2015). Remarks by the President on Community Policing. Retrieved from https://obamawhitehouse.archives.gov/the-press-office/2015/05/18/remarks-president-community-policing

CHAPTER 13

1. Rubin, J., & Winton, R. (2009, January 1). Crime continues to fall in Los Angeles despite bad economy. *Los Angeles Times*. Retrieved from https://www.latimes.com/archives/la-xpm-2009-jan-01-me-socalcrime1-story.html
2. Vito, G., Reed, J., & Walsh, W. (2017). Police executives' and managers' perspectives on Compstat. *Police Practice and Research*, *18*(1), 15–25. DOI: 10.1080/15614263.2016.1205986
3. Kovandzic, T., Schaffer, V., Vieraitis, M., Orrick, E., & Piquero, L. (2016). Police, crime and the problem of weak instruments: Revisiting the "more police, less crime" thesis. *Journal of Quantitative Criminology*, *32*(1), 133–158. doi: 10.1007/s10940-015-9257-6
4. Rubin, J., & Winton, R. (January 1, 2009). Crime continues to fall in Los Angeles despite bad economy. *Los Angeles Times*. Retrieved from https://www.latimes.com/archives/la-xpm-2009-jan-01-me-socalcrime1-story.html
5. Ahlstrom, R. (n.d.). Predicting and surviving a no-confidence vote. *Best practices guide*. International Association of Chiefs of Police. Project supported by a grant from Bureau of Justice Assistance. Retrieved from https://www.theiacp.org/sites/default/files/2018-08/BP-NoConfidenceVote.pdf
6. The Plain View Project. (n.d.). https://www.plainviewproject.org/
7. Garcia, U. J. (October 22, 2019). Phoenix officer in Facebook post scandal fired, chief says. AZCentral. Retrieved from https://www.azcentral.com/story/news/local/phoenix/2019/10/22/phoenix-police-officer-david-swick-fired-after-facebook-post-scandal/4066312002/
8. Garcia, U. J. (October 22, 2019). Phoenix officer in Facebook post scandal fired, chief says. AZCentral. Retrieved from https://www.azcentral.com/story/news/local/phoenix/2019/10/22/phoenix-police-officer-david-swick-fired-after-facebook-post-scandal/4066312002/
9. FBI 2018 crime in the United States. Full-time law enforcement employees (Table 78). Retrieved from https://ucr.fbi.gov/crime-in-the.u.s/2018/crime-in-the.u.s.-2018/tables/table-78/table-78-state-cuts/arizona.xls
10. Phoenix police union discussing possible no-confidence vote in Chief Jeri Williams after officers fired. (October 25, 2019). *12 News*. Retrieved from https://www.12news.com/article/news/local/valley/phoenix-police-union-discussing-possible-no-confidence-vote-in-chief-jeri-williams-after-officers-fired/75-e3749ad6-654f-4874-9c61-f2efff65a13c
11. Bratton, W. J. (September 16, 2016). William J. Bratton: How to reform policing from within. Op-ed contributor in *The New York Times*. Retrieved from https://www.nytimes.com/2016/09/16/opinion/william-j-bratton-how-to-reform-policing-from-within.html
12. Ahlstrom, R. (n.d.). Predicting and surviving a no-confidence vote. *Best Practices Guide*. International Association of Chiefs of Police. Project supported by a grant from Bureau of Justice Assistance. Retrieved from https://www.theiacp.org/sites/default/files/2018-08/BP-NoConfidenceVote.pdf
13. Guzman, Z. (July 13, 2018). Bill Bratton reveals what his "biggest mistake" taught him about ambition. *CNBC Make It*. Retrieved from https://www.cnbc.com/2018/07/12/bill-bratton-reveals-what-his-biggest-mistake-taught-him-about-manag.html

14. Connors, E., & Webster B. (June 24, 2003). *Transforming the law enforcement organization to community policing*. Final Monograph. NCJRS funded grant 95-IJ-CX-0091. Retrieved from https://www.ncjrs.gov/pdffiles1/nij/grants/200610.pdf
15. Wexler, C. (February 2015). Defining moments for police chiefs. Introduction in *Critical Issues in Policing Series*. Police Executive Research Forum. Retrieved from https://www.policeforum.org/assets/definingmoments.pdf
16. International Association of Chiefs of Police. (1971). Police unions and other police organizations. (Bulletin (International Association of Chiefs of Police); 4. New York: Arno Press: *New York Times*.
17. Bruinius, H. (August 30, 2016). Why police are pushing back on body cameras. *The Christian Science Monitor*. Retrieved from https://www.csmonitor.com/USA/Justice/2016/0830/Why-police-are-pushing-back-on-body-cameras
18. Kadleck, C. & Travis, III, L. (March 2008). *Police department and police officer association leaders' perceptions of community policing: Describing the nature and extent of agreement*. Final Report. This project was supported by Grant No. 98-IJ-CX-0005 awarded by the National Institute of Justice Programs, U.S. Department of Justice. Retrieved from https://www.ncjrs.gov/pdffiles1/nij/grants/226315.pdf
19. Garrity v. New Jersey, 385 U.S. 493 (1967)
20. Levinson, R. (January 13, 2017). Protecting the blue, across the U.S., police contracts shield officers from scrutiny and discipline. *Reuters Investigates*. Retrieved from https://www.reuters.com/investigates/special-report/usa-police-unions/
21. Calams, S. (2020). Quick take: How police leaders can improve officer safety during the COVID-19 crisis. PoliceOne.Com. Retrieved from https://www.policeone.com/police-products/firstnet-products-service/articles/quick-take-how-police-leaders-can-improve-officer-safety-during-the-COVID-19-crisis-jba5oNIB9I5AJ5uT/
22. Connors, E., & Webster B. (June 24, 2003). *Transforming the law enforcement organization to community policing*. Final Monograph. NCJRS funded grant 95-IJ-CX-0091. Retrieved from https://www.ncjrs.gov/pdffiles1/nij/grants/200610.pdf
23. Connors, E., & Webster B. (June 24, 2003). Transforming the Law Enforcement Organization to Community Policing, Final Monograph. NCJRS funded grant 95-IJ-CX-0091. Retrieved from https://www.ncjrs.gov/pdffiles1/nij/grants/200610.pdf
24. Gallup. (July 10, 2017). Confidence in police back at historical average. Retrieved from https://news.gallup.com/poll/213869/confidence-police-back-historical-average.aspx
25. Federal Bureau of Investigation, *Uniform crime reports, 1960–1999*. Retrieved from https://www.bjs.gov/ucrdata/Search/Crime/State/RunCrimeStatebyState.cfm
26. Gramlich, J. (October 17, 2019). 5 facts about crime in the U.S. Pew Research Center. Retrieved from https://www.pewresearch.org/fact-tank/2019/10/17/facts-about-crime-in-the-u-s/
27. Gramlich, J. (October 17, 2019). 5 facts about crime in the U.S. Pew Research Center. Retrieved from https://www.pewresearch.org/fact-tank/2019/10/17/facts-about-crime-in-the-u-s/
28. Dolliver, M., Kenney, J., Reid, L., Prohaska, A., & Henson, B. (2018). Examining the relationship between media consumption, fear of crime, and support for controversial criminal justice policies using a nationally representative sample. *Journal of Contemporary Criminal Justice*, *34*(4), 399–420. doi: 10.1177/1043986218787734
29. The United States Department of Justice. Justice news retrieved from https://www.justice.gov/opa/pr/department-justice-awards-119-million-hire-community-policing-officers
30. Williams, E. J. (2003). Structuring in community policing: Institutionalizing innovative change. *Police Practice & Research*, *4*(2), 119–130. doi: 10.1080/1561426032000084909
31. Maguire, E., Shin, Y., "Solomon" Zhao, J., & Hassell, K. (2003). Structural change in large police agencies during the 1990s. *Policing: An International Journal*, *26*(2), 251–275. https://doi.org/10.1108/13639510310475750
32. Walker, S. (1993). Does anyone remember team policing? Lessons of team policing experiences for community policing. *American Journal of Police*, *12*(1): 33–55.
33. Stanley, D., Meyer, J., & Topolnytsky, J. (2005). Employee cynicism and resistance to organizational change. *Journal of Business and Psychology*, *19*(4), 429–459.
34. Van Maanen, J. (1978). The asshole. In P. K. Manning & J. Van Maanen (Eds.). *Policing: A view from the street* (pp. 221–238). Santa Monica, CA: Goodyear.
35. Connors, E., & Webster B. (June 24, 2003). *Transforming the law enforcement organization to community policing*. Final monograph. NCJRS funded grant 95-IJ-CX-0091. Retrieved from https://www.ncjrs.gov/pdffiles1/nij/grants/200610.pdf
36. Williams, E. J. (2003). Structuring in community policing: Institutionalizing innovative change. *Police Practice & Research*, *4*(2), 119–130. doi: 10.1080/1561426032000084909
37. Weichselbaum, S. & Schwartzapfel, B. (2017) When warriors put on the badge. The Marshall Project, in collaboration with *USA Today* Network. Retrieved from https://www.themarshallproject.org/2017/03/30/when-warriors-put-on-the-badge

38. Pannell, I. (March 17, 2018). Baltimore: "This is what poverty in the US looks like." *BBC News*. Retrieved from YouTube: https://youtu.be/tCgqIN-6A20
39. Carr, P. J., Napolitano, L., & Keating, J. (2007). We never call the cops and here is why: A qualitative examination of legal cynicism in three Philadelphia neighborhoods. *Criminology, 45*(2), 445–480.
40. Holmes, M. D., Smith, B. W., Freng, A. B., & Munoz E. A. (2008). Minority threat, crime control, and police resource allocation in the Southwestern United States. *Crime & Delinquency, 54*(1), 128–152.
41. The Center for Popular Democracy (2017). *Freedom to thrive. Reimagining safety and security in our communities*. Retrieved from https://populardemocracy.org/sites/default/files/Freedom%20To%20Thrive%2C%20Higher%20Res%20Version.pdf
42. Police Executive Research Forum (2010). *Is the economic downturn fundamentally changing how we police? Critical issues in policing series*. Retrieved from https://www.policeforum.org/assets/docs/Critical_Issues_Series/is%20the%20economic%20downturn%20fundamentally%20changing%20how%20we%20police%202010.pdf
43. Police Executive Research Forum (2010). *Is the economic downturn fundamentally changing how we police? Critical issues in policing series*. Page 5. Retrieved from https://www.policeforum.org/assets/docs/Critical_Issues_Series/is%20the%20economic%20downturn%20fundamentally%20changing%20how%20we%20police%202010.pdf
44. Police Executive Research Forum (2010). *Is the economic downturn fundamentally changing how we police? Critical issues in policing series*. Retrieved from https://www.policeforum.org/assets/docs/Critical_Issues_Series/is%20the%20economic%20downturn%20fundamentally%20changing%20how%20we%20police%202010.pdf
45. LeDuff, C. (February 25, 2018). Inside a broken police department in Flint, Michigan. *The New Times*. Retrieved from https://www.newyorker.com/culture/photo-booth/inside-a-broken-police-department-in-flint-michigan
46. LeDuff, C. (February 25, 2018). Inside a broken police department in Flint, Michigan. *The New Times*. Retrieved from https://www.newyorker.com/culture/photo-booth/inside-a-broken-police-department-in-flint-michigan
47. Rainguet, F. W., & Dodge, M. (2001). The problems of police chiefs: An examination of the issues in tenure and turnover. *Police Quarterly, 4*(3), 268–288. Retrieved from file:///C:/Users/000002508/Downloads/RainguetandDodge.pdf
48. City of Lincoln Nebraska, Tom Casady, Public Safety Director. (n.d.). Website for Lincoln Nebraska Police: http://www.lincoln.ne.gov/city/police/cbp.htm
49. Holmes, M. D., Smith, B. W., Freng, A. B., & Munoz E. A. (2008). Minority threat, crime control, and police resource allocation in the Southwestern United States. *Crime & Delinquency, 54*(1), 128–152.
50. Reaves, B. A. (May 2015). Local police departments, 2013: Personnel, policies, and practices. U.S. Department of Justice office of Justice Programs Bureau of Justice Statistics. NCJ 248677 https://www.bjs.gov/content/pub/pdf/lpd13ppp.pdf
51. Collaborative Reform Initiative. Website https://cops.usdoj.gov/collaborativereform
52. Reaves, B. A. (May 2015). Local police departments, 2013: Personnel, policies, and practices. U.S. Department of Justice office of Justice Programs Bureau of Justice Statistics. NCJ 248677 https://www.bjs.gov/content/pub/pdf/lpd13ppp.pdf
53. Marino, P. (September 16, 2017). Salinas Police rec-commit to community policing after DOJ cancels program. *Monterey County Weekly*. Retrieved from http://www.montereycountyweekly.com/blogs/crime_blog/salinas-police-re-commit-to-community-policing-after-doj-cancels/article_7bb350f2-9a6e-11e7-a4fe-132f2f7433c5.html
54. Wexler, C. (February 2015). Defining moments for police chiefs. Introduction in *Critical issues in policing series*. Police Executive Research Forum. Retrieved from https://www.policeforum.org/assets/definingmoments.pdf

CHAPTER 14

1. Arcega-Dunn, M. (August 25, 2019). Family wants answers after man is shot, killed by police. FOX 5 News. Retrieved from https://fox5sandiego.com/2019/08/25/family-wants-answers-after-man-is-shot-killed-by-police/
2. Salonga, R. (November 5, 2019). San Jose: Man shot by police had history of mental-health struggles. *The Mercury News*. Retrieved from https://www.mercurynews.com/2019/11/05/san-jose-man-fatally-shot-by-police-had-history-of-mental-health-struggles/
3. *Graham v. Connor*, 490 U.S. 386, 109 S.Ct. 1865, 104 L.Ed.2d 443 (1989).
4. Sharp, C. (2018). Jails: America's biggest mental health facilities. *Kennedy School Review, 18*, 48–56.
5. Seattle Police Department. Community Policing. About MCPP. Retrieved from https://www.seattle.gov/police/community-policing/about-mcpp
6. Community Policing Defined. (Published 2012, revised 2014). Community Oriented Policing Services. U.S. Department of Justice. Pamphlet retrieved from https://cops.usdoj.gov/RIC/Publications/cops-p157-pub.pdf

7. Community Policing Defined. (Published 2012, revised 2014). Community Oriented Policing Services. U.S. Department of Justice. Pamphlet retrieved from https://cops.usdoj.gov/RIC/Publications/cops-p157-pub.pdf
8. COPS Training & Technical Assistance. See website: https://cops.usdoj.gov/training-technical-assistance
9. CRI-TAC Collaborative Reform Initiative Technical Assistance Center. See website: https://cops.usdoj.gov/collaborativereform
10. Chavis, D. M., & Lee, K. (May 12, 2015) What is community anyway? *Stanford Social Innovation Review*. Retrieved from https://ssir.org/articles/entry/what_is_community_anyway#
11. Chavis, D. M., & Lee, K. (2015, May 12). What is community anyway? Informing and inspiring leaders of social change. *Stanford Social Innovation Review*. Retrieved from https://ssir.org/articles/entry/what_is_community_anyway#
12. Stein, R., & Griffith, C. (2017). Resident and police perceptions of the neighborhood: Implications for community policing. *Criminal Justice Policy Review, 28*(2), 139–154. doi: 10.1177/0887403415570630
13. Clinical and Translational Science Awards Consortium. Community Engagement Key Function Committee Task Force on the Principles of Community Engagement. (June 2011) *Principles of community engagement* (2nd ed.). NIH Publication No. 11-7782. Book retrieved from https://www.atsdr.cdc.gov/communityengagement/pdf/PCE_Report_508_FINAL.pdf
14. Scott, W., & Lazar, D. (October 3, 2018). Community policing strategic plans. *Police Chief*. Retrieved from https://www.policechiefmagazine.org/community-policing-strategic-plan/
15. Community Policing Defined. (Published 2012, revised 2014). Community Oriented Policing Services. U.S. Department of Justice. Pamphlet retrieved from https://cops.usdoj.gov/RIC/Publications/cops-p157-pub.pdf
16. Van Maanen, J. (1995). The Asshole. In Kappeler, V. (Ed.). *The Police & society: Touchstone readings* (pp. 307–328). Prospect Heights, IL: Waveland.
17. National Neighborhood Watch, a Division of the National Sheriff's Association. Retrieved from https://www.nnw.org/our-history
18. Luscombe, R. (July 2013). George Zimmerman acquittal leads to protests across US cities. *The Guardian*. Retrieved from https://www.theguardian.com/world/2013/jul/15/trayvon-martin-protests-streets-acquittal
19. Alcindor, Y. (June 25, 2013). Trial turns to Zimmerman's neighborhood-watch role. *USA Today*. Retrieved from https://www.usatoday.com/story/news/2013/06/25/zimmerman-trial-trayvon-neighborhood-watch/2455163/
20. Rosenbaum, D. (1988). Community crime prevention: A review and synthesis of the literature. *Justice Quarterly, 5*(3), 323–395. DOI: 10.1080/07418828800089781
21. Liberation Staff. (April 13, 2012). The ugly history and repressive role of Neighborhood Watch. *Liberation Newspaper*. Retrieved from https://www.liberationnews.org/the-ugly-history-and-html/
22. Ross, N. (November 21, 2013). 8 tips for designing your Citizen Police Academy. PoliceOne.com. Retrieved from https://www.policeone.com/citizen-police-academies/articles/8-tips-for-designing-your-citizen-police-academy-tbZKuPIZ5qAvOdn7/
23. Brewster, J., Stoloff, M., & Sanders, N. (2005). Effectiveness of citizen police academies in changing the attitudes, beliefs, and behavior of citizen participants. *American Journal of Criminal Justice, 30*(1), 21–34. doi: 10.1007/BF02885879
24. Jordan, W. (2000). Citizen police academies: Community policing or community politics? *American Journal of Criminal Justice, 25*(1), 93–105. DOI: 10.1007/BF02886813
25. Cohn, E. (1996). The Citizen Police Academy: A recipe for improving police-community relations. *Journal of Criminal Justice, 24*(3), 265–271. DOI: 10.1016/0047-2352(96)00011-6
26. Brewster, J., Stoloff, M., & Sanders, N. (2005). Effectiveness of citizen police academies in changing the attitudes, beliefs, and behavior of citizen participants. *American Journal of Criminal Justice, 30*(1), 21–34. doi: 10.1007/BF02885879
27. Brown, C. (July 31, 2019). Next-gen citizen's police academies. *Police Chief*. Retrieved from https://www.policechiefmagazine.org/next-gen-citizens-police-academies/
28. Fox, J.A., & Piquero, A. (2003). Deadly demographics: Population characteristics and forecasting homicide trends. *Crime and Delinquency, 49*(3), 339–359. doi: 10.1177/00111287030490030
29. Males, M. (2015). Age, poverty, homicide, and gun homicide: Is young age or poverty level the key issue? *SAGE Open, 5*(1). doi: 10.1177/2158244015573359
30. Allen, A., & Lo, C. (2012). Drugs, guns, and disadvantaged youths: Co-occurring behavior and the code of the street. *Crime & Delinquency, 58*(6), 932–953. doi: 10.1177/0011128709359652
31. *Los Angeles police department youth programs manual*. (Updated August 2017). Retrieved from http://assets.lapdonline.org/assets/pdf/2017_Youth_Programs_manual.pdf
32. TAPS Center Teen and Police Service Academy. Retrieved from http://www.tapsacademy.org/Programs
33. TAPS Center Teen and Police Service Academy. Retrieved from http://www.tapsacademy.org/events

34. Blake, M. (June 2015). California seniors police patrol. *Reuters/The Wider Image.* Retrieved from https://widerimage.reuters.com/story/california-seniors-police-patrol
35. Garcia, S. & ABC7.com staff. (January 13, 2020). Long Beach retired veterans patrol streets like regular officers, integral part of police department. *ABC7 Eyewitness News.* Retrieved from https://abc7.com/5845886/
36. Hetey, R. C., Monin, B., Maitreyi, A., & Eberhardt, J. L. (2016). Data for change: A statistical analysis of police stops, searches, handcuffings, and arrests in Oakland, Calif., 2013–2014. Stanford University, SPARQ: Social Psychological Answers to Real-World Questions. Report retrieved from https://stanford.app.box.com/v/Data-for-Change
37. Spencer, K., Charbonneau, A., & Glaser, J. (2016). Implicit bias and policing. *Social and Personality Psychology Compass, 10*(1), 50–63. doi: 10.1111/spc3.12210
38. Kramer, R., Remster, B., & Charles, C. Z. (2017). Black lives and police tactics matter. *Contexts, 16*(3), 20–25. https://doi.org/10.1177/1536504217732048
39. Quillian, L., & Pager, D. (2001). Black neighbors, higher crime? The role of racial stereotypes in evaluations of neighborhood crime 1. *American Journal of Sociology, 107*(3), 717–767. doi: 10.1086/338938
40. Diehr, A., & Mcdaniel, J. (2018). Lack of community-oriented policing practices partially mediates the relationship between racial residential segregation and "black-on-black" homicide rates. *Preventive Medicine, 112,* 179–184. doi: 10.1016/j.ypmed.2018.04.032
41. Jones, J. M. (June 19, 2015). In U.S., confidence in police lowest in 22 years. Gallup. Retrieved from https://news.gallup.com/poll/183704/confidence-police-lowest-years.aspx
42. Carbado, D. W., & Richardson, L. S. (May 10, 2018). *The Black police: Policing our own.* Book review. *Harvard Review.* Retrieved from https://harvardlawreview.org/2018/05/the-black-police-policing-our-own/
43. Importance of police–community relationships and resources for further reading. (n.d.). U.S. Department of Justice. Retrieved from: https://www.justice.gov/crs/file/836486/download
44. Holtzman, M. (n.d.) The barbershop: where real conversations take place. National Police Foundation. Retrieved from https://www.policefoundation.org/the-barbershop-where-real-conversations-take-place/
45. Hartmann, D. (2001). Notes on midnight basketball and the cultural politics of recreation, race, and at-risk urban youth. *Journal of Sport and Social Issues, 25*(4), 339–371. doi: 10.1177/0193723501254002
46. Basketball Cop Foundation. Website: https://www.basketballcop.net/
47. Solomon, D., Jawetz, T., & Malik, S. (March 21, 2017). The negative consequences of entangling local policing and immigration enforcement. *Center for American Progress.* Retrieved from https://cdn.americanprogress.org/content/uploads/2017/03/20140134/LawEnforcementSanctuary-brief.pdf
48. The Associated Press. (July 12, 2019). Trump Administration gets court victory in sanctuary cities case. *The New York Times.* Retrieved from https://www.nytimes.com/2019/07/12/us/sanctuary-cities-ruling.html
49. The Associated Press. (July 12, 2019). Trump Administration gets court victory in sanctuary cities case. *The New York Times.* Retrieved from https://www.nytimes.com/2019/07/12/us/sanctuary-cities-ruling.html
50. Radford, J. & Noe-Bustamante. (June 3, 2019). Facts on U.S. immigrants, 2017. Hispanic trends. Pew Research Center. Retrieved from https://www.pewresearch.org/hispanic/2019/06/03/facts-on-u-s-immigrants/
51. Radford, J. & Noe-Bustamante. (June 3, 2019). Facts on U.S. immigrants, 2017, *Hispanic trends. Pew Research Center.* Retrieved from https://www.pewresearch.org/hispanic/2019/06/03/facts-on-u-s-immigrants/
52. Bever, L. (May 12, 2017). Hispanics 'are going further into the shadows' amid chilling immigration debate, police say. *The Washington Post.* Retrieved from https://www.washingtonpost.com/news/post-nation/wp/2017/05/12/immigration-debate-might-be-having-a-chilling-effect-on-crime-reporting-in-hispanic-communities-police-say/
53. Weber, N. (July 15, 2018). Aurora embraces role as the top landing spot for immigrants in Colorado. *The Denver Post.* Retrieved from https://www.denverpost.com/2018/07/15/aurora-colorado-immigrants-home/
54. Police Executive Research Forum. (2019). *Community policing in immigrant neighborhoods: Stories of success.* Publication supported by Carnegie Corporation of New York. Police Executive Research Forum. Washington, DC. https://www.policeforum.org/assets/CommunityPolicingImmigrantNeighborhoods.pdf
55. State of Homelessness. (2020). National Alliance to End Homelessness website: https://endhomelessness.org/homelessness-in-america/homelessness-statistics/state-of-homelessness-report/
56. Homeless Victory: Santa Ana law criminalizes status. (1994). *ABA Journal, 80*(4), 102–103.
57. Kelly, D. (March 18, 2008). "Ontario residents only" at Tent City. *Los Angeles Times.* Retrieved from https://www.latimes.com/local/la-me-tents18mar18-story.html
58. Kelly, D. (March 18, 2008). "Ontario residents only" at Tent City. *Los Angeles Times.* Retrieved from https://www.latimes.com/local/la-me-tents18mar18-story.html

59. Do, A. (September 4, 2016). Santa Ana struggles with increasing homeless population around Civic Center. *Los Angeles Times*. Retrieved from https://www.latimes.com/local/lanow/la-me-ln-santa-ana-homeless-20160904-snap-story.html
60. The City of Santa Ana, Police Department, Homeless Engagement. Retrieved from https://www.ci.santa-ana.ca.us/pd/homeless-engagement
61. Salonga, R. (May 14, 2018). Report: Mentally ill are in nearly 40 percent of South Bay Police shootings. *The Mercury News*. Retrieved from https://www.mercurynews.com/2018/05/11/report-mentally-ill-are-in-nearly-40-percent-of-south-bay-police-shootings/
62. Fuller, D. A., Lamb, H. R., Biasotti, M., & Snook, J. (2015, December). Overlooked in the undercounted, the role of mental illness in fatal law enforcement encounters. Report can be found online: https://www.treatmentadvocacycenter.org/storage/documents/overlooked-in-the-undercounted.pdf
63. Clifford, K. (2013). Mental health crisis and interventions and the politics of police use of deadly force. In D. Chappell (Ed.), *Policing and the mentally ill: International perspectives* (pp. 171–195). Boca Raton, FL: CRC Press.
64. Sweeney, K. (1999, January 10). No crisis here untrained Memphis Police used to kill seven mentally ill people a year. But that was before the police got smart. They cooperated with mental health experts on an action plan that has saved lives -- and Memphis is now a national model. Might there be lessons for Jacksonville? *The Florida Times Union*. Retrieved from https://www.questia.com/newspaper/1G1-57527308/no-crisis-here-untrained-memphis-police-used-to-kill
65. Watson, A., & Fulambarker, A. (2012). The crisis intervention team model of police response to mental health crises: A primer for mental health practitioners. *Best Practices in Mental Health*, 8(2), 71–81.
66. Watson, A. (2010). Research in the real world: Studying Chicago Police Department's crisis intervention team program. *Research on Social Work Practice*, 20(5), 536–543. doi: 10.1177/104973151037420
67. Jany, L. (January 30, 2020). Minn. police to hand out vouchers instead of citations for equipment violations. *PoliceOne.com*. Retrieved from https://www.policeone.com/vehicle-incidents/articles/minn-police-to-hand-out-vouchers-instead-of-citations-for-equipment-violations-SE42yZ8g4G9yrEEg/
68. Warth, G. (January 5, 2020). Police program clears homeless infractions in exchange for shelter stays. PoliceOne.com. Retrieved from https://www.policeone.com/community-policing/articles/police-program-clears-homeless-infractions-in-exchange-for-shelter-stays-qwTRJCgYjRns9c8p/
69. Marrero, T. (June 7, 2019). Hillsborough deputies will be walking and talking with public under new policing effort. *Tampa Bay Times*. Retrieved from https://www.tampabay.com/news/publicsafety/hillsborough-deputies-will-be-walking-and-talking-with-public-under-new-policing-effort-20190606/
70. Dukmasova, M. (August 25, 2016). Abolish the police? Organizers say it's less crazy than it sounds. *Reader News & Politics*. Retrieved from https://www.chicagoreader.com/chicago/police-abolitionist-movement-alternatives-cops-chicago/Content?oid=23289710
71. Dukmasova, M. (2016, August 25). Abolish the police? Organizers say it's less crazy than it sounds. *Reader News & Politics*. Retrieved from https://www.chicagoreader.com/chicago/police-abolitionist-movement-alternatives-cops-chicago/Content?oid=23289710
72. https://blacklivesmatter.com/what-defunding-the-police-really-means/
73. For a world without police. Website is http://aworldwithoutpolice.org/about-us/
74. The Abolition Research Group. (October 8, 2017). The problem with community policing. For a world without police. Retrieved from http://aworldwithoutpolice.org/2017/10/08/the-problem-with-community-policing/

CHAPTER 15

1. Corman, D. (October 26–29, 2019). *Best ways to engage communities through outreach*. Paper presented at the International Conference of Police Chiefs in Chicago, IL.
2. Bradsher, K. (November 22, 1997). G.M. to close car factory, delivering big blow to Flint. *The New York Times*. Retrieved from https://www.nytimes.com/1997/11/22/business/gm-to-close-car-factory-delivering-big-blow-to-flint.html
3. Federal Bureau of Investigation (2019). *Crime in the U.S.* U.S. Department of Justice. Retrieved from https://ucr.fbi.gov/crime-in-the-u.s
4. Murembya, L. (2016). *Demographic and labor market profile: City of Flint*. State of Michigan. Department of Technology, Management, and Budget. Retrieved from https://milmi.org/Portals/198/publications/Flint_City_Demographic_and_Labor_Mkt_Profile.pdf
5. Adams, D. *See how Flint's police staffing fares against the nation's most-violent cities*. MLive Michigan. Retrieved December 11, 2019 from https://www.mlive.com/news/flint/2017/09/see_how_flint_police_staffing.html

6. Parsons, J. Policing. Vera Institute of Justice. Retrieved December 14, 2019 from https://www.vera.org/centers/policing
7. Nickel, D. S. (2018). *Critical factors in public satisfaction with police services*. Major paper submitted in partial fulfillment of the requirements for the degree of Masters of Arts (Criminal Justice). Fraser Valley University. Retrieved from https://arcabc.ca/islandora/object/ufv:18286/datastream/PDF/file.pdf
8. Nickel, D. S. (2018). *Critical factors in public satisfaction with police services*. Major paper submitted in partial fulfillment of the requirements for the degree of Masters of Arts (Criminal Justice). Fraser Valley University. Retrieved from https://arcabc.ca/islandora/object/ufv:18286/datastream/PDF/file.pdf
9. Tyler, T. R. (1990). *Why people obey the law*. New Haven, CT: Yale University Press.
10. Morin, R., Parker, K., Stepler, R., & Mercer, A. (2017). *Behind the badge: Police views, public views*. PEW Research Center. Retrieved from https://www.pewsocialtrends.org/2017/01/11/police-views-public-views/
11. Morin, R., Parker, K., Stepler, R., & Mercer, A. (2017). *Behind the badge: Police views, public views*. PEW Research Center. Retrieved from https://www.pewsocialtrends.org/2017/01/11/police-views-public-views/
12. Nickel, D. S. (2018). *Critical factors in public satisfaction with police services*. Major paper submitted in partial fulfillment of the requirements for the degree of Masters of Arts (Criminal Justice). Fraser Valley University. Retrieved from https://arcabc.ca/islandora/object/ufv:18286/datastream/PDF/file.pdf
13. Weitzer, R., & Tuch, S. (2005). Determinants of public satisfaction with police. *Police Quarterly*, *8*(3), 279–297.
14. Morin, R., Parker, K., Stepler, R., & Mercer, A. (2017). *Behind the badge: Police views, public views*. PEW Research Center. Retrieved from https://www.pewsocialtrends.org/2017/01/11/police-views-public-views/
15. Tate, J., Jenkins, J., and Rich, S. (December 19, 2019). Fatal force. *Washington Post*. Retrieved from https://www.washingtonpost.com/graphics/2019/national/police-shootings-2019/
16. Neusteter, R., O'Toole, M., & Doyle, L. (2018). *Why a Michigan law enforcement agency employs formerly incarcerated people to bridge the police-community divide*. Vera Institute of Justice. Retrieved from https://www.vera.org/downloads/Publications/washtenaw-county-sheriffs-office-case-study/legacy_downloads/Washtenaw-County-Sheriffs-Office-Case-Study_final.pdf
17. U.S. Census Bureau (2018). *Quick facts: United States*. Retrieved from https://www.census.gov/quickfacts/fact/table/US/PST045218
18. Parks, W. (2000). *Community policing as a foundation for restorative justice*. International Institute for Restorative Practices.
19. Parks, W. (2000). *Community policing as a foundation for restorative justice*. International Institute for Restorative Practices.
20. Center for Justice and Reconciliation. *Restorative Justice*. Retrieved 2019 from http://restorativejustice.org/restorative-justice/#sthash.bfm5l1jA.dpbs.
21. Bazemore, G., & Griffith, C. (2003). Police reform, restorative justice and restorative policing. *Police Practice and Research*, *4*(4), 335–346.
22. Demand, J. (2015). *How a citizens' academy can be a positive approach to public relations*. PoliceOnce.com. Retrieved from https://www.policeone.com/community-policing/articles/how-a-citizens-police-academy-can-be-a-positive-approach-to-public-relations-H3ekAqVXz9TH1vO5/
23. Bazemore, G., & Griffith, C. (2003). Police reform, restorative justice and restorative policing. *Police Practice and Research*, *4*(4), 335–346.
24. Parks, W. (2000). *Community policing as a foundation for restorative justice*. International Institute for Restorative Practices.
25. Bazemore, G., & Griffith, C. (2003). Police reform, restorative justice and restorative policing. *Police Practice and Research*, *4*(4), 335–346.
26. Bazemore, G. and Griffith, C. (2003). Police reform, restorative justice and restorative policing. *Police Practice and Research*, *4*(4), 335–346.
27. Walsh, W. D. (October 26–29, 2019). *Citizen academies: Building trust, increasing legitimacy, correcting misconceptions*. Paper Presented at the International Conference of Police Chiefs in Chicago, IL.
28. Demand, J. (2015). *How a citizens' academy can be a positive approach to public relations*. PoliceOnce.com. Retrieved from https://www.policeone.com/community-policing/articles/how-a-citizens-police-academy-can-be-a-positive-approach-to-public-relations-H3ekAqVXz9TH1vO5/
29. Walsh, W. D. (October 26–29, 2019). *Citizen academies: building trust, increasing legitimacy, correcting misconceptions*. Paper Presented at the International Conference of Police Chiefs in Chicago, IL.
30. Demand, J. (2015). *How a citizens' academy can be a positive approach to public relations*. PoliceOnce.com. Retrieved from https://www.policeone.com/community-policing/articles/how-a-citizens-police-academy-can-be-a-positive-approach-to-public-relations-H3ekAqVXz9TH1vO5/

31. Demand, J. (2015). *How a citizens' academy can be a positive approach to public relations*. PoliceOnce.com. Retrieved from https://www.policeone.com/community-policing/articles/how-a-citizens-police-academy-can-be-a-positive-approach-to-public-relations-H3ekAqVXz9TH1vO5/
32. Walsh, W. D. (October 26–29, 2019). *Citizen academies: building trust, increasing legitimacy, correcting misconceptions.* Paper Presented at the International Conference of Police Chiefs in Chicago, IL.
33. Bazemore, G., & Griffith, C. (2003). Police reform, restorative justice and restorative policing. *Police Practice and Research, 4*(4), 335–346.
34. Nicholls, C.G. (1998). *Community policing, community justice, and restorative justice: Exploring the links for the delivery of a balanced approach to public safety.* US Department of Justice, Office of Community Oriented Policing Services. Retrieved from https://cops.usdoj.gov/RIC/Publications/cops-w0033-pub.pdf
35. Bazemore, G., & Griffith, C. (2003). Police reform, restorative justice and restorative policing. *Police Practice and Research, 4*(4), 335–346.
36. Bazemore, G., & Griffith, C. (2003). Police reform, restorative justice and restorative policing. *Police Practice and Research, 4*(4), 335–346.
37. Bazemore, G., & Griffith, C. (2003). Police reform, restorative justice and restorative policing. *Police Practice and Research, 4*(4), 335–346.
38. Bazemore, G., & Griffith, C. (2003). Police reform, restorative justice and restorative policing. *Police Practice and Research, 4*(4), 335–346.
39. Umbreit, M. (1994). *Victim meets offender: The impact of restorative justice in mediation.* Monsey, NY: Criminal Justice Press.
40. Bazemore, G., & Umbreit, M. (2001). A comparison of four restorative conferencing programs. *Juvenile Justice Bulletin.* Office of Juvenile Justice and Delinquency Prevention. Retrieved from https://www.ncjrs.gov/pdffiles1/ojjdp/184738.pdf
41. Bazemore, G., & Umbreit, M. (2001). A comparison of four restorative conferencing programs. *Juvenile Justice Bulletin.* Office of Juvenile Justice and Delinquency Prevention. Retrieved from https://www.ncjrs.gov/pdffiles1/ojjdp/184738.pdf
42. Bazemore, G., & Umbreit, M. (2001). A comparison of four restorative conferencing programs. *Juvenile Justice Bulletin.* Office of Juvenile Justice and Delinquency Prevention. Retrieved from https://www.ncjrs.gov/pdffiles1/ojjdp/184738.pdf
43. Stuart, B. (1996). Circle sentencing—Turning swords into ploughshares. In B. Galaway & J. Hudson (Eds.). *Restorative justice: International perspectives.* Boulder, CO: Lynne Reimer.
44. Corman, D. (October 26–29, 2019). *Best ways to engage communities through outreach.* Paper Presented at the International Conference of Police Chiefs in Chicago, IL.
45. Federal Bureau of Investigation (2017). *2017 crime in the United States.* Retrieved from https://ucr.fbi.gov/crime-in-the-u.s/2017/crime-in-the-u.s.-2017/tables/table-8/table-8-state-cuts/louisiana.xls
46. Moore, Hillar C., III. (2018). *State of crime, 2018: A look at crime and crime response needs in East Baton Rouge Parish.* Baton Rouge, LA: East Baton Rouge District Attorney's Office.
47. Paul, M. (October 26–29, 2019). *Best ways to engage communities through outreach (PSP).* Paper presented at the International Conference of Police Chiefs in Chicago, IL.
48. Hutchins, M. (October 26–29, 2019). *Best ways to engage communities through outreach (PSP).* Paper presented at the International Conference of Police Chiefs in Chicago, IL.
49. Calhoun, J. *How law enforcement and faith-based groups can work together for cities.* National League of Cities. Retrieved 2017 from https://www.nlc.org/article/how-law-enforcement-and-the-faith-community-can-work-together-for-cities
50. Calhoun, J. *How law enforcement and faith-based groups can work together for cities.* National League of Cities. Retrieved 2017 from https://www.nlc.org/article/how-law-enforcement-and-the-faith-community-can-work-together-for-cities
51. PoliceOneStaff (2015). Video: Anti-police protester undergoes use of force scenario training. PoliceOne.Com. Retrieved from https://www.policeone.com/use-of-force/articles/video-anti-police-protester-undergoes-use-of-force-scenario-training-ucSt7LklFGBrQ8F9/
52. Congressional Research Service (2018). *Public trust and law enforcement—A discussion for policymakers.* Retrieved from https://fas.org/sgp/crs/misc/R43904.pdf
53. Deliso, M. (2020). Timeline: *The impact of George Floyd's death in Minneapolis and beyond,* June 10, ABC News, https://abcnews.go.com/US/timeline-impact-george-floyds-death-minneapolis/story?id=70999322
54. SB20-217 Enhance Law Enforcement Integrity (2020). *Concerning measures to enhance law enforcement integrity, and, in connection therewith, making an appropriation.* Colorado General Assembly. Available online: https://leg.colorado.gov/bills/sb20-217

55. Perry, N. (June 22, 2018). Looking back, looking forward: predicting the state of policing in the future. PoliceOne. Retrieved from https://www.policeone.com/police-products/body-cameras/articles/looking-back-looking-forward-predicting-the-state-of-policing-in-2043-P8e1UuSz03F8hbHU/
56. Perry, N. (June 22, 2018). Looking back, looking forward: predicting the state of policing in the future. PoliceOne, Retrieved from https://www.policeone.com/police-products/body-cameras/articles/looking-back-looking-forward-predicting-the-state-of-policing-in-2043-P8e1UuSz03F8hbHU/
57. Federal Bureau of Investigation. *The augmented reality technology*. Retrieved April 2, 2020 https://www.fbi.gov/file-repository/stats-services-publications-police-augmented-reality-technology-pdf/view
58. Federal Bureau of Investigation. *The augmented reality technology*. Retrieved April 2, 2020 from https://www.fbi.gov/file-repository/stats-services-publications-police-augmented-reality-technology-pdf/view
59. Federal Bureau of Investigation. *The augmented reality technology*. Retrieved April 2, 2020 from https://www.fbi.gov/file-repository/stats-services-publications-police-augmented-reality-technology-pdf/view
60. Griffith, D. (2017). *SWAT in the 21st century*. Police Law Enforcement Solutions. Retrieved from https://www.policemag.com/342295/swat-in-the-21st-century
61. Federal Bureau of Investigation. *The augmented reality technology*. Retrieved April 2, 2020 from https://www.fbi.gov/file-repository/stats-services-publications-police-augmented-reality-technology-pdf/view
62. Federal Bureau of Investigation. *The augmented reality technology*. Retrieved April 2, 2020 from https://www.fbi.gov/file-repository/stats-services-publications-police-augmented-reality-technology-pdf/view
63. Federal Bureau of Investigation. *The augmented reality technology*. Retrieved April 2, 2020 from https://www.fbi.gov/file-repository/stats-services-publications-police-augmented-reality-technology-pdf/view
64. Fleming, C. Remote drone dispatch: Law enforcement's future? *Police Chief Magazine*. Retrieved April 4, 2020 from https://www.policechiefmagazine.org/remote-drone-dispatch/
65. Salle, V. (March, 2020). Drone as a first responder. The new paradigm in public safety. *Police Chief Magazine*. Retrieved from https://www.policechiefmagazine.org/wp-content/uploads/Police_Chief_March_2020_web.pdf

INDEX

Abolition Research Group, 279
Abortion extremists, 68, 302
Actiq, 148
Adam, S., 172
Adult diversion programs, 172
Afterschool program, 264
Aggression, 45, 302
Alcohol and Drug Abuse Institute, 154
Alcohol-related crime and victimization
    alcohol outlet density reduction, 47–48
    binge drinking, 44
    heavy drinking, 44
    incidents in college towns, 44–46
    intervention strategies, 47–49
    NSDUH reports, 44
    Oregon State University, 47
    pharmacological effects, 46
All-or-nothing approach, 253
Al-Qaeda, 67
American Association for Access, Equity and Diversity (AAAED), 110
American Civil Liberties Union (ACLU), 130, 156
American law enforcement, 26
American policing, 252
Andy Griffith Show, 193
Anger, 110
Animal Liberation Front (ALF), 68
Animal rights/environmental extremism, 68, 302
Anti-Defamation League, 81, 117
Anti-Drug Abuse Act, 130
Antigovernment/antiauthority extremism, 67–68, 302
Anti-swatting bill, 127
Arrest activity
    and convictions, 174
    and force incidents, 198
Asian gangs, 99
Asset forfeiture
    civil asset forfeiture laws, 178
    defense, 182
    funds offering to compensate victims, 180
    goals, 178, 180
    incapacitation, 180
    socially desirable causes, 180
    stop-and-seize practices, 181
Augmented reality (AR), 299
Aurora (CO) Police Department, 274
Aurora Police Explorer Program and Global Teen Academy, 275
Austin, J., 172
Automated license plate readers (ALPR), 21

Basketball Cop Foundation, 273
Baton Rouge Police Department, 293–294
Battered Women Criminal Justice Center (BWJP), 59
Battle dress uniforms (BDUs), 302
    advantage, 139
    color, 136, 138
    no-knock warrants, 136, 137
    shootouts and deaths, 137
Bike patrol strategy, 278
Binge drinking, 44, 302
Biometrics, 21
Black Asphalt Electronic Networking and Notification System, 181
Black Lives Matter (BLM) Movement, 212, 279
Body Cam, 207
Bonds, 83, 302
Border security, 143
Boston Marathon Bombings, 69
Boston Police Department, 156
Bratton, W., 240, 241, 243, 249
Breitbart News, 111
Bridges, 82, 302
Broken windows theory, 37, 38, 48, 175, 240, 302
Budgets allocations for policing, 253–254
Bureau of Justice Statistics (BJS), 111

Cadet Program, 269
California Personality Inventory (CPI), 235
Call subjects, 51, 302
Care Quality Commission, 54
Center for Problem-Oriented Policing, 264
Center for the Study of Hate and Extremism, 113
Centers for Disease Control (CDC), 263
Centralized intake system, 170
Chain-of-command, 6, 302

345

Change management, 240, 302
Chaplinsky v. New Hampshire, 112
Chicago decree agreement, 4
Chinese gang, 99
Chula Vista Police Department, 20
Circle sentencing, 293
Citizen Contact Patrol program, 39
Citizen Police Academy (CPA), 266, 267–269
Citizens' arrest, 266, 302
Civil asset forfeiture, 178, 302
Civilian population, 66
Civil Rights Act, 230
Civil rights riots, 128
Civil Rights Under Law, 110
Civil unrest, 248
Clearance rates, 10, 302
Closed system of policing, 252–253, 302
Cocaine, 160
Code of Silence, 56, 228
Coercive change motivations, 248
Coercive pressure, 241
Coffee-with-a-Cop programs, 279
Cognitive interference, alcohol, 46
Cokely, C. L., 120
Collaborative Reform Initiative for Technical Assistance Center (CRI-TAC), 257
Collective efficacy, 82, 166, 302
Colorado's police reform bill, 298
Colombian drug cartels, 101
Community Based Crime Reduction initiative, 153
Community-based programs, 170
Community building, 288, 302
Community, definition, 262–263
Community engagement
   community policing, 261
   definition, 263, 302
   three-pronged approach, 261
Community intelligence, terrorism
   Boston Marathon Bombings, 69, 70
   intelligence-led policing, 72
   operational intelligence, 71–72
   strategic intelligence, 70
   tactical intelligence, 72
Community mobilization, OJJDP, 104
Community Organizing Response Team (CORT), 39
Community-Oriented Policing Services (COPS), 13–14, 65, 74–75, 84, 122–123, 147, 225, 250, 262
Community outreach, 302
   Black American communities, 272
   citizen police academies, 267–269
   and citizens involvement, 266
   historical citizen participation, 266–267
   homeless, 275–276
   honorable mentions, 278–279
   immigrant communities, 273–275
   mental health outreach, 276–278
   program development, 264–265
   program planning, 264–265
   scale and focus, 264
   senior programs, 270
   youth programs, 269–270
Community partnerships, 14–15, 262, 302
Community policing, 303
   adoption, incentives, 247–250
   broken windows theory, 36, 38
   and community issues, 23
   community partnerships, 14–15, 262
   crime and budgetary restraints, 253–255
   cultural challenges, 251–253
   definition, 13, 262
   fear of crime, 19
   Flint Town story, 281–283
   foot patrol, 34–36
   military-style training of police officers, 139–141
   Obama's report, 225
   officer-centered challenges, 227–228
   officer resistance, 250–251
   organizational transformation, 15–16, 262
   principles, 16–18
   problem solving, 16, 262
   recruitment and hiring, 230–234
   Sir Robert Peel's Nine Principles, 32
   technological developments, 19
   technology, 298–301
   vs. traditional models, 18–20
   U.S. law enforcement history, 31
Community Policing Advisory Team (CPAT), 60, 275
Community policing implementation
   federal agencies, 225
   split force concept, 226–227
Community reparative boards, 292
Community stakeholders, 263
Compstat, 241, 303
Conducted energy device (CED), 197
Conflict theory, 256
Consent decree agreement, 4, 302
Continuum and ladder force models, 196
Controlled items, 133, 302

Conversations for Change program, 155
Coordinated partnerships, domestic violence, 59
Counterdrug, 143
Counterproductive work habits (CWH), 235
Counterterrorism tactics, 81, 143
Covid-19 crisis, 24, 246
Crack cocaine market, 164
Crime, 110
    budget cuts, 255
    budgets allocations, 253–254
    crime and disorder challenges, 253
    and criminal behavior, 25
    and disorder challenges, 253
    inducements, 249
    investigation, 217–218, 221–222
    unintended consequences, 254–255
Criminal Interference with Right to Fair Housing statute, 116
Criminal justice programing, 206
Criminal justice system, 252–253, 261
Crisis intervention teams (CITs), 52–53, 278, 303

Damage to Religious Property, Church Arson Prevention Act, 116
Danger Assessment Tool, 55
Data-driven policing, 240
Decentralized policing, 26, 303
Decker, S., 103
Defense Logistics Agency (DLA), 133, 303
Defund policing, 3, 303
Delinquent behavior, 170
Department of Defense (DOD) Excess Property Program, 133–134
Deterrence theory, 176–177
Discontinuation symptoms, 156
Diversion programs, 303
    goals, 171–172
    Group Stigma, 174–175
    juvenile justice system, 172
    loss of human capital, 172–175
Domain Awareness System, 21
Domestic Abuse Intervention Program, 61
Domestic abuse situations, 236
Domestic terrorism
    abortion extremists, 68
    Al-Qaeda, 66, 67
    animal rights/environmental extremism, 68
    antigovernment/antiauthority extremism, 67–68
    definition, 66, 303
    hate crime, 67
    violent extremists, 67
Domestic violence
    community policing practices, 58–61
    definition, 53
    and intimate partner homicide, 56–57
    officer-involved domestic violence, 57–58
    and officer safety, 55–56
    services and units, 55
Domestic Violence Liaison Officers (DVLOs), 59
Domestic violent extremists, 303
Draconian methods, 234
Drill instructor approach, 140
Drones, 21, 300
Drug courts, 173–174
Drug Enforcement Agency (DEA), 157
Drug market intervention model, 166
Drug tolerance, 149, 303
Drug trafficking and gangs
    children's as drug runners, 101–102
    Colombian drug cartels, 101
    drug distribution, 102
    Mexican drug cartels, 100–101
    Mexican drug trafficking organizations, 101
    MS-13 members, 100
    street-level drug sales, 100
Drug use and abuse
    Cocaine, 160
    fentanyl, 148–149, 151
    heroin, 148–149
    marijuana, 158–159
    methamphetamine, 160–161
    open-air drug markets, 161–167
    opioids. *See* Opioid drug use
    overdose death crisis, 156–161
    in United States, 150–156
Due care, 11, 303
Duragesic, 148

Earth Liberation Front (ELF), 68
Elected officials, 263
Elliott, L., 177
Emergency medical technicians (EMTs), 246
Emotional intelligence (EI), 236–237
Emotional quotient (EQ), 236–237
Entertainment media
    criminal justice programming, 206
    media cops, 206
    police dramas, 206–207

prime-time portrayals, 206
reality and infotainment shows, 207
television police programs, 207
EpiPen, 149
Explorer and Cadet Programs, 266
Explorer Programs, 269

Fair Housing Act, 118
Family group conferencing, 292
Family Reunification Program, 276
Fear of crime, 18, 210–211, 249, 303
Federal Bureau of Investigation (FBI), 29, 66, 110
Federal Drug Administration, 149
Federal hate crime legislation
   civil rights worker murder, 115–116
   Criminal Interference with Right to Fair Housing statute, 116
   Damage to Religious Property, Church Arson Prevention Act, 116
   Hate Crime Prevention Act, 116
   Violent Interference with Federally Protected Rights Statute, 116
Fentanyl, 148–149, 151
Ferguson effect, 231
Field training, 234
Fleeing felon rule, 192
Flint Foot Patrol Study, 34, 35
Flint Neighborhood Foot Patrol Experiment, 249
Foot patrol, 34–38
Force escalation model, 196
Force option models, 303
Free Speech Week, 111
Fund for Leadership, Equity, Access and Diversity (LEAD Fund), 110

Gang resistance education and training (G.R.E.A.T.) program, 107, 303
Gangs and gang activity, 303
   Asian gangs, 99–100
   community policing strategy, 103
   crimes, 90
   definition, 86
   and drug trafficking, 100–102
   education and training, 106–107
   hospital-based violence intervention programs, 105
   juvenile members, 96–97
   law enforcement, 90–93
   motorcycle gangs, 88–89
   prevalence, 89–93
   prison gangs, 87–88
   street gangs, 87
   tattoos as identifiers, 88*f*
   White street gangs, 98–99
General Accounting Office (GAO) Study, 134
Goofball, 154
Graffiti, 93
Graham v. Connor, 192, 260
Group Stigma, 174–175

Habitual alcohol use, 48
Hate crime
   community policing strategies, 121–124
   data collection, 111–113
   definition, 110, 303
   and first amendment, 112
   laws, 110, 303
   recording statutes, 117
   types of laws, 117–118
   in United States, 113–115
Hate crime legislation
   civil lawsuits success, 121
   critics, 119
   federal hate crime legislation, 115–117
   jury decision-making, 120
   low-level crimes, 118
   minority rights groups, 121
   offender motive, 119
   prosecutor, 120
   State hate crime laws, 117
   violence, 118
Hate Crime Prevention Act, 124
Hate Crime Statistics Act, 111
Hate, definition, 110
Hate incidents, 110, 303
Hate speech, 110, 303
Heavy drinking, 44, 303
Hells Angels motorcycle gang, 89, 90*f*
Herman Goldstein's concept of problem-oriented policing, 43
Heroin, 148–149
   price, 157
   trafficking route, 157
Heroin withdrawal syndrome, 156
Highway interdiction techniques, 181
Highway Patrol, 207
Hiring process, 225
   prescreening tests, 229–230
Homegrown terrorism in Europe, 82
Homeland Security, 181
Homeless, community outreach

hands-off approaches, 275
H.E.A.R.T. Program, 276
homeless camp, Ontario, 275
Homeless Evaluation Assessment Response Team (H.E.A.R.T.), 276
Hospital-based violence intervention programs, 105
Human capital, 171–174, 303
Hypoxia, 148, 303

Illinois drug raid, 129
Illinois Police Department, 254
Immigrant outreach, 273–275
Immigration and Customs Enforcement (ICE), 86
Immigration and Nationality Act, 273
Incentives, community policing adoption
    coercive change motivations, 248
    crime-related inducements, 249
    criminal and civil liability, 247
    financial incentives, 249–250
    focus groups and case studies, 247
    social disorganization inducements, 247, 249
    state and federal funds, 248
Inconvenience policing, 164, 303
Industrial Revolution, 27–28, 28f
Informal social control, 288, 303
In-service training, 234
Institutional vandalism statutes, 118
Integrity testing, 235–236
Intelligence, 69, 303
Intelligence-led policing, 72, 303
Intent, 119, 304
International Association of Chiefs of Police (IACP), 244, 264
International terrorism, 304
Interrupted threat detection, alcohol, 46
Intimate partner homicide, 55–56
Investigative reporting, 208, 304
Inwald Personality Inventory (IPI), 235
Islamic Revolutionary Guard Corps-Qods Force, 66
Islamic State of Iraq and the Levant (ISIS), 66

Jackson, A., 27
Job assignments and force, 198
Juvenile gangs
    Josephina's story, 94–96
    motivating factors, 96
    MS-13 members, 94
    Office of Juvenile Justice and Delinquency Prevention, 103–104
    recruiting strategies, 93
    risk factors, 94
    social media platforms, 93
    Trei Hernandez's story, 97
Juvenile justice system, 172

Kansas City Preventive Patrol Study, 8

Laser Village, 195
Law enforcement, 252
    agencies, 262
    dangers of, 223
    gang violence, 90–93
    Pew Research Center and survey, 283–284
    police influences, 296–297
    police legitimacy and procedural fairness, 283
    political influences, 297
    principles, 75
    public expectations and satisfaction, 283–285
Law Enforcement Assistance Administration (LEAA), 250
Law Enforcement Management and Administrative Statistics (LEMAS) Survey, 257
Law Enforcement Officers Safety Act (LEOSA), 193
Law Enforcement Support Office (LESO) Program, 133, 134, 304
    benefits, 142, 145
    Department of Defense Excess Controlled Property, 133–134
    General Accounting Office (GAO) Study, 134
Law enforcement, United States, 26
Legal case precedent, 192–193
Legitimacy, 213
Less than lethal weapons, 197, 304
Linking, 83, 304
Live PD, 207
Lone actors, 82
Long Beach (CA) Police Department, 270
Loss of human capital, 171–174
Lost in transit, 152, 304

Maricopa County Police Department, 136
Marijuana
    adverse effects, 158
    butane, 159
    edibles, 158–159
    legalization, 158
    legal uncertainties, 159
    plant extract, 159
    seizure, 159
    U.S. Customs and Border Protection, 158
Marin Abused Women's Services, 60

Mayor's Office on Domestic Violence (MODV), 60
Media and police
    entertainment media, 207
    fear of crime, 210–211
    news media, 208–2210
    police portrayals, 206–208
    social constructionism, 205
    social media. *See* Social media
    symbiotic relationship, 204–205
    war on police, 212
Memorandum of Understanding (MOU), 276, 304
Memphis model, 278
Mental disorders
    adverse outcomes, 53
    area knowledge, 51
    call subjects, 51–52
    chronic vulnerability and solutions, 51
    City of Miami and Miami-Dade County, 53
    core features, 40
    crisis intervention teams (CITs), 52–53
    deinstitutionalization movement, 50
    incarceration rate, 50
    police response, 49–53
    vulnerabilities, 49
Mental health outreach
    addictions and mental health issues, 277
    civil grand jury study, 277
    community engagement models, 277
    crisis intervention teams, 278
    depression, 277
    Research and Public Affairs, 277
Mentally ill and police, 260
Methamphetamine
    production, 158, 159
    transportation, 159
    withdrawal syndrome, 157
Mexican drug trafficking organizations (MDTOs), 101
Mexican Mafia, 87
Micro Community Policing Plans (MCPP), 261
Militarization of police
    battle dress uniforms, 136–139
    Black and Brown communities, 128
    military equipment, 141
    military-style training of police officers, 139–141
    politics, 146
    Specialized Weapons and Tactics developments, 132–133
    war on drugs, 128–131
    war on terrorism, 131
    weapons arsenal, 136

Military equipment
    controlled items, 141
    LESO program, 141–145
    uncontrolled items, 141
    vehicles, 141–142, 142*f*
Minnesota Multiphasic Personality Inventory (MMPI), 235
Minority threat theory, 257
Misconduct, 235
Mongolian Boys Society, 99
Mothers, Sons, and Police program, 270
Motive, 119, 304
Motorcycle gangs, 88–89
MS-13 Marks Gang Territory, 91f–92
Muslim and Arab American communities
    community-based intelligence gathering, 79
    community engagement, 79–80
    community policing strategies, 79
    consultation and community input, 81
    discrimination, 76
    orthodox groups, 80
    poor coordination among police, 77
    social capital, 78
    9/11 terrorist attacks, 77

Naloxone, 149
Nashville Police Department, 292–293
National Alliance on Mental Illness (NAMI), 50, 52
National Center for Women and Policing, 58
National Coalition Against Hate Violence, 121
National Crime Victimization Survey (NCVS), 55, 112, 114
National Drug Threat Assessment (NDTA), 150
National Gang Center, 99
National Gang Intelligence Center, 89
National Gang Threat Assessment, 87
National Institute of Justice, 264
National Law Enforcement Officers Memorial Fund, 212
National Longitudinal Survey of Youth (NLSY), 98
National Network for Safe Communities, 264
National Security Agency (NSA), 72
National Security Decision Directive, 129
National Sheriffs' Association (NSA), 266
Nationwide Suspicious Activity Reporting (SAR) Initiative, 67
Natural disasters, 144
Neighborhood and Block Watch programs, 266
Neighborhood-based organizing and problem solving, 60
Neighborhood Watch Programs, 266–267
Net widening and social control, 304
    asset forfeiture, 178–183
    definition, 171

diversion programs, 171–172, 174
  in New York City, 178
  quality-of-life policing, 175–178
Newark Foot Patrol Experiment, 34, 35
New nets, 171, 304
New Orleans Police Department, 234
News media, 208–210
New York City Police Department (NYPD), 179
New York Times, 66, 128
Next-gen citizen's police academies, 268
Nguyen, T. Q., 154
No-knock raids, 129
North County Regional Gang Task Force, 102

Oak Creek Police Department, 123
Obama's Task Force, 225
Occupational values, 238
Oehme, K., 58
Off-duty cop, 194
Off-duty law enforcement officer, 193
Office of Drug Abuse Law Enforcement (ODALE) team, 129
Office of Juvenile Justice and Delinquency Prevention (OJJDP), 103
Office of Postsecondary Education (OPE), 113
Officer-centered challenges, 227
Officer discretion and use of force, 197–198
Officer-involved domestic violence (OIDV), 57, 304
Officer resistance, 250–251
Omnibus Crime Control and Safe Streets Act, 128, 250
One-percenter motor-cycle clubs, 89
On-the-job fatality rate, 223
Open-air crack cocaine markets, 164
Open-air drug markets, 173, 304
  adverse effects, 158, 163
  basic enforcement strategies, 164–165
  characteristics, 161
  crack cocaine market, 164
  diffusion, 165
  injection process, 165
  physical displacement of drugs, 165
  police strategies, 164–167
  problem-oriented policing strategies, 166
  Violent Crime Initiative, 161
Open system, 253, 304
Operational intelligence, 71–72, 304
Opioid drug use, 148–149
  adverse economic impact, 153
  community–police partnerships, 153–156
  family issues, 155
  federal government funds, 153

fentanyl, 148–149, 151
HIV and hepatitis, 153
opioid epidemic, 151
overdose deaths, 150–153
and police role, 156–161
prescription drug abusers, 151
syringe exchange programs (SEP), 154–155
Opportunities provision, OJJDP, 104
Orange County Health Care Agency, 276
Oregon State University, 47
Organizational change and development, OJJDP, 104
Organizational culture
  all-or-nothing approach, 253
  characteristics, 251
  closed system, 252–253, 302
  institutionalized values, norms, and beliefs, 252
  training and work experience, 252
Organizational structure of traditional policing, 6–7
Organizational transformation, 15–16, 262, 304
Outlaw motorcycle gangs (OMGs), 87–89, 304
Oxycontin, 151

Pain system alteration, alcohol, 46
Paradigm shift, 25, 304
Paramilitary organization, traditional policing, 6, 7
Peace Officer Standards and Training (POST), 229
Peelian Principles, 32
Pepper spray, 197
Perishable skills, 195, 304
16 Personality Factor Questionnaire (16PF), 235
Personality tests, 235
Pew Research Center, 283
Phoenix Arizona Chief, 242, 243
Pill Mills, 151
Plain View Project, 243
Plakas v. Drinski, 192
Police
  and barbers, 272
  brutality, 3, 189–190
  and community programs, 261
  and mentally ill individuals, 260–261
Police abolition movement, 272, 279
Police Area Representatives (PAR), 275
Police-based restorative conferencing programs
  circle sentencing, 293
  community reparative boards, 292
  family group conferencing, 292–293
  victim–offender mediation programs, 291–292
Police community relations (PCR)
  Kerner Report, 32–33

objectives, 32–33
    Report of the Presidents' Crime Commission, 33
    and team policing, 33–34
Police corruption, 235
Police Executive Research Forum (PERF), 75, 157, 264
Police leadership
    financial challenges, 253–255
    incentives, community policing adoption, 247–250
    innovations and technologies, 258
    motivations, inducements, 241–242
    officer resistance, 250–251
    organizational culture, 251–253
    top cop, 241–246
Police legitimacy, 187, 304
Police misconduct, 235
Police Night School, 268
Police Officer Bill of Right, 245
Police officer resistance, 250–251
Police organization, 25
Police portrayals, media, 206–208
Police recruitment
    educated officers, 233–234
    Ferguson effect, 231
    message, 231–232
    recommendation, 230
    selecting in process, 230
    self-selection, 231
    transformation, 230
    women and minority recruitment, 233
Police subculture, 252
Police training
    field training, 234
    in-service training, 234
    soft skills, 236–237
Police union, 244–245
Political era, police service
    adjuncts to local political machines, 27
    Industrial Revolution, 27–28, 28f
    police departments, 28
    politically controlled police, 28–29
    Spoils System, 27
Posse Comitatus Act, 130, 266, 304
Post-traumatic stress disorder (PTSD), 53
Precinct commanders, 240
Predictive policing, 19
Preemployment screening, 230
Prescription drug abusers, 150–151
Pretextual stop, 253, 304
Preventive patrol, 8–9, 304
Prison gangs, 87–88, 304

Problematic split force, 227
Problem-oriented policing strategies, 166
Problem-solving strategies, 19, 262, 304
Procedural fairness, 283, 304
Procedural justice, 186–187, 304
Pseudoephedrine, 162
Psychological Emergency Response Team (P.E.R.T.), 276
Psychomotor stimulant effects, alcohol, 46
Public health model, 103, 104t
Public information officers (PIOs), 212–213

Quality-of-life policing, 305
    broken windows theory, 175
    The Cut, 178
    deterrence theory, 176–178
    disorder and crime rates, 175–176
    marijuana possession and use, 177
    in New York City, 179
Quasi-military organizational structure, 33–34

Race relations, 248
Racial disparities
    net widening, 174–175
Rational choice theory, 256
Reasonable force, 191
Recontacting Victims Program, 39
Recruitment. See Police recruitment
Red Cup Campaign, 47
Reform era, police service
    classical theory, 30
    division of labor, 30
    Hoover's strategies, 29
    police and citizen relationship, 30
    social changes, 31
    Uniform Crime Report (UCR), 30
    unity of control, 30
Regional clubs, 89
Resource allocation to police departments, 256
Restorative conferencing, 291–293, 305
Restorative justice, 287
Restorative policing
    Baton Rouge Police Department, 293–294
    citizens academies, 289–290
    community building, 288
    vs. community policing, 288
    implementation, 291
    Nashville Police Department, 293–294
    police and faith-based community relationship, 295–296
    restorative conferencing, 291–293
Retired law enforcement officer, 193

Retired Senior Volunteer Patrol (RSVP), 270
Revolutionary War, 26
Revolving door of incarceration, 50, 305
Rodney King beating trial, 224
Rogers Police Department, 93

Salt River Pima-Maricopa Indian Reservation, 92
San Bernardino Inland Regional Center (IRC), 64–65
San Bernardino terrorist attack, 141
San Diego program, 270
Santa Ana Police Department, 275–276
Satellite clubs, 89
Scanning, analysis, response, and assessment (SARA) model, 43, 61, 72, 103, , 104*t*, 262, 305
Scared Straight program, 170–171
Seattle Police Department, 246, 261
Senior Police Partners (SPP), 270, 271*f*
Serve and protect theme, 136
Shoot–Don't Shoot Simulators, 196
Shooting galleries, 165, 305
Shop-with-a-Cop program, 278
Simon City Royals, 99
Skogan, W.G., 175
Social capital, 78
Social constructionism, 205, 305
Social disorganization inducements, 249
Social intervention, OJJDP, 104
Social media, 21
    beating of Rodney King, 215–216
    body cams and dash cams, 213–214
    crime investigation, 218–219, 221–222*f*
    and news media merge, 216–218, 218*f*, 219*f*
Soft skills, 236–237
Soft Skills Training Conference, 236
Specialized Weapons and Tactics (SWAT) team, 7, 70, 305
    augmented reality, 299–300
    death by swatting, 127
    officer safety, 132
    public protest and civil unrest, 132
    and suicide attempt, 132
    survey questions, 132–133
    team raids, 136
    training, 128
    weaponry, 132
Split force concept, 226–227
Spoils System, 27, 305
Springfield (MO) Police Department, 223
Stakeholders, 305
State hate crime laws, 117
State Homeland Security Program (SHSP), 131

Stop and Frisk searches, 192, 305
Stop-and-seize practices, 181
Strategic intelligence, 70, 305
Street gangs, 87, 305
Stronger nets, 171, 305
Subculture of police, 228
Sublimaze, 148
Super speedballs, 160
Support clubs, 89
Suppression, OJJDP, 104
Swatting, 127, 305
Syringe Exchange Programs (SEP), 154–155

Tactical intelligence, 72, 305
Tasers, 197
Tattoos, gangs and gang activity, 88*f*
Team policing, 33–34
Teen Academy, 270
Teen and Police Service Academy (TAPS), 270
Teen Police Academies, 266
Tennessee v. Garner, 192
Terrorism, 305
    community intelligence, 68–72
    community resilience and social connections, 82, 83
    definition, 65–66
    domestic terrorism, 66–68
    in Minnesota, 72
    Muslim and Arab American communities, 73–81
9/11 terrorist attacks, 65, 68, 181
Terry Stop, 192
Terry v. Ohio, 192
The Center for Popular Democracy, 253
The Cut, 178, 303
The Police Foundation, 264
There Is No Room for Domestic Violence in This Neighbourhood campaign, 60
The Signs of Crime Program, 39
The 20/20 vision of hindsight, 260
The Washington Post, 181, 188, 192, 285
Timbs v. Indiana, 182, 183
Top cop
    career path, 241
    community leader, 244
    and elected officials, 244
    police chief's role, 242–243
    police officer rights, 245–246
    police union, 244–245
    public figure, 244
    responsibilities, 243
    as a visionary, 246

vote of no confidence, 242
Tough on crime approach, 128
Tough on drugs policy, 128
Traditional policing model
    corrupt practices, 6
    crime-fighting, 248
    crime-fighting focus, 7–8
    investigations, 11–12
    organizational structure, 6–7
    preventive patrol, 8–9
    professionalization of policing, 6
    public safety, 10
    rapid response, 9–10
Transformative change, 25
Trusting relationships, 74, 305
Tuch, S., 284

Unconscious biases, 120, 305
Uncontrolled items, 133, 305
Uniform Crime Report (UCR), 4, 11, 30, 111, 114
Unity of command, 6, 304
Unpaid reserve program, 268
Urban Areas Security Initiative (UASI), 131
Urban minority criminal street gangs, 98
U.S. Charter Citizen Police Academies, 268
U.S. Department of Justice, 172
Use of force, 305
    data collections, 188–192
    force options, 192
    force rate reduction, 187
    law enforcement tactical gear, 193
    misconceptions, 199
    and officer discretion, 197–199
    options models, 196–197
    police brutality, 189–190
    police–citizen contacts, 186
    police legitimacy, 187
    policies and legal case precedent, 191–193
    prevalence rates, 188–190
    procedural justice, 186–187
    public distrust, 186
    public outrage, 185
    and race, 189
    research limitations, 186
    risk reduction, 199–200
    standards, 200
    training, 195–196
    types, 188
U.S. Immigration and Customs Enforcement (ICE), 273

U.S. Marshals Service, 26
U.S. Postal Service, 151
Us-*versus*-them mentality, 139, 141, 228

Vallejo Alcohol Policy Coalition (VPAC), 49
Vicarious liability, 224
Victim–offender mediation programs, 291
Victim Recontact Program, 39
Vietnamese gang, 99
Vigilante justice, 266, 305
Violence, 46, 305
    consent decree agreement, 4
    domestic violence. *See* Domestic violence
    Uniform Crime Report, 4
Violent Crime Initiative (VCI), 161
Violent Interference with Federally Protected Rights Statute, 116
Virginia v. Black, 112
Visionary leadership, 241
Vote of no confidence, 242, 305

Walk and Talk programs, 278
War on drugs
    Illinois drug raid, 129
    joint task forces, 130
    Law Enforcement Act, 130
    Nixon's statement, 128
    Office of Drug Abuse Law Enforcement (ODALE) team, 129
War on police, 212
War on terrorism, 131
Washington Post, 66
Washtenaw County Sherriff's Office (WCSO), 286
Watts v. U.S., 112
Weitzer, R., 284
Wheel force model, 196–197
White street gangs
    arrest data, 98
    law enforcement data, 98
    National Youth Survey, 98
    royals, 98–99
Wide nets, 171, 305
Withdrawal syndromes, 156
Wolff, K. B., 120

Xanax pills, 152

Youth ALIVE! organization, 105–106
Youth programs, 269–271